Nanomaterials for Sustainable Tribology

With the advent of nanotechnology, the properties offered by nano-sized particles in various engineering applications have revolutionized the area of material science. Furthermore, due to the use of nanomaterials in various engineering components, particularly in moving parts, it is imperative to understand the behaviour of these nanomaterials under sliding conditions. Therefore, an augmented approach of nanotechnology and tribology has been addressed in this book. It presents recent advancements on the topics related to:

- Mechanical and tribological behaviours of nanocomposites
- Nanomaterials in lubricating oils
- Synergetic effects of nanomaterials
- Surface texturing at nano-scale
- Nanocoatings for various applications
- Biotribological applications of nanomaterials

Nanomaterials for Sustainable Tribology covers major aspects of tribology of nanomaterials, and its current status and future directions. This book will provide the readers an insight on several aspects of tribology of nanomaterials. It will act as a strong stimulant for readers to appreciate and initiate further advancements in the field of tribology, particularly at nano-scale.

Nanomaterials for Sustainable Tribology

Edited by
Ankush Raina
School of Mechanical Engineering,
Shri Mata Vaishno Devi University
Mir Irfan Ul Haq
School of Mechanical Engineering,
Shri Mata Vaishno Devi University
Patricia Iglesias Victoria
Kate Gleason College of Engineering at the Rochester
Institute of Technology
Sudan Raj Jegan Mohan
ELTE University Center Savaria (ELTE Savaria Egyetemi Központ)
Ankush Anand
School of Mechanical Engineering,
Shri Mata Vaishno Devi University

CRC Press is an imprint of the
Taylor & Francis Group, an **informa** business

Designed cover image: Neon_dust/Shutterstock

First edition published 2023
by CRC Press
6000 Broken Sound Parkway NW, Suite 300, Boca Raton, FL 33487-2742

and by CRC Press
4 Park Square, Milton Park, Abingdon, Oxon, OX14 4RN

CRC Press is an imprint of Taylor & Francis Group, LLC

© 2023 selection and editorial matter, Ankush Raina, Mir Irfan Ul Haq, Patricia Iglesias Victoria, Sudan Raj Jegan Mohan, Ankush Anand; individual chapters, the contributors

First edition published by CRC Press 2023

Reasonable efforts have been made to publish reliable data and information, but the author and publisher cannot assume responsibility for the validity of all materials or the consequences of their use. The authors and publishers have attempted to trace the copyright holders of all material reproduced in this publication and apologize to copyright holders if permission to publish in this form has not been obtained. If any copyright material has not been acknowledged please write and let us know so we may rectify in any future reprint.

Except as permitted under U.S. Copyright Law, no part of this book may be reprinted, reproduced, transmitted, or utilized in any form by any electronic, mechanical, or other means, now known or hereafter invented, including photocopying, microfilming, and recording, or in any information storage or retrieval system, without written permission from the publishers.

For permission to photocopy or use material electronically from this work, access www. copyright.com or contact the Copyright Clearance Center, Inc. (CCC), 222 Rosewood Drive, Danvers, MA 01923, 978-750-8400. For works that are not available on CCC please contact mpkbookspermissions@tandf.co.uk

Trademark notice: Product or corporate names may be trademarks or registered trademarks and are used only for identification and explanation without intent to infringe.

ISBN: 978-1-032-30690-2 (hbk)
ISBN: 978-1-032-30691-9 (pbk)
ISBN: 978-1-003-30627-6 (ebk)

DOI: 10.1201/9781003306276

Typeset in Sabon
by codeMantra

Contents

Preface ix
Editors xi
Contributors xv

1 Nanomaterials and tribology: An introduction 1
FAROOQ AHMAD NAJAR, JUNAID H. MASOODI, SUMMERA BANDAY, AND NADEEM AKBAR NAJAR

2 Nanocomposites and tribology: Overview, sustainability aspects, and challenges 25
ABRAR MALIK, NITISH SINGH JAMMORIA, ASRAR RAFIQ BHAT, PRATEEK SAXENA, MIR IRFAN UL HAQ, AND ANKUSH RAINA

3 Tribological behaviour of aluminium metal composites reinforced with nanoparticles 53
PAWANDEEP SINGH, VIVUDH GUPTA, AND MD IRFAN UL HAQUE SIDDIQUI

4 Tribological performance of RGO and Al_2O_3 nanodispersions in synthetic lubricant 65
PRANAV DEV SRIVYAS, M. S. CHAROO, SOUNDHAR ARUMUGAM, AND TANMOY MEDHI

5 Synergism of the hybrid lubricants to enhanced tribological performance 75
PRANAV DEV SRIVYAS, M. S. CHAROO, SOUNDHAR ARUMUGAM, AND TANMOY MEDHI

6 Recent progress on the application of nano-cutting fluid in turning process 87
MOHD BILAL NAIM SHAIKH AND MOHAMMED ALI

7 Dispersion stability of nanoparticles and stability measurement techniques 101
PRANAV DEV SRIVYAS, SANJAY KUMAR, SOUNDHAR ARUMUGAM, AND MANOJ KUMAR

8 Natural fiber–reinforced polymer nanocomposites 115
ASRAR RAFIQ BHAT AND PRATEEK SAXENA

9 Epoxy-based nanocomposites: Tribological characteristics and challenges 143
VIVUDH GUPTA, PAWANDEEP SINGH, AND NIDA NAVEED

10 Tribo-response of multifunctional polymer nanocomposites 155
MEGHASHREE PADHAN

11 Synthesis of nanomaterial coatings for various applications 167
MOHD RAFIQ PARRAY, HIMANSHU KUMAR, AMRITBIR SINGH, BUNTY TOMAR, AND S. SHIVA

12 Corrosion mitigation using polymeric nanocomposite coatings 191
AVI GUPTA, JAYA VERMA, AND DEEPAK KUMAR

13 Sustainable nanomaterial coatings for anticorrosion: A review 203
SANGEETA DAS, PREETAM BEZBARUA, AND SHUBHAJIT DAS

14 Effect of nano-additives lubricant on the dynamic performance of textured journal bearing 215
DEEPAK BYOTRA, SANJAY SHARMA, ARUN BANGOTRA, AND RAJEEV KUMAR AWASTHI

15 Impact of nano-lubricants on the dynamic performance of journal bearings with surface waviness 237
ARUN BANGOTRA, SANJAY SHARMA, DEEPAK BYOTRA, AND RAJEEV KUMAR AWASTHI

16 Effect of nano-hydroxyapatite and post heat treatment on
   biomedical implants by sol-gel and HVOF spraying   257
   KHUSHNEET SINGH, SANJAY MOHAN, SERGEY KONOVALOV,
   AND MARCEL GRAF

Index   287

Preface

Tribology, being a phenomenon of interacting surfaces and an interdisciplinary field, finds applications in different areas. The properties offered by nano-sized particles in various engineering components have revolutionized the area of material science. The results obtained from different findings suggest that there is a lot of improvement in the tribological properties with the use of nanomaterials. Therefore, it is imperative to understand the behaviour of these nanomaterials under sliding conditions at nano-scale. Furthermore, owing to the interdisciplinary nature of the subject, there is a great scope in nanotechnology with regard to tribology.

In this context, some of the initiatives taken by tribologists include the use of nanomaterials in lubricating oils, use of nano-additives in green lubricants, nano-additives as self-lubricating materials, surface texturing at nano-scale, use of nanocoatings for improved surface characteristics, synergetic effects of nanomaterials and the use of nanomaterials for biomedical applications. All these approaches have led to substantial improvements in the tribological behaviour of different materials under different operating conditions. Keeping in view these aspects, in this book, an attempt has been made to address the role of nanotechnology in the field of tribology.

The main aim of this book is to highlight the contribution of nanomaterials in the field of tribology. This book covers major aspects of tribology of nanomaterials, and its current status and future directions. It presents the recent advancement on the topics related to tribology of nanolubricants, and special topics related to nanomaterial coatings and nano-biotribology. The book starts with chapters focused on introduction to nanotechnology and its role in the field of tribology followed by the role of nanomaterials in metal matrix composites and polymer composites. Furthermore, chapters also cover topics related to liquid lubrication and the importance of different nano-additives in lubricating oils. The later part of the book includes chapters on nanomaterial coatings for various applications and the role of surface texturing in tribology followed by a chapter focused on the role of nanotechnology in biotribology.

Editors believe that this book will provide the readers with various insights on several aspects of tribology of nanomaterials and shall be helpful to material scientists, industry personnel, researchers working the field of tribology and students pursuing various courses related to material science, tribology and nanotechnology. Furthermore, this book will act as a strong stimulant for readers to augment the concepts of nanotechnology with tribology for development of systems and materials which are greener, cost-effective and efficient.

Editors

Dr. Ankush Raina is an Assistant Professor at the School of Mechanical Engineering, Shri Mata Vaishno Devi University, Jammu and Kashmir, India. His areas of interest include wear and lubrication, additive manufacturing and rheological properties of lubricating oils. He has been awarded a PhD in the field of industrial lubrication. He completed his M.Tech in Mechanical System Design from NIT Srinagar, Jammu and Kashmir. He was awarded a Gold Medal for securing the first position in M.Tech. He has extensive research experience in the field of nanolubrication and has carried out various studies using nano-additives in different lubricating oils and composite materials. Furthermore, he has also explored the tribological characteristics of different types of 3D-printed polymeric materials. He has published more than 50 articles in the SCI/SCIE/Scopus-indexed journals with more than 1400 citations, and an h-index of 19 and i-10 index of 34. He has also presented several papers in national and international conferences in India and abroad. He is the reviewer of several reputed SCI/SCIE-indexed journals and has also coordinated several events at university level.

Dr. Mir Irfan Ul Haq is currently an Assistant Professor at the School of Mechanical Engineering, Shri Mata Vaishno Devi University. He has previously worked with the R&D wing of Mahindra and Mahindra. He has obtained a Bachelor of Technology in Mechanical Engineering from SMVD University and a Master of Engineering in Mechanical System Design (Gold Medallist) from National Institute of Technology, Srinagar. He has done his PhD in the area of tribology of lightweight materials. He is actively involved in teaching and research for the past 10 years in the field of materials, tribology and 3D printing. His research interests include lightweight materials, new product development and additive manufacturing, development of green lubricants, self-lubricating materials and cutting fluids, mechanical testing of composites, friction and wear of materials and surface engineering. He has published around 50 research papers in SCI- and Scopus-indexed journals. He has edited three books and has served as a guest editor in reputed journals.

Moreover, he has been awarded various research grants for various projects from various agencies like DST and NPIU AICTE. He has attended numerous conferences and workshops both in India and abroad (The United States and Singapore). He is a member of reviewer boards of various international journals apart from chairing technical sessions in various international and national conferences. Dr. Haq is actively involved in organizing various conferences and workshops. He has also coordinated various student events such as SAEBAJA, ECOKART and TEDX. Moreover, he has supervised around 20 Master- and Bachelor-level projects. Dr. Haq has served as a member of various committees at national and state levels.

Dr. Patricia Iglesias Victoria earned her PhD in Mechanical Engineering with focus on tribology from the Polytechnic University of Cartagena (Spain). Dr. Iglesias also worked as a post-doctoral researcher in the Center for Materials Processing and Tribology at Purdue University. Currently, she is an Associate Professor and the Director of the Tribology Laboratory in the Kate Gleason College of Engineering at the Rochester Institute of Technology. Her research focuses on wear and friction of materials, ionic liquids as lubricants and additives of lubricants, bio-lubricants, nanostructured materials and textured surfaces. Dr. Iglesias has extensive experience working on tribology and has published 40 peer-reviewed articles and more than 40 conference proceedings in the field of tribology.

Dr. Sudan Raj Jegan Mohan is currently an Assistant Professor in the Faculty of Informatics, Eötvös Loránd University, Szombathely, Hungary. He has an industrial experience of 3 years at Bharath Electronics Limited, Bengaluru as Contract Engineer in the Production Department. He has an additional 2 years of experience at Wipro Technologies, Chennai as Design Engineer in the Design and Development Team. He has obtained his Bachelor of Engineering in Mechanical Engineering from Anna University, Chennai and Master of Engineering in Engineering Design from Anna University, Coimbatore. He has done his PhD in the field of industrial tribology in automotive disk brake pads. He is actively involved in teaching and research for a decade in the field of industrial tribology, composite materials and 3D printing. His research interests include noise, vibration and harshness studies in automotive brakes, yaw brakes, biomedical materials, eco-friendly polymers, material development and characterization. To his credit, he has published various research papers in SCI- and Scopus-indexed journals and presented various papers in national and international conferences.

Dr. Ankush Anand is currently an Associate Professor in the School of Mechanical Engineering, at Shri Mata Vaishno Devi University, Katra, India. His areas of interest include sustainability and product life-cycle design, tribology of metal matrix composites, polymer composites and biotribology. He is actively involved in teaching and research activities for more than 17 years. He has successfully guided 6 PhD and supervised more than 15 MTech dissertations. He has published more than 50 articles in the SCI/SCIE/Scopus-indexed journals with more than 1,000 citations and an h-index of 18. He has also attended various international conferences in India and abroad, has presented research articles and has also chaired various technical sessions in the United States, the United Kingdom, Europe, etc. He is an active member of various professional bodies including the American Society for Mechanical Engineers.

Contributors

Mohammed Ali
Department of Mechanical
   Engineering
Aligarh Muslim University
Aligarh, India

Soundhar Arumugam
Mechanical Engineering
   Department
Indian Institute of Technology
Guwahati, India

Rajeev Kumar Awasthi
Department of Mechanical
   Engineering
Sardar Beant Singh State University
Gurdaspur, India

Summera Banday
Department of Mechanical
   Engineering
University of Kashmir
Srinagar, India

Arun Bangotra
Govt. Polytechnic College, Kathua
Badala, India
and
School of Mechanical Engineering
Shri Mata Vaishno Devi University
Katra, India

Preetam Bezbarua
Department of Mechanical
   Engineering
Girijananda Chowdhury Institute
   of Management and Technology
Guwahati, India

Asrar Rafiq Bhat
School of Mechanical and
   Materials Engineering
Indian Institute of Technology
   Mandi
Kamand, India

Deepak Byotra
School of Mechanical Engineering
Shri Mata Vaishno Devi University
Katra, India

M. S. Charoo
Department of Mechanical
   Engineering
National Institute of Technology
Srinagar, India

Sangeeta Das
Department of Mechanical
   Engineering
Girijananda Chowdhury Institute
   of Management and Technology
Guwahati, India

Shubhajit Das
Department of Mechanical
 Engineering
National Institute of Technology,
 Arunachal Pradesh
Jote, India

Marcel Graf
Institute of material science and
 engineering
Technische Universität Chemnitz
Chemnitz, Germany

Avi Gupta
Centre for Automotive Research
 and Tribology
Indian Institute of Technology,
 Delhi
New Delhi, India

Vivudh Gupta
School of Mechanical Engineering
Shri Mata Vaishno Devi University
Katra, India

Mir Irfan Ul Haq
Shri Mata Vaishno Devi University
Katra, India

Nitish Singh Jammoria
School of Mechanical Engineering
Shri Mata Vaishno Devi University
Katra, India

Sergey Konovalov
Research and Innovation
Siberian State Industrial University
Novokuznetsk, Russia

Deepak Kumar
Centre for Automotive Research
 and Tribology
Indian Institute of Technology,
 Delhi
New Delhi, India

Himanshu Kumar
Laboratory for Advanced
 Manufacturing and Processing
Indian Institute of Technology
 Jammu
Jammu & Kashmir, India

Manoj Kumar
Department of Metallurgical and
 Materials Engineering
National Institute of Technology
Srinagar, India

Sanjay Kumar
Department of Mechanical
 Engineering
National Institute of Technology
Srinagar, India

Abrar Malik
School of Mechanical Engineering
Shri Mata Vaishno Devi University
Katra, India

Junaid H. Masoodi
Department of Mechanical
 Engineering
University of Kashmir
Srinagar, India

Tanmoy Medhi
Mechanical Engineering
 Department
Indian Institute of Technology
Guwahati, India

Sanjay Mohan
School of mechanical engineering
Shri Mata Vaishno Devi University
Katra, India

Farooq Ahmad Najar
Department of Mechanical Engineering
University of Kashmir
Srinagar, India

Nadeem Akbar Najar
Centre for Policy Studies, IIT Bombay
Mumbai, India

Nida Naveed
School of Engineering
University of Sunderland
Sunderland, United Kingdom

Meghashree Padhan
Centre for Automotive Research and Tribology
Indian Institute of Technology, Delhi
New Delhi, India

Mohd Rafiq Parray
Laboratory for Advanced Manufacturing and Processing
Indian Institute of Technology Jammu
Jammu & Kashmir, India

Ankush Raina
School of Mechanical Engineering
Shri Mata Vaishno Devi University
Katra, India

Prateek Saxena
School of Mechanical and Materials Engineering
Indian Institute of Technology Mandi
Kamand, India

Mohd Bilal Naim Shaikh
Department of Mechanical Engineering
Aligarh Muslim University
Aligarh, India

Sanjay Sharma
School of Mechanical Engineering
Shri Mata Vaishno Devi University
Katra, India

S. Shiva
Laboratory for Advanced Manufacturing and Processing
Indian Institute of Technology Jammu
Jammu & Kashmir, India

Md Irfan ul Haque Siddiqui
Department of Mechanical Engineering
King Saud University
Riyadh, Saudi Arabia

Amritbir Singh
Laboratory for Advanced Manufacturing and Processing
Indian Institute of Technology Jammu
Jammu & Kashmir, India

Khushneet Singh
School of mechanical engineering
Shri Mata Vaishno Devi University
Katra, India

Pawandeep Singh
School of Mechanical Engineering
Shri Mata Vaishno Devi University
Katra, India

Pranav Dev Srivyas
Mechanical Engineering
  Department
Indian Institute of Technology
Guwahati, India

Bunty Tomar
Laboratory for Advanced
  Manufacturing and Processing
Indian Institute of Technology
  Jammu
Jammu & Kashmir, India

Jaya Verma
Centre for Automotive Eesearch
  and Tribology
Indian Institute of Technology,
  Delhi
New Delhi, India

Chapter 1

Nanomaterials and tribology
An introduction

Farooq Ahmad Najar, Junaid H. Masoodi, and Summera Banday
University of Kashmir

Nadeem Akbar Najar
Centre for Policy Studies
Indian Institute of Technology, Bombay

CONTENTS

1.1 Introduction to nanomaterials	1
1.2 Types of nanomaterials	3
1.2.1 Polymeric-based nanomaterials	5
1.2.2 Metal-based nanomaterials	5
1.2.3 Ceramic-based nanomaterials	7
1.3 Introduction to sustainable tribology	7
1.4 Nanomaterials in tribological systems	9
1.4.1 Bulk materials	9
1.4.2 Lubricants	11
1.4.3 Semisolid materials	13
1.4.4 Coatings	15
1.5 Applications	16
1.5.1 Industrial applications	16
1.5.2 Biomedical applications	18
1.6 Conclusion	18
References	19

1.1 INTRODUCTION TO NANOMATERIALS

The word nanomaterial(s) itself portrays that it is something associated with very small sizes and dimensions of a material of the order of nanometre, or in other words, nanomaterials are colloidal particles with size ranging from 10 to 1,000 nm. In the last few years, they have caught the attention of researchers across all scientific fields, and the products based on this particular cutting-edge technology have flooded the world markets at large. Therefore, they have achieved significant commercial importance, which will definitely increase furthermore in the near future. However, materials containing particles of nanodimension have a serious concern for human

DOI: 10.1201/9781003306276-1

health and the environment, and the risks of these developed nanomaterials need to be studied meticulously.

Nanomaterials are generally different from the already existing materials, as they possess some excellent optical, magnetic, and electrical properties, which make these materials so smart and they can be used for purposes where conventional materials cannot perform well. Nanomaterials are resources at the molecular or tiny level as these materials have such properties which are not visible at bulk scale. Materials at the nanoscale exhibit different properties due to the relatively higher surface area and are altogether different (quantum size effects), which is desirable. So, nanomaterials have utmost chemical reactivity as compared to the bulk-size materials. Some of the nanomaterials occur naturally, some are being engineered (EN), which are later on customized for specific applications. EN nanomaterials are extensively used in industries for the production of sunscreens, cosmetics, lubricants, wear-resistant materials, sports goods, clothing, tyres, etc. Nanomaterials possess a wider range of applications in the medical field for several purposes like diagnosis, imaging, and drug delivery (Figure 1.1).

Nanocoatings and nanocomposites are the other applications of nanomaterials, and they are used in diverse consumer products like windows, sports equipment, bicycles, and automobiles. Some ultraviolet-resistant coatings on glass bottles, which protect the edibles or food stuff from continuous sunlight exposure, and sports items like long-lasting balls (tennis) use composites such as butyl-rubber/nanoclay. Nanosized titanium dioxide, which is from the family of nanomaterials, is finding an application in self-cleaning windows, and also nanosized silica is nowadays being used as filler material in a variety of products, including cosmetics and particularly dental fillings. Over the past few decades, nanomaterials have attracted great attention for their promising friction and wear reduction performance and also some important tribological applications.

Nowadays, it is an explicitly known fact that energy and its crisis are one of the most challenging concerns in scientific progress in totality. As per the literature, it has been observed that almost one-third of the fuel gets consumed to combat friction, particularly in internal combustion engines. With the adoption of green energy concept and seeking environmental benefits, a drastic decrease is noticed in friction and subsequently reduction in wear as well. Tribological optimizations and associated improvements, such as reduction in friction and wear magnitude, are increasingly

Figure 1.1 New direction and development in tribology.

recognized as paramount strategies for energy saving not only in the progress of macro-scaled turbo-machinery units, such as heavy pumps, compressors, and steam turbines, but also optimizations in tribology, which has brought about technology-driven changes enabling the industries to incorporate micro- or nanotechnologies, such as micro- or nanoelectromechanical systems, which are notably known as (MEMS/NEMS) devices and which have assumed dramatic roles in the modern industries.

For the last few years, carbon materials have been used in the fields of optics, electrochemistry, mechanics, and tribology as well, because of the variety in their morphologies, like they are corrosion resistive, have excellent mechanical behaviour, and have high thermal conductivity. Carbon nanomaterials (CNMs) having zero-dimensional fullerenes, one-dimensional carbon nanotubes (CNTs), two-dimensional graphene and three-dimensional nanodiamonds exhibit extraordinary electrochemical properties, making them ideal materials for use in related areas. In other words, CNMs have received massive attention in the area of nanotechnology because of their certain properties and flexible dimensional structure. CNMs have electrical, thermal, and optical properties, which make them the best media for drug delivery, in particular, biosensing, bioimaging, etc.

1.2 TYPES OF NANOMATERIALS

Nanomaterials are incredibly tiny, with at least one of the dimensions being 100 nm or less. Nanoscale materials come in one or more dimensions, such as surface coatings, fibres, strands, or three dimensions (e.g. particles). Their typical geometries are spherical, tubular, and irregular shapes, and they can be found alone, fused, aggregated, or agglomerated. Some typical examples of nanomaterials include fullerenes, quantum dots, dendrimers, and nanotubes. Nanomaterials, such as CNTs, fullerene, silver, and silica, have been used in the field of nanotechnology.

Zero-dimensional nanomaterials: In such materials, all of the three dimensions (x, y, and z) are found at the nanoscale, that is, none of the dimensions exceed 100 nm and nanospheres and nanoclusters are parts of them.

One-dimensional nanomaterials: In this case, two dimensions (x, y) are present at the nanoscale range, while the third dimension is not. Nanomaterials with a needle form result from this. Nanowires, nanorods, nanotubes, and nanofibres all belong to this category.

Two-dimensional nanomaterials: In this case, one dimension (x) is on the nanoscale, although the other two dimensions are not. The 2D nanomaterials have forms like plates. They consist of nanometre-thick nanofilms, nanocoatings, and nanolayers.

4 Nanomaterials for Sustainable Tribology

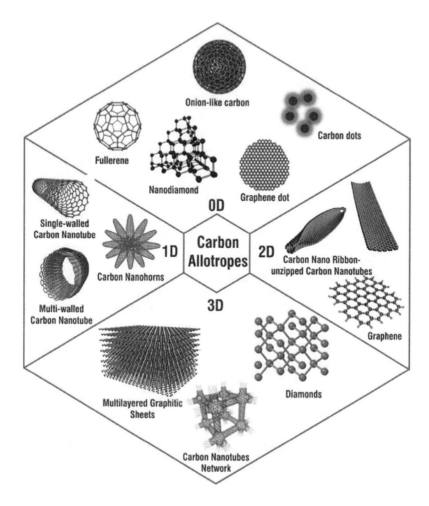

Figure 1.2 Carbon nanomaterial allotropes with examples for multi-dimensions in carbon nanostructure [1].

Three-dimensional nanomaterials: In the case of three-dimensional nanomaterials, these are not restricted to the nanoscale level in any dimension. They are always above 100 nm, and these materials have three dimensions (arbitrary). There are a variety of nanoscale crystals, and they are arranged in such a way that they make up most of the (3D) nanomaterials. 3D nanomaterials possess multi-nanolayers in which the dimensions (0D, 1D, and 2D) and structural elements therein are in proximate contact with one another, and thus, they form sort of interfaces, as well as nanoparticle dispersions, nanotubes, and the bundles of nanowires (Figure 1.2).

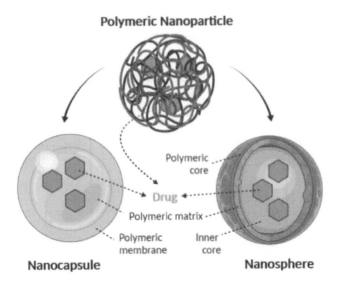

Figure 1.3 Structures of nanocapsules and nanospheres [2].

1.2.1 Polymeric-based nanomaterials

Polymeric nanoparticles (NPs) have caught the attention of researchers over the last several years because of certain properties as a result of their small and finer sizes. There are various advantages for polymeric NPs when they are used in pharmaceutical companies/industries. Polymeric-based nanomaterials as drug carrier(s) include their breakthrough and potential usage for controlled releasing and the capability to shield the drugs and some other fine molecules with bioactivity, which is highly detrimental to the environment, thus improving their bioavailability and therapeutic index.

The term nanoparticle is composed of nanocapsules, nanoshots, and nanospheres, which differ completely from one another with their distinct morphologies. Nanocapsules are basically oil-based substances, which possess the drugs in a dissolved form covered by a polymeric shell, which later on controls the release distribution of the drug from its core. Nanospheres are the continuous polymeric matrix and the drug is retained inside or adsorbed on their surfaces. There are certain types of polymeric NPs which are recognized as a reservoir system (nanocapsule type) and the matrix system (nanosphere type), both of which are shown in Figure 1.3.

1.2.2 Metal-based nanomaterials

Biomedical sciences and the engineering field have become the platform for the utilization of metal-based nanomaterials, and they have diverted the

6 Nanomaterials for Sustainable Tribology

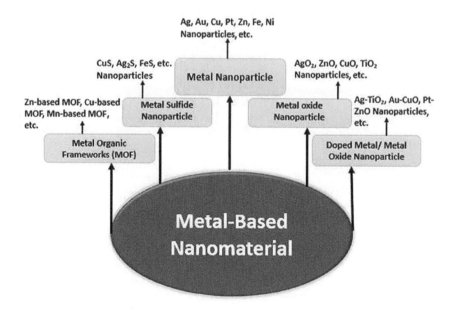

Figure 1.4 Different types of metal-based nanomaterials [3].

attention of researchers from the conventional methods of diagnosis to this breakthrough scanning. In short, nanomaterials play an indispensable role in modern medical science, and they occur commonly in basic patterns and forms, as shown in Figure 1.4. The first common category of nanomaterials is composed of the pure form of metal-based NPs. These are termed as metal NPs, for example, silver, copper, gold, titanium, platinum, zinc, magnesium, iron, and aligned NPs. The rest of the nanomaterials are the metal oxide– based NPs, for example, titanium dioxide, silver oxide, zinc oxide, etc. Metal, or metal oxide, or metal NPs in a doped form are considered as another kind of classification among nanomaterials [3,4].

In the field of medical sciences, different imaging techniques have been introduced in the last two decades, for example, ultrasound, positron emission tomography (PET), Magnetic resonance imaging (MRI), Surface-Enhanced Raman Spectroscopy (SERS), Computed Tomography (CT), and other optical imaging as tools to study the kind of disease and the damage levels. Imaging characteristics vary in terms of both technologies and instrumentations involved, and more importantly, they require an agent (contract) and should have unique physiochemical properties. In addition to this, metal sulphides, metal organic frameworks (MOFs), and nanomaterials have a paramount role because of their wide applications in various biological sciences. For example, handling of drug delivery and antimicrobial activities is one of the prominent medical applications in which the following metal base NPs /nanometals are used.

1.2.3 Ceramic-based nanomaterials

Nanoceramics were discovered in the early 1980s, and they were composed of ceramics and advanced ceramics, which are made from materials like alumina and silica carbide. They are entirely different from the traditional pots or utensils (clay based). Nanoceramics have their own unique properties and characteristics in terms of physical, chemical, mechanical, and magnetic attributes, which are altogether different from metal-based, plastic-based, and conventional-type bulk ceramics. These utmost and indispensable properties like ferroelectric, dielectric, pyroelectric, piezoelectric, magneto-resistant, ferromagnetic, and superconductivity also depends on material type and amount. Nanoceramics also have exceptional properties in terms of mechanical and surface texture characteristics including machinability, super plasticity, bioactivity, strength, and toughness. All of them wholly and solely depend on the particle size that is used to create them.

Ceramic nanomaterials and their utilization in medical science have increased extensively, particularly in the area of drug delivery. Numerous types of nanosystems have been introduced and designed, for example, nanoclays, scaffolds, and nanotubes. Ceramic nanosystems have applications in areas such as drug loading, bioassay, target cell uptake, and imaging. Nanoceramics are extensively being used in biomedicine, particularly in bone repair. Bioactive ceramics closely match with the properties of bone materials, and they can act as a nano-scaffold to help the bones to rejuvenate. Ceramic materials have got high mechanical strength and least biodegradability. Furthermore, some aspects concerned with ceramic NPs are advantageous over some other systems, and their cellular uptake and toxicity concerns are appreciably minimal.

1.3 INTRODUCTION TO SUSTAINABLE TRIBOLOGY

Sustainable tribology has gained utmost interest towards further progress and development, as environmental protection and energy generation are the main concerns at the global level. In light of this, tribology targets have steadily grown, and several additional concepts were introduced with the passing of time, including energy conservation tribology and eco-tribology. Particularly as we entered the new millennium, it was anticipated that tribology would play an even bigger role and shape the world's resource and energy, as the environment is heavily losing its quality. In addition to its original goals of controlling friction, wear reduction, and enhancing lubrication performance, it has shown a supportive extension in saving energy and minimizing material wastage, emission control, noise pollution reduction, development of biolubricants and ecolubricants, and thereby improvement in the quality of life at large. Environmentally friendly tribology, also

Figure 1.5 The impact of sustainable tribology on sustainable development [6].

known as "Green Tribology," underlines the tribological aspects of ecological balance, and the effects of interacting surfaces on the environment and biological systems are being studied in this area [5].

Optimizing friction and its effects, wear rate assessment and its reduction, and enhancing of lubrication performance are the three main basic concerns of tribology. Reducing power loss, saving energy and reducing materials are the important targets of tribology. So, tribology is considered one of the better solutions to address the needs of a sustainable society. Therefore, green tribology keeps on developing in the way that sustainable advancements occur in the society without jeopardizing the environment. In other words, "green" means a novel way of understanding that what predominately contemplates on ecology and environmental conservation. In short, it encompasses the sustainable growth of nature and society as well. In recent researches (Figure 1.5), sustainable tribology mentions the rudimentary aspects of sustainable growth and developments by addressing the practicability of an engineering problem in order to improve its efficiency and durability at large, which is being done while optimizing the cost of the overall system to benefit the end users and by minimizing material consumption and energy savings for the sake of environment. Eventually, with the help of sustainable tribology or green tribology concepts, social progress can be achieved.

1.4 NANOMATERIALS IN TRIBOLOGICAL SYSTEMS

In tribological systems where there are surface interactions at large, it is imperative to understand the concept of roughness at nanoscale level. Nanoscale surfaces alter the tribology properties to either enhance or diminish at a greater scale than the properties provided by the macroscopic level of lubrication and adhesion. It is common practice to use (scanning probe microscopy) techniques such as the atomic force microscope in order to assess tribological characteristics at the nanoscale. Properties of surfaces like friction, adhesion, and wear are observed using the tip of the atomic force microscope. In the same manner, nanomaterials have an indispensable role in bulk materials, semisolids, and coatings. The whole information provided gives an insight and further understanding of nanotribology, and it helps a designer to predict the influences of the nanomaterial surface interactions.

1.4.1 Bulk materials

With regard to tribological characteristics like friction coefficient and wear rate in bulk materials from various experimental studies, it has been noticed that bulk materials come in three categories, namely, metals, polymers, and ionic solids, and they are different in terms of tribological behaviour. The research community has a deep insight and understanding of the correlations in between the parameters of environment, properties of materials, and tribological characteristics in the overall system. Concurrently, the demand has increased for the incorporation of novel synthesis and *in situ* experimentation using start-of-the-art techniques. In the recent past, anisotropy in friction response with anisotropy in wear behaviour of crystalline ionic solids was reported as a function of crystallographic orientation. The tribological behaviours in polymers, metals, and ionic solids vary completely. It seems clear that the mechanical failures, for example, corrosion, fatigue, adhesion, and abrasion, are almost equal. Eventually, *in situ* experimental techniques play an indispensable role, and they serve as a conduit between the basic atomic molecular structures and the tribological contact.

Several researchers have incorporated various kinds of nanofillers to increase the composite materials' properties, as the properties of such materials improve sufficiently when the matrix is under low-load condition. Amongst the three, one of the exterior dimensions in the nanometre range of 1, 103, or 109 nm is present in the nanofiller [7,8]. Polymer matrix nanocomposite (PMNC) materials are made up of various thermosetting and thermoplastic polymer matrices such as polystyrene, polyetherimide, epoxy, polyester, and others [9–12]. Polymer matrix (epoxy resin) is frequently employed in the manufacture of PMNC [13]. In general, nanofillers are classed as 1D (nanorods, nanofibres, nanowires, nanotubes),

2D (nanocoatings, nanofilms), and 3D (NPs) based on their dimensions. CNTs, carbon nanofibres (CNFs), and halloysite nanotubes (HNTs) are well-known types of 1D fibrous materials [10,11,14]. Furthermore, the 2D nanoclay has a plate-like structure with four groups, kaolin-serpentine, smectite, illite, and vermiculite with a high aspect ratio (30–1,000) [10,15]. 3D NPs include silica particles, silica oxide, carbon black (CB), titanium oxide, and fullerene [16,17]. In general, the polymer matrix and nanofiller are appropriately combined using a variety of processes, including *in situ* polymerization, *in situ* embedding kind of polymerization, polymer-particle direct mixing, melt intercalation, and the sol-gel process [11,18]. Various approaches, among them solution intercalation method, polymerization method, *in situ* intercalation, melt intercalation method, and direct *in situ* synthesis method [19], are prevalent for creating polymer/nanomaterial mixes.

Researchers classified dispersions into four types depending on the size of nanofiller: surface volume fraction, aspect ratio, mixing method of polymer, and degree of mixing [11,18,20]. These are phase separated or tactoid, intercalated disordered, and exfoliated. In reality, the dispersion speed determines how well nanocomposites withstand. A researcher [21] used Nomex fabric blend packed with nano-SiO_2 and Polyfluo 150 wax (PFW) to create a nanocomposite. Phenolic resin was first used to coat Nomex fibres. Tribological investigations show that the addition of PFW and NPs leads laminates to wear more quickly and have a lower coefficient of friction (COF). This kind of research has enhanced the tribological characteristics of nanocomposites at large, as they are capable enough to resist wear and minimize friction. In addition, the optimal concentration of nano-SiO_2 in Nomex fabric composites significantly elevated the tribology performance. A test was carried in which microscale and nanoscale indentation and scratch resistance of epoxy or silica nanocomposites coatings using the (Hysitron Tribo Indenter) system was performed and it was reported as to when silica particles reduce scratch depth and friction coefficient [22]. The tribological characteristics of the nanocoatings were looked at in another study on friction and wear. Epoxy resin and hydrophilic nano-silica particles were combined to create the coatings. The test findings show that the tribological characteristics were interestingly changed by the nano-silica particles. In addition, the filler-loaded coatings greatly outperformed the basal epoxy coating in terms of surface roughness and water contact angle properties [23].

The characteristics (tribological) of vinyl ester–based nanocomposites embedded with three separate, same sized micron or submicron particles were evaluated in the study [24]. The Taguchi design approach was used for the research's design of experiment (DOE). As seen below, the Taguchi design was produced using an L27 (36) orthogonal array with six components and three distinct levels. In order to assess the effects of factors on the friction coefficient and wear resistance (sliding) under dry sliding circumstances, analysis of variance (ANOVA) was utilized. The tribological test revealed that the submicron-sized particles (filler based) significantly shows the better performance by 21.18% and 11.58% more than that of micro-sized particles in terms of wear characteristics. The DOE confirms that among the other parameters that affected the COF by 68.33% and 9.81%, and wear resistance by 63.89% and 13.39%, the load and particle content had been the primary restrictions. In general, as the cenosphere's content increased, laminates' wear characteristics improved. Investigations at microscopic level resulted that the worn-out surfaces bears the homogeneous and strong adherence to the counter face, resulted in the improved properties of composite). The results also showed that the six weight % nanoparticle-filled composites exhibited substantial wear resistance. Hard abrasive wear is replaced by moderate abrasive wear as the primary wear in the mechanism of panels.

Nanocomposites created by compression moulding silica (SiO_2) particles of nanoscale size mixed with nylon-6 polymer were examined for their friction and wear characteristics. During the test, a steel flat pin was moving towards a composite disc in the pin-on-disk test. Friction can be decreased by 0.5–0.18 when 2% by weight of SiO_2 particles are added as compared to pure nylon 6. Due to the low silica concentration, the wear rate was 140 times lower. Over the steel pin, a transfer layer has formed due to the material's adherence to it and interlocking with metal asperities. The test also turned up two other oddities. The adhesion in between the transfer film and the surface (mating) is the prime factor that determines the wear rate. The second method is to cover the polymer surface with a transfer film to shield it from metal abrasion [25]. The abrasive and delamination were the two main contributors to wear processes, according to a wear resistance investigation on the influence of SiC particles in laminated polyimide/SiC laminates produced by the hot dynamic compaction method. In addition, when the quantity of SiC particles increases, the wear resistance of the materials containing these NPs declines noticeably [17]. Nanoscale materials are employed as particles (such as NPs), sheets (such as nanoclay), threads (such as carbon nanofibres), and tubes depending on their various forms (e.g. CNTs). Various mechanical properties, i.e., inter-laminar properties, strength like ILSS, tribological properties like the COF, fracture toughness, chemical resistance, and dielectric properties, were all improved by the incorporation of small nanofiller materials.

1.4.2 Lubricants

To diminish the friction coefficient and power loss in tribo-pairs or any bearing system, lubricants support to minimize wear in contacting surfaces. The incorporation of nanoparticle-based lubricating oil will improve a certain lubricating oil property, such as it boosts load-bearing capacity, improving friction, wear resistance, etc. In the past decade, scientists have been closely following the fast emerging field of nanoparticle lubrication technology. The production of NPs had a lot of limitations given the technologies available then, but at present, some synthetic methods have evolved to such an extent that the bulk quantities of these materials can be produced at very meagre costs and the efficiency of these manufacturing processes is paramount. Nanolubrication offers an ultimate solution to certain problems as linked with the lubricants containing phosphorus and Sulphur. When NPs are suspended in a lubricant, NPs have the ability to pass through tiny gaps between abrasive surfaces in contact and change the overall tribological performance between the contacts. Therefore, NPs provide a different method of lubricating by bringing third-body entities into the interface [26].

The nanoparticle features, such as shape, size, and concentration, determine the frictional reduction and anti-wear behaviours in general. In order to increase tribological characteristics, only a small amount of NPs (1%) is needed [27]. Each nanophase and lubricating oil has a different optimal concentration. In order to generate anti-wear surface protection coatings, a smaller size particle is more suitable for (asperity or peaks) interactions of the sliding material's surfaces [28]. Due to unusual structure and exceptional capabilities of monoatomic, graphene, two-dimensional (sp2) carbon atom lattice, has received an ample attention [29] in the recent past. In order to distribute graphene in non-polar lubricating oils, it is imperative to alter it using ionic liquid (IL) molecules. ILs are molecules that are typically generated by bulk organic cations and organic or inorganic anions. Molten salts (ILs) are in the form of liquids at room temperature [30]. ILs have a special set of characteristics. Some of the most important characteristics from a tribological perspective are their low inflammability and volatility as well as their great thermal stability.

The rolling action in between the mating or friction surfaces, using composite graphene/ILs nanostructures, development of a tribo-film with a significant load-carrying capacity, and the separation in between the mating surfaces are all responsible for increase in performance. Metal oxides and heavy metals are added as nanoadditives, and they serve as lubricant additives. Spherical metal oxide NPs are present in the lubricant, namely, polyalphaolefines, like Al_2O_3, CuO, ZnO, and TiO_2 [30]. This kind of addition also helps in yielding better performance. Tribo-sinterization processes have been criticized for the decreases in friction and wear. ILs with imidazolium and ammonium cations that stabilize CuO nanorods showed good friction and wear reduction (approximately up to 43%) when compared with engine oil [31]. The CuO nanorods' effective lubricating properties in ILs have been attributed to their excellent rolling action mechanism and dispersion stability. ZnO NPs are likewise excellent in reducing wear and friction [32], but their tribological properties were enhanced when an IL was used to change the surface composition and shape of ZnO NPs [33]. Although balanced formulations have yet to be created, nanomaterials offer the potential to improve several lubricant characteristics.

IF-WS2-based NPs are used in lubricants to achieve load capacity in hydrodynamic bearings. Hydrodynamic lubrication bearings have been used extensively in heavy industries (like hydropower plants) to support shafts or elements (rotating) due to their utmost durability at the cost of least maintenance and high load-carrying capacity (LCC). Fullerene-type tungsten disulphide NPs (IF-WS2) are the most known additives used for the purpose of lubrication because of their characteristics (mechanical). A research [34] carried out using numerical approach with discrete phase modelling of NPs (suspended) in order to enhance the load capacity of hydrodynamic journal bearings, wherein it is evident over here that the normal load on these bearing setups are typically very much high of the order of hundreds of tons. To get numerical simulation done precisely, nanofluid viscosity index was figured out using some basic empirical viscosity models, taking into account the effect of suspension (IF-WS2-based) NPs. Eventually, pressure distribution achieved was later on integrated throughout the oil film, in between the mating surfaces of the bearing, to compute the load capacity. From this particular numerical effort, around 20% enhancement in LCC has been noticed at around 5% weight fraction of the nanofluid considered.

Nanoparticle-based lubricants influence the performance of thin-film elasto-hydrodynamic lubrication (EHDL). It was clearly assessed that NPs-based lubricants diminish the coefficient of friction, when some basic friction tests and evaluations of a nanolubricant in the regime of thin-film EHDL were carried out [28]. On the basis of surface analyses techniques and molecular dynamics simulations, it was noticed that the effectiveness of various interactions in the lubricant (suspended particle based) and the associated tribo system, and the interactions in between the NPs with the lubricant or surfaces. On the basis of results, the friction reduction in the NPs induces an obstructed flow called plug flow in the oil film between the mating surfaces. Hence, plug flow reduces friction by forcing some of the layers of lubricant molecules to slip with one another.

1.4.3 Semisolid materials

Semisolid materials (with NPs) and grease lubricants immersed with nano-sized particles are the best examples of semisolid nanomaterials, which are being used in a variety of machines. Therefore, it is mandatory to introduce high-efficiency lubricants to increase machine performance and its lifespan and efficiency as well as quality of the product. It is a well-known fact that with the incorporation of NPs in semisolid lubricant greases and their additives, the lubricity of the nanolubricant becomes high, and hence, it benefits the overall lubrication performance scenario in machines. If surface-to-volume ratio of NPs seems high, properties and parameters like thermal conductivity, thermal stability, heat transfer coefficient, high mobility, efficient interaction between sliding surfaces, and also this kind of nano-based grease are advantageous in high-temperature and high-pressure operating conditions.

Incorporation of NPs has recently become a valid means for enhancing the tribological properties of lubricants and greases, and it guarantees the efficient functioning of the machinery. Certain laboratories have carried out some preliminary tests, and it was observed that such greases possess anti-wear and anti-friction qualities in addition to increased performance under the conditions of severe pressure and temperature [35]. Metal powders (Cu, Ni, and Fe) [36], Al_2O_3 NPs [37,38], SiO_2 NPs [39,40], and ZrO_2/SiO_2 nanocomposites [41] have all been extensively used in lubricating oils, as they operate on the surface of the tribo-pair, have greater specific surface area and high diffusion, are simple to sinter, and diminish the melting point. Lubricating grease incorporated with NPs has various benefits. In the recent researches [42], SiO_2 NPs distributed in paraffin grease resulted in friction and wear rate reductions of up to 20% and 42%, respectively.

In the past investigations [43,44], it was tested how addition of NPs to lithium grease influences the friction coefficient. To disperse the lithium grease, nanotitanium, silicon, and hybrid nano-oxides have been used and tested individually at five different concentration levels as reported [45]. It was found that when 6 wt.% hybrid NPs of TiO_2/SiO_2 were applied, as opposed to either 6 wt.% TiO_2 (27%) or 8 wt.% SiO_2 (46%) alone, the friction coefficient decreased by 50%. Recent research has reported [46] that the tribological characteristics of lithium grease with additions to the graphene layer were also examined, and it was demonstrated that the high specific surface area, greater quantity of active groups, and characteristic composed layered structure of lithium grease were primarily responsible for lowering the friction coefficient and wear caused by its use. Al_2O_3 nanorods were used as an addition in lithium grease at various concentrations, and the tribological and rheological behaviour of these nanorods was studied [47]. This study also presented the relationship between the rheological characteristics of nanorod-Al_2O_3 grease and its anti-wear and friction-reducing mechanism. Lithium grease with an Al_2O_3 nano-anti-friction additive and surface reconditioning characteristics was studied [48]. The tribological characteristics of different types of NPs used as lithium grease additives were investigated using an MRH-3 friction and wear tester; the results reported that the addition of Al_2O_3 NPs enhanced the anti-friction properties of the base grease's surface reconditioning ability [49].

With the help of four-ball and thrust-ring tester, the vibration and coefficient friction of modified Al_2O_3/SiO_2 (composite NPs) were investigated [50,51]. The use of Co, Fe, and Cu NPs reduced the COF and wear between tribo-pairs drastically in accordance with the tribological study of oils with nanoadditives [52]. In order to notice the vibration behaviour of ball bearings, CNTs have been doped as NPs to lithium grease. With higher CNT concentrations, the vibration amplitude got enhanced [53]. By using a four-ball tester, the tribological characteristics of lithium grease when mixed with CNTs were assessed. The findings indicated that the wear magnitude and friction coefficients have got reduced by around 63% and 81.5%, respectively [54].

Researchers tested the anti-wear capabilities of NPs suspended in polyalphaolefin (PAO$_6$) [55,56] using ZrO$_2$, CuO, ZnO, and NPs dispersed at 0.5 wt.%, 1.0 wt.%, and 2.0 wt.% in (PAO$_6$). The SAE-30 motor oil's lubricating properties were unaffected by the copper nanopowder addition, which has changed the topology of the worn-out surface [32]. Copper NPs with modified surfaces used as 50 CC oil additives have been tested for their wear and friction characteristics [57]. The function of fullerene NPs distributed in a mineral oil–based lubricant was studied using a disc-on-disc-type tribometer [58].

1.4.4 Coatings

The intended function determines the mass percentage of a nanomaterial in a coating system. For instance, the mass fraction of NPs used to increase the wear resistance of a lacquer is between 3% and 7%. Functional coatings based on nanotechnology frequently include the following nanomaterials, depending on the intended function: TIO$_2$, CB, silicon dioxide, iron oxide, zinc oxide, and silver. In actual practice, different particle sizes are constantly present. The usual range of the particle size distribution is at least one order of magnitude. The pigments' particle sizes before processing range from a few tens to a few thousand nanometres. Binders that may form tiny particles in water with the aid of a stabilizer have a diameter between 50 and 500 nm. The particles combine when the coating dries. The use of particles having a restricted particle size distribution (monodispersed) is recommended for applications where specific technical features, such as scratch resistance, are required. Higher opacity, greater surface-coating interaction, and increased coating endurance are all made possible by the introduction of nanomaterials. Some nanomaterials are appropriate for use in transparent coating systems because they include particles that are 100 nm or less in size. In addition, because these NPs are transparent in visible light (as TiO$_2$), it is possible to develop new additives that give previously opaque coatings new characteristics. Nano-based coatings frequently fit the concept of "nanocomposite" or "nanohybrid" materials, depending on their structure. A mixture of several materials is called a nanocomposite or nanohybrid. Its physical characteristics are different from those of each of its constituent parts. As a result, it is feasible to mix traits that are incompatible with one another, such as hardness and elasticity in place of hardness and brittleness. In comparison to conventional coatings, coatings incorporating nanomaterials have much improved material and processing characteristics (such as quick drying, higher resistance for indentation, high elasticity behaviour, least expansion once in contact with moisture, and high water vapour permeability). For instance, these characteristics are employed to create nanocomposite-based wood stains that cure more quickly and have more flexibility. TiO_2-based photocatalytic coatings provide antibacterial and self-cleaning qualities. These surfaces are appropriate, for instance, for the following applications since they repel water: flat glass, mirrors, automatic windows, window frames, bricks, wall paint, and tiles. It is anticipated that environmentally harmful halogenated flame retardants would be replaced with nanomaterial additions in coatings. Glass, wood, metal, plastic, and concrete surfaces that need to be protected are treated directly with nanocoatings. Within seconds after a fire, a ceramic layer is created. It offers heat insulation and significantly lowers the quantity of smoke produced. In debates on fire-retardant coatings, CNTs are becoming more prevalent. Other goods that require specialty coatings include those that must adhere to rigorous transmittance and antireflection specifications, such as glass coverings for solar water heaters and photovoltaic modules, or certain architectural and greenhouse applications. Antireflective coatings can boost the energy transmission by glass by 6%. Steel surfaces coated with nanomaterials are anticipated to provide better corrosion protection and replace the environmentally toxic chromate.

1.5 APPLICATIONS

Nanomaterials have multiple applications in the fields of industrial tribology and biomedicine.

1.5.1 Industrial applications

In the modern industries, anti-wear enhancement and friction reduction works in the field of tribology are at their peak. A remarkable development and speedy growth in tribological characteristics and performance of tribo materials has been observed in the recent past. In the recent past, one of the major developments is surface coatings, which has shown a massive change in terms of tribological behaviours for bulk materials. Ceramics, metals, and polymers are forms of bulk materials, and they have witnessed multiple tribological applications in the last few decades.

Similarly, in the field of coating, a huge volume of research work has been carried out to model in such a way that it should yield breakthrough tribological characteristics for bulk materials reinforced by carbon nanomaterials, including carbon nanotubes, graphene and nanodiamonds. Embodiment of carbon nanotubes is the best way to build up the tribological, mechanical, and electrical properties of the ceramic polymer and metal matrices. Scatter of CNTs in the polymeric materials, a drastic reduction of friction and wear characteristics is witnessed in CNTs-reinforced composites. Such a massive reduction in wear and friction was attributed to the doping mechanisms including the impediments of the cutting and ploughing during the contact formation.

In the same manner, graphene is another kind of additive material possessing lubrication behaviour, which is reinforced on bulk materials (polymer); a homogeneously distributed graphene in composite form enhances the anti-wear properties. Graphene is properly dispersed in polymeric matrix in order to develop graphene/polymer composites, which are being studied extensively since long. This particular type of doping leads to the decrease of COF and wear. Fluorinated graphene-based composites possess excellent properties (mechanical) and they are biocompatible, which is a suitable material for the purpose of joint replacement (artificially). It has been observed that composites of graphene increase resistance for wear and minimize the COF, because of the lubricating property of graphene and the formation of transfer film on the counterpart. Thermal conductivity of graphene can be helpful for heat transfer (frictional), and hence it reduces the decomposition of the phenolic resin.

Lubricant additives are the most common practice for contact mating surfaces, and scrolls, bearings, gears, and seals are lubricated with lubricants in order to keep friction and wear at an optimal level. Lubrication (hydrodynamic) theory and extensive researches have been carried out in the field for more than a century, and it is still continuously developing. Some basic parameters that control the friction and wear magnitudes are the lubricant viscosity indexes and possible additives. Thorough researches on the NPs and additives have shown utmost tribological characteristics as compared with solid lubricants and their associated additives. With regard to NPs in lubricants and in their additives, there are various mechanisms by which the reduction of COF and wear rate is achieved such as colloidal effect, protective film, size effect, and third-body effect.

Fullerene NPs in lubricating oils have shown a bigger volume of development in the tribological performances. Based on several investigations, the tribological behaviour of the lubricant (oil) with the embodiment of different volume percentages of fullerene-based NPs (0.01, 0.05, 0.1, and 0.5 volume %) using disc-on-disc-type tribo-meter set-up has been investigated. It was observed from the tests carried out on the tribometer that the variation of the quantity of fullerene NPs in the base oil has a pivotal role to maintain the COF and the level of wear between the mating surfaces.

The concept of superlubricity over the last decade and the level of research interest towards superlubricity have substantially shown an elevation across the world, because of chaotic energy crisis. Various liquid superlubricity systems that work under ambient conditions have been employed. So, it seems a unique and promising solution to cope up with the upcoming alarming energy demand and to reduce material losses. During the process of sliding, lubricants and their additives have progressively shown an increase in the LCC (up to 600 MPa); therefore, there is significant reduction in wear and acceleration of running in a certain period (1,000 seconds) of superlubricity system. Recently, a study of two-dimensional (2D) nanoadditives with ultrathin longitudinal dimensions has shown that it can minimize the shear resistance in between sliding solid surfaces; hence, it can be further optimized for the applied conditions [59].

1.5.2 Biomedical applications

Nanotechnology is one of the prominent scientific fields, which is emerging in almost all the key areas of research and enables mankind to visualize the environment at minute levels ranging from molecular to atomic sizes. It has changed and revolutionized all the scientific fields due to its vibrant and breakthrough features. Amongst all fields, medicine seems to be one of the prominent fields that has witnessed altogether a massive nanotechnology-based revolution that has embarked medical scholars and scientists towards the new paradigm, which has helped them to investigate the pathologies and search out some possible therapeutic stratagems by the incorporation of nanotechnology, which can operate specifically at molecular targets and intricacies, to minimize the risks/side effects that conventional methods had imposed on patients at large. One can easily understand the influence of nanotechnology on medicine on the basis of results of this technology, and it shall be counted in terms of its potential outcomes. It has drastically reduced the expenses related with human life and health services, as it produces the exact location of pathologic conditions; it also diminishes the seriousness of treatment.

In general, nanomaterials have a major influence on biomedical applications, due to intensive utilization in diagnostic and therapeutic applications and have a tendency to revolutionize the medicine industry. Use of nanomaterials in biomedical applications is ever since growing due to their ultrasmall size and some excellent mechanical properties, and super paramagnetic behaviour. Besides, these properties of nanomaterials are biocompatible and are manufactured quite easily. Nanosized devices can pass through the whole human body with quite ease and safely [60].

Thus, these devices serve as excellent means of detecting diseases and delivering targeted medicine. Nanomaterials can be used in targeted drug delivery, biosensors, bioimaging, hyperthermia, biomolecules [61], tissue engineering [62], etc. Owing to the excellent magnetic properties of nanomaterials, magnetic resonance imaging technique is highly improved. Nanomaterials with property like fluorescence can both detect and kill the harmful pathogens.

1.6 CONCLUSION

Certain important cases were discussed thoroughly, and it has given a clear picture as to how the sustainable or green tribology is helping the mankind in terms of energy saving or energy conservation. In particular, it has helped the world through health care advancements. If we see in actual practice, there have been diseases since the origin of mankind. However, because of the poor diagnosis system and the low level of technological infrastructure, people were dying cheap. With the advent of sustainable tribology, a new chapter of nanomaterials has come into prominence. This in combination with conventional tribology has resulted in a paradigm shift of research, from macro–micro–nanoscale, which in turn has helped medical science to diagnose and learn the problems with intricate details and seek a viable solution.

Also, from various studies' point of view, it has been noticed that the wide range of contributions in the field of tribology for sustainability are in line for progress towards a better future for sustainable civilization. The relationship and linkage of tribology and sustainability is an inextricable combination for the tribological targets meant to prolong lifecycle with least wear rate, energy savings, usage of such lubricants and additives, which will not deteriorate the environment at large, and the creation of such bearing materials which will withstand heavy pressures with minimal material waste. These targets and benefits shall really contribute towards the wider range of socioeconomic development without jeopardizing the environment.

REFERENCES

[1] M. Gaur, C. Misra, A. B. Yadav, S. Swaroop, F. Ó. Maolmhuaidh, M. Bechelany, and A. Barhoum, "Biomedical applications of carbon nanomaterials: Fullerenes, quantum dots, nanotubes, nanofibers, and graphene," *Materials*, vol. 14, no. 20, p. 5978, 2021.

[2] A. Zielińska, F. Carreifo, A. M. Oliveira, A. Neves, B. Pires, D. N. Venkatesh, A. Du-razzo, M. Lucarini, P. Eder, A. M. Silva, et al., "Polymeric nanoparticles: Production, characterization, toxicology and ecotoxicology," *Molecules*, vol. 25, no. 16, p. 3731, 2020.

[3] A. A. Yaqoob, H. Ahmad, T. Parveen, A. Ahmad, M. Oves, I. M. Ismail, H. A. Qari, K. Umar, and M. N. Mohamad Ibrahim, "Recent advances in metal decorated nanomaterials and their various biological applications: A review," *Frontiers in Chemistry*, vol. 8, p. 341, 2020.

[4] K. Umar, M. Haque, N. A. Mir, M. Muneer, and I. Farooqi, "Titanium dioxide-mediated photocatalysed mineralization of two selected organic pollutants in aqueous suspensions," *Journal of Advanced Oxidation Technologies*, vol. 16, no. 2, pp. 252–260, 2013.

[5] S. Zhang, "Current industrial activities of tribology in China plenary lecture to the China Int. in Symp. on Tribology (CIST 2008) (Beijing)," 2008.

[6] I. Tzanakis, M. Hadfield, B. Thomas, S. Noya, I. Henshaw, and S. Austen, "Future perspectives on sustainable tribology," *Renewable and Sustainable Energy Reviews*, vol. 16, no. 6, pp. 4126–4140, 2012.

[7] S. Komarneni, "Nanocomposites," *Journal of Materials Chemistry*, vol. 2, no. 12, pp. 1219–1230, 1992.

[8] E. P. Giannelis, "Polymer layered silicate nanocomposites," *Advanced Materials*, vol. 8, no. 1, pp. 29–35, 1996.

[9] S. Kumar, S. Raju, N. Mohana, P. Sampath, and L. Jayakumari, "Effects of nanomaterials on polymer composites—An expatiate view," *Reviews on Advanced Materials Science*, vol. 38, no. 1, pp. 40–54, 2014.

[10] R. Boujmal, C. A. Kakou, S. Nekhlaoui, H. Essabir, M.-O. Bensalah, D. Rodrigue, R. Bouhfid, and A. E. K. Qaiss, "Alfa fibers/clay hybrid composites based on polypropylene: Mechanical, thermal, and structural properties," *Journal of Thermoplastic Composite Materials*, vol. 31, no. 7, pp. 974–991, 2018.

[11] Y. Zhu, J. O. Iroh, R. Rajagopolan, A. Aykanat, and R. Vaia, "Optimizing the synthesis and thermal properties of conducting polymer–montmorillonite clay nanocomposites," *Energies*, vol. 15, no. 4, p. 1291, 2022.

[12] K. P. Rajan, A. Al-Ghamdi, S. P. Thomas, A. Gopanna, and M. Chavali, "Dielectric analysis of polypropylene (PP) and polylactic acid (PLA) blends reinforced with halloysite nanotubes," *Journal of Thermoplastic Composite Materials*, vol. 31, no. 8, pp. 1042–1053, 2018.

[13] D. Miracle, "Metal matrix composites—From science to technological significance," *Composites Science and Technology*, vol. 65, no. 15–16, pp. 2526–2540, 2005.

[14] S. Ghasemi, R. Behrooz, I. Ghasemi, R. S. Yassar, and F. Long, "Development of nanocellulose-reinforced PLA nanocomposite by using maleated PLA (PLA-g-MA)," *Journal of Thermoplastic Composite Materials*, vol. 31, no. 8, pp. 1090–1101, 2018.

[15] M. Alexandre and P. Dubois, "Polymer-layered silicate nanocomposites: Preparation, properties and uses of a new class of materials," *Materials Science and Engineering: R: Reports*, vol. 28, no. 1–2, pp. 1–63, 2000.

[16] D. Schmidt, D. Shah, and E. P. Giannelis, "New advances in polymer/layered silicate nanocomposites," *Current Opinion in Solid State and Materials Science*, vol. 6, no. 3, pp. 205–212, 2002.

[17] R. He, F. Niu, and Q. Chang, "Tribological properties of PI-SiC nanocomposite prepared by hot dynamic compaction," *Journal of Thermoplastic Composite Materials*, vol. 31, no. 8, pp. 1066–1077, 2018.

[18] K. Müller, E. Bugnicourt, M. Latorre, M. Jorda, Y. Echegoyen Sanz, J. M. Lagaron, O. Miesbauer, A. Bianchin, S. Hankin, U. Bölz, et al., "Review on the processing and properties of polymer nanocomposites and nanocoatings and their applications in the packaging, automotive and solar energy fields," *Nanomaterials*, vol. 7, no. 4, p. 74, 2017.

[19] L. Mrah and R. Meghabar, "In situ polymerization of styrene–clay nanocomposites and their properties," *Polymer Bulletin*, vol. 78, no. 7, pp. 3509–3526, 2021.

[20] C. I. Park, O. O. Park, J. G. Lim, and H. J. Kim, "The fabrication of syndiotactic polystyrene/organophilic clay nanocomposites and their properties," *Polymer*, vol. 42, no. 17, pp. 7465–7475, 2001.

[21] F.-H. Su, Z.-Z. Zhang, and W.-M. Liu, "Tribological and mechanical properties of Nomex fabric composites filled with polyfluo 150 wax and nano-SiO$_2$," *Composites Science and Technology*, vol. 67, no. 1, pp. 102–110, 2007.

[22] Z. Z. Wang, P. Gu, Z. Zhang, L. Gu, and Y. Z. Xu, "Mechanical and tribological behavior of epoxy/silica nanocomposites at the micro/nano scale," *Tribology Letters*, vol. 42, no. 2, pp. 185–191, 2011.
[23] Y. Kang, X. Chen, S. Song, L. Yu, and P. Zhang, "Friction and wear behavior of nanosilica-filled epoxy resin composite coatings," *Applied Surface Science*, vol. 258, no. 17, pp. 6384–6390, 2012.
[24] S. Thakur and S. Chauhan, "Friction and sliding wear characteristics study of submicron size cenosphere particles filled vinylester composites using Taguchi design of experimental technique," *Journal of Composite Materials*, vol. 48, no. 23, pp. 2831–2842, 2014.
[25] M. García, M. De Rooij, L. Winnubst, W. E. van Zyl, and H. Verweij, "Friction and wear studies on nylon-6/SiO$_2$ nanocomposites," *Journal of Applied Polymer Science*, vol. 92, no. 3, pp. 1855–1862, 2004.
[26] R. Greenberg, G. Halperin, I. Etsion, and R. Tenne, "The effect of WS2 nanoparticles on friction reduction in various lubrication regimes," *Tribology Letters*, vol. 17, no. 2, pp. 179–186, 2004.
[27] H. Ghaednia, R. L. Jackson, and J. M. Khodadadi, "Experimental analysis of stable cuo nanoparticle enhanced lubricants," *Journal of Experimental Nanoscience*, vol. 10, no. 1, pp. 1–18, 2015.
[28] H. Ghaednia, H. Babaei, R. L. Jackson, M. J. Bozack, and J. Khodadadi, "The effect of nanoparticles on thin film elasto-hydrodynamic lubrication," *Applied Physics Letters*, vol. 103, no. 26, p. 263111, 2013.
[29] I. Suarez-Martinez, N. Grobert, and C. Ewels, "Nomenclature of sp2 carbon nanoforms," *Carbon*, vol. 50, pp. 741–747, 2011.
[30] P. Walden, "Molecular weights and electrical conductivity of several fused salts," *Bulletin de l'Académie impériale des sciences de (St. Petersburg)*, vol. 1800, pp. 405–422, 1914.
[31] R. Gusain and O. P. Khatri, "Ultrasound assisted shape regulation of CuO nanorods in ionic liquids and their use as energy efficient lubricant additives," *Journal of Materials Chemistry A*, vol. 1, no. 18, pp. 5612–5619, 2013.
[32] A. H. Battez, J. F. Rico, A. N. Arias, J. V. Rodriguez, R. C. Rodriguez, and J. D. Fernandez, "The tribological behaviour of ZnO nanoparticles as an additive to PAO6," *Wear*, vol. 261, no. 3–4, pp. 256–263, 2006.
[33] J. Sanes, F.-J. Carrión, and M.-D. Bermúdez, "Zno–ionic liquid nanostructures," *Applied Surface Science*, vol. 255, no. 9, pp. 4859–4862, 2009.
[34] H. Sadabadi and A. S. Nezhad, "Application of IF-WS2 nanoparticles in lubricants to enhance load carrying capacity of plain bearings," *AIP Conference Proceedings*, vol. 2380, 2021, p. 050002.
[35] E. Ahmed, A. Nabhan, N. M. Ghazaly, and G. Abd El Jaber, "Tribological behavior of adding nano oxides materials to lithium grease: A review," *American Journal of Nano-Materials*, vol. 8, no. 1, pp. 1–9, 2020.
[36] M. Astakhov, V. Moskovtsev, and V. Muratov, "Ultrafine powders and properties of lubricating oil," *Automobile Promst*, vol. 2, pp. 4–23, 1994.
[37] S. Radice and S. Mischler, "Effect of electrochemical and mechanical parameters on the lubrication behaviour of Al$_2$O$_3$ nanoparticles in aqueous suspensions," *Wear*, vol. 261, no. 9, pp. 1032–1041, 2006.
[38] G. Shi, M. Q. Zhang, M. Z. Rong, B. Wetzel, and K. Friedrich, "Sliding wear behavior of epoxy containing nano-Al$_2$O$_3$ particles with different pretreatments," *Wear*, vol. 256, no. 11–12, pp. 1072–1081, 2004.

[39] X. Li, Z. Cao, Z. Zhang, and H. Dang, "Surface-modification in situ of nano-SiO$_2$ and its structure and tribological properties," *Applied Surface Science*, vol. 252, no. 22, pp. 7856–7861, 2006.

[40] D. Peng, Y. Kang, R. Hwang, S. Shyr, and Y. Chang, "Tribological properties of diamond and SiO$_2$ nanoparticles added in paraffin," *Tribology International*, vol. 42, no. 6, pp. 911–917, 2009.

[41] S. Y. Ma, S. H. Zheng, H. Y. Ding, and W. Li, "Anti-wear and reduce-friction ability of ZrO$_2$/SiO$_2$ self-lubricating composites," *Advanced Materials Research*, vol. 79, pp. 1863–1866, 2009.

[42] S. S. Rawat, A. Harsha, and A. P. Deepak, "Tribological performance of paraffin grease with silica nanoparticles as an additive," *Applied Nanoscience*, vol. 9, no. 3, pp. 305–315, 2019.

[43] E. Ahmed, A. Nabhan, M. Nouby, and G. Abd El Jaber, "Influence of adding contaminants particles to lithium grease on the frictional coefficient," *Journal of the Egyptian Society of Tribology (EGTRIB Journal)*, vol. 14, no. 1, pp. 15–29, 2017.

[44] A. Samy and W. Ali, "Developing the tribological properties of lithium greases to withstand abrasion of machine elements in dusty environment," *International Journal of Scientific & Engineering Research*, vol. 4, no. 10, pp. 1176–1181, 2013.

[45] E. Ahmed, A. Nabhan, M. Nouby, and G. Abd El Jaber, "Influence of dispersing lithium grease by hybrid nano titanium and silicon oxides on friction coefficient," *Journal of the Egyptian Society of Tribology*, vol. 15, pp. 61–71, 2018.

[46] Z. Li, Q. He, S. Du, and Y. Zhang, "Effect of few layer graphene additive on the tribological properties of lithium grease," *Lubrication Science*, vol. 32, no. 7, pp. 333–343, 2020.

[47] H. Qiang, T. Wang, H. Qu, Y. Zhang, A. Li, and L. Kong, "Tribological and rheological properties of nanorods–Al$_2$O$_3$ as additives in grease," *Proceedings of the Institution of Mechanical Engineers, Part J: Journal of Engineering Tribology*, vol. 233, no. 4, pp. 605–614, 2019.

[48] L. Baoliang, L. Zhigang, and L. Gaozhi, "Experimental study on tribological performance of lithium grease containing various nano-particles as reconditioned additives," *Lubrication Engineering Huangpu*, vol. 2, no. 186, p. 150, 2007.

[49] B. Li, Z. Li, G. Luo, and Q. Jiang, "Experimental study of tribological properties of various nanoparticles as additive in lubrication of lithium grease," *Journal of Dalian Railway Institute*, vol. 3, no. 9, 2006.

[50] D. Jiao, S. Zheng, Y. Wang, R. Guan, and B. Cao, "The tribology properties of alumina/silica composite nanoparticles as lubricant additives," *Applied Surface Science*, vol. 257, no. 13, pp. 5720–5725, 2011.

[51] T. Luo, X. Wei, X. Huang, L. Huang, and F. Yang, "Tribological properties of Al$_2$O$_3$ nanoparticles as lubricating oil additives," *Ceramics International*, vol. 40, no. 5, pp. 7143–7149, 2014.

[52] J. Padgurskas, R. Rukuiza, I. Prosyčevas, and R. Kreivaitis, "Tribological properties of lubricant additives of Fe, Cu and Co nanoparticles," *Tribology International*, vol. 60, pp. 224–232, 2013.

[53] A. Nabhan, "Vibration analysis of adding contaminants particles and carbon nanotubes to lithium grease of ball bearing," *Vibroengineering Procedia*, vol. 8, pp. 28–32, 2016.

[54] A. Mohamed, T. Osman, A. Khattab, and M. Zaki, "Tribological behavior of carbon nanotubes as an additive on lithium grease," *Journal of Tribology*, vol. 137, no. 1, p. 011801, 2015.

[55] J. Dong and Z. Hu, "A study of the anti-wear and friction-reducing properties of the lubricant additive, nanometer zinc borate," *Tribology International*, vol. 31, no. 5, pp. 219–223, 1998.

[56] A. H. Battez, R. González, J. Viesca, J. Fernández, J. D. Fernández, A. Machado, R. Chou, and J. Riba, "Cuo, ZrO_2 and ZnO nanoparticles as anti-wear additive in oil lubricants," *Wear*, vol. 265, no. 3–4, pp. 422–428, 2008.

[57] S. Tarasov, A. Kolubaev, S. Belyaev, M. Lerner, and F. Tepper, "Study of friction reduction by nanocopper additives to motor oil," *Wear*, vol. 252, no. 1–2, pp. 63–69, 2002.

[58] K. Lee, Y. Hwang, S. Cheong, Y. Choi, L. Kwon, J. Lee, and S. H. Kim, "Understanding the role of nanoparticles in nano-oil lubrication," *Tribology Letters*, vol. 35, no. 2, pp. 127–131, 2009.

[59] H. Wang and Y. Liu, "Superlubricity achieved with two-dimensional nano-additives to liquid lubricants," *Friction*, vol. 8, no. 6, pp. 1007–1024, 2020.

[60] C. R. Patra, R. Bhattacharya, D. Mukhopadhyay, and P. Mukherjee, "Application of gold nanoparticles for targeted therapy in cancer," *Journal of Biomedical Nanotechnology*, vol. 4, no. 2, pp. 99–132, 2008.

[61] T. Liu, J. Tang, and L. Jiang, "The enhancement effect of gold nanoparticles as a surface modifier on DNA sensor sensitivity," *Biochemical and Biophysical Research Communications*, vol. 313, no. 1, pp. 3–7, 2004.

[62] B. S. Harrison and A. Atala, "Carbon nanotube applications for tissue engineering," *Biomaterials*, vol. 28, no. 2, pp. 344–353, 2007.

Chapter 2

Nanocomposites and tribology

Overview, sustainability aspects, and challenges

Abrar Malik and Nitish Singh Jammoria
Shri Mata Vaishno Devi University

Asrar Rafiq Bhat and Prateek Saxena
Indian Institute of Technology, Mandi

Mir Irfan Ul Haq and Ankush Raina
Shri Mata Vaishno Devi University

CONTENTS

2.1	Introduction	26
2.2	Fabrication methods	28
	2.2.1 Stir-casting	28
	2.2.2 Disintegrated melt deposition (DMD)	29
	2.2.3 Semi-solid casting	29
	2.2.4 Powder metallurgy	30
	2.2.5 Friction stir processing	31
	2.2.6 Accumulative roll bonding	31
2.3	Role of nano-reinforcements in metallic materials	31
2.4	Sustainability aspects of nanomaterials in metal matrix composites (MMCs)	32
2.5	Nanoadditives as self-lubricating materials	34
	2.5.1 Aluminium alloys	35
	2.5.2 Copper composites	35
	2.5.3 Magnesium alloys	35
	2.5.4 Nickel-graphite composites	36
2.6	Nanomaterials used in different composites	36
	2.6.1 Studies related to carbon-related materials	38
	2.6.2 Studies related to other nanomaterials	39
	2.6.3 Studies related to hybrid nanomaterials	40
2.7	Challenges and future scope	41
2.8	Conclusion	44
References		45

DOI: 10.1201/9781003306276-2

2.1 INTRODUCTION

Nanoparticle-reinforced metal matrix composites (MMCs), also known as metal matrix nanocomposites (MMnCs), have attracted a lot of research attention in the past decade due to the various advantages offered by them in various structural and functional applications (Singh et al., 2018). When combined with additional strengthening effects generally present in traditional MMCs, the nanoscale reduction in the reinforcement phase's size makes the interaction of molecules with dislocations more important. This results in a significant improvement in the mechanical characteristics (Sanaty-Zadeh, 2012).

The major challenge in manufacturing MMnCs is the poor wettability of metal melt matrix with ceramic nanoparticles, which prevents the use of traditional casting methods. Small powder aggregates are actually more likely to group together and lose their ability to be evenly distributed in the matrix for the best use of the strengthening capabilities. As a result, various alternate approaches have been suggested as a solution to this issue (Casati & Vedani, 2014).

There are two main kinds of production methods: *ex situ* and *in situ*. *Ex situ* method involves the addition of nano-reinforcements to powdered or liquid metal, whereas in the *in situ* method, ceramic nanocompounds are formed by reactions while processing, generally by employing reactive gases (Singh et al., 2020). For *ex situ* production of MMnCs, different methods have been developed, which involved several powder metallurgy techniques. Ultrasound-assisted casting has yielded promising results in terms of high productivity. The techniques used to characterize MMnCs are those which are used to characterize traditional MMCs and alloys. Naturally, the reduction in reinforcement calls for the adoption of greater resolution techniques to characterize the local chemistry and morphology of the components.

The studies carried out in this regard include various types of metals as matrix in combination with different nanomaterials. Al, Mg, Cu, and other metals and alloys were reinforced using ceramic compounds (SiC, Al_2O_3, etc.; Jammoria et al., 2022a), intermetallic materials, and carbon allotropes (Jammoria et al., 2022b). Carbon nanotubes (CNT) have a special importance owing to their remarkable strength, electrical conductivity, and stiffness. These characteristics enhance the parent material's mechanical strength while also enhancing its electrical and thermal qualities (Bakshi et al., 2010). Furthermore, MMnCs have proved to enhance other important properties such as damping capacity (Deng et al., 2007), creep behaviour (Shehata et al., 2009), and wear resistance (Ferkel & Mordike, 2001). Tensile and compressive properties, ductility, high-temperature mechanical properties, dynamic mechanical properties, ignition temperature, and corrosion resistance have also improved by the use of nano-reinforcements (Malaki et al., 2019; Farooq et al., 2022). The main properties that are enhanced by the use of nano-reinforcements are shown in Figure 2.1.

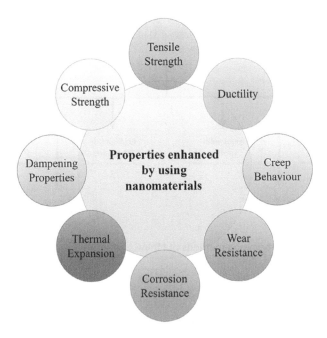

Figure 2.1 Properties enhanced by using nano-reinforcements.

The enhancement in properties in MMCs is obtained from a combination of different factors. The three main factors include careful selection of matrix and reinforcement, proper choice of primary and secondary processing methods, and heat treatment. However, in MMnCs, another factor important for enhancement in properties is minimizing the particle-based damage methods. These are primarily found in micro-scale ceramic reinforcements. They include particle breakage under stress, which leads to crack initiation and debounding at matrix-reinforcement interface under stress.

These particle-based damage mechanisms degrade the mechanical properties of MMCs. In particular, fracture toughness and the elongation till failure of MMCs are compromised by particle-based damage mechanisms, which are undesirable for engineering and biomedical applications. Also, nano-MMCs have several advantages in comparison to micro-MMCs owing to the better combination of properties. Apart from the type of matrix and reinforcement selected, the processing method and parameters, which may give a homogenous distribution of reinforcement and a suitable interface at reinforcement-matrix bounding, play a critical role in the improvement of characteristics of MMCs (Ul Haq & Anand, 2018). It is also difficult to combine reinforcement in molten metal in large production processes like casting without subjecting the reinforcement to excessive temperature exposure.

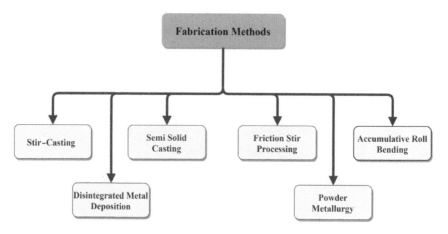

Figure 2.2 Different fabrication methods of MMnCs.

2.2 FABRICATION METHODS

The major problem in the synthesis of MMnCs is the formation of clusters that results in non-uniform distribution of nanoparticles in the matrix. As a result, a number of methods have been used for the fabrication. Figure 2.2 shows the different fabrication methods of MMnCs, and these methods are discussed briefly below.

2.2.1 Stir-casting

Stir-casting is one of the more popular methods of fabricating MMnCs. In this method, MMnCs are made using the widespread, affordable, and relatively easy stir-casting technique, in which the reinforcement particles are mixed with the molten matrix metal (in this case, aluminium). An impeller is used to provide stirring, which gives an even mix of the particles in the molten metal. This method has successfully been employed for ceramic, graphene, and various metal oxide powders in aluminium and magnesium matrices (Hamedan & Shahmiri, 2012).

Despite its common use, there are certain challenges, which are encountered while using this method. One of the main challenges is the formation of clusters of nanoparticles because of their large surface area. Less wettability between nanoparticles and metal melt also presents a challenge. Another challenge faced in this method is the entrapment of air generated by rotating stirrer, which increases the porosity in the matrix (Akbari et al., 2013). These challenges are presented in Figure 2.3.

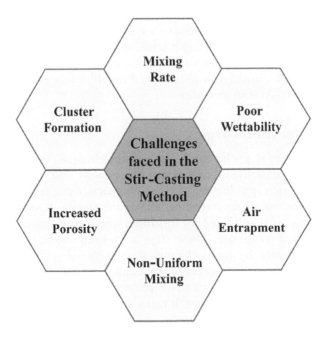

Figure 2.3 Challenges faced in stir-casting.

2.2.2 Disintegrated melt deposition (DMD)

This method has been derived from stir-casting and is particularly used for magnesium-based nanocomposites in which stir-casting principles are followed to produce a composite slurry (Ceschini et al., 2017). An inert gas jet (Argon) is used to pass the developed slurry through a pouring nozzle at superheated temperature (750°C) after which the slurry is put on a metallic substrate. The DMD method yields an ingot as its final product, which can subsequently be hot extruded to produce the required shape. Magnesium nanocomposites are mainly derived using this process, as they eliminate all the main shortcomings of the stir-casting method such as high density of oxides in nanocomposite, and holding back of reinforcement such as Al_2O_3, SiC, Y_2O_3, B_4C, BN, ZrO_2, ZnO, and CNT due to difference in density between Mg alloys and particles (Ceschini et al., 2017).

2.2.3 Semi-solid casting

Semi-solid casting (SSC) is a slightly elevated die casting method in which a semi-liquid/solid metal is fed into a mould cavity having "slushy" appearance. This method provides various advantages over high-pressure die casting, including easier handling and lesser post-moulding machining.

However, the globular microstructure (consisting of solid matter suspended in a liquid phase), which is necessary for SCC, is challenging to produce. De Cicco (2009) observed that cast semi-solid zinc alloy (AC43A) having 0.5 wt. percentage of SiC with size 20–30 nm nanoparticles had enhanced ductility, lowered shrinkage, and improved mould filling in comparison to liquid monolithic AC43A zinc alloy.

2.2.4 Powder metallurgy

Powder metallurgy is a solid-form fabrication technique and involves three basic things sequentially: mixing of metallic powder and reinforcement in proper composition, production of green compacts using compaction, and sintering by the use of resistance or microwave energy. Post sintering, an additional process is employed for finishing, which confirms consolidation of the fabricated composite. These steps have been shown in Figure 2.4. This is usually done by equal channel angular pressing (ECAP). Powder metallurgy gives various advantages to other manufacturing methods. It can be used to produce near-net shape parts and has the ability to process more volume of the reinforcement. It can also be used to manufacture large batches of different automotive parts.

A subset of powder metallurgy approach for making MMnCs uses mechanical alloying, where powder grains are continuously cold weld, fractured, and re-weld using a high-energy ball milling device (Suryanarayana & Al-Aqeeli, 2013). This technique is most useful in the case of ceramic reinforcement in nanocomposites due to its ability to successfully split ceramic clusters and thus promotes even distribution of ceramic reinforcement in the matrix (Ye et al., 2006). The ball-to-powder ratio, milling time, and rotating speed are the primary processing variables in mechanical alloying process. A process control agent (stearic acid, methanol, etc.) is occasionally used during the ball milling process to avoid the creation of big clusters of powder. After the ball milling process is completed, the remaining steps are similar to those in the Powder Metallurgy (PM), which include green compact and sintering.

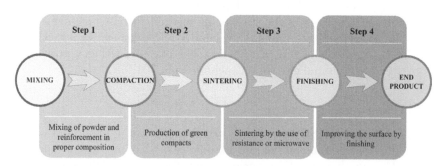

Figure 2.4 Steps followed in powder metallurgy.

Reaction milling is used to generate a thin layer of carbides, nitrides, and oxides in a light (non-ferrous) alloy matrix. However, in this case, the process control agent isn't provided to start the reaction.

2.2.5 Friction stir processing

Friction stir processing (FSP), a version of friction stir welding, is employed by various industries and enterprises to integrate nanoparticles into a metal matrix to create surface nanocomposites. The base metallic substance first generates a groove, which is subsequently packed with the necessary volume proportion of nanoparticles. The groove is sealed off, encasing the particles therein, by employing a non-consumable tool that is placed on the surface of the work piece. The particles are then stirred with the majority of the metallic substrate in the following step using a pin attached to a tool. The surface quality of the fabricated part and homogenous nanoparticle dispersion inside the matrix are this method's key challenges. This technique is currently in the development stage and has been utilized to create nanocomposites with aluminium and magnesium as the main components (Bauri et al., 2011).

2.2.6 Accumulative roll bonding

Accumulative roll bonding was introduced by Saito et al. in 1998, and it works upon extreme plastic deformations produced by rolls, which rotate on metallic sheets. It consists of the following four steps: removal of contaminants, oxide layers on sheet layers by using a wire brush, stacking sheets one over the other, and roll bonding the sheets collectively to at least half reduction in thickness and cutting the sheets into pieces. This process is repeated for every set of samples to attain the required number of pieces. This method has effectively been used in the production of several metal matrix nanocomposites including Al/SiC (Amirkhanlou et al., 2011), Al/Al$_2$O$_3$ (Jamaati & Toroghinejad, 2010), Al/B$_4$C (Yazdani & Salahinejad, 2011), Al/CNT (Karimi et al., 2016), Al/SiO$_2$ (Soltani et al., 2012), Al/TiO$_2$ (Hashemi et al., 2012), Al/W (Amirkhanlou et al., 2013), Al/WC (Liu et al., 2012), and Mg/CNT (Lv et al., 2017).

2.3 ROLE OF NANO-REINFORCEMENTS IN METALLIC MATERIALS

The main objective of adding nano-reinforcements/particles to metallic materials is to achieve enhanced material properties. The key objectives are:

1. To increase the yield strength and tensile strength at room temperature as well as at higher temperatures without compromising on ductility of the material

2. To increase creep resistance at higher temperatures
3. To increase fatigue strength
4. To increase Young's modulus
5. To improve thermal shock resistance
6. To decrease coefficient of thermal expansion
7. To increase wear resistance

In order to avoid the particles from dissolving in the melt while also maintaining adequate adherence to the matrix, chemical compatibility is crucial at the onset. As is generally the case with ceramic particles, the melting temperature of the particles should be considerably greater than that of the matrix.

2.4 SUSTAINABILITY ASPECTS OF NANOMATERIALS IN METAL MATRIX COMPOSITES (MMCs)

Every human endeavour is primarily driven by energy. These days, energy plays a significant role primarily in three facets of human development, namely, social, environmental, and economic. Energy services enhance living standards and environmental quality while serving as a vital input to economic activity. However, excessive energy consumption can raise carbon footprints and greenhouse gas emissions, and on the other hand, inefficient energy resource management can destroy the existing ecosystems (Malik et al., 2022). As a result, there are many intricate connections between energy use and human growth (Tzanakis et al., 2012).

It is the responsibility of the designer to make an effort to minimize the gap between demand for energy and its negative impacts by creating technologies that can meet societal needs while significantly lowering our reliance on fossil fuels. Reduced efficiency from friction and wear lowers the performance of these machines and is a key component to achieve sustainability. Thus, it is essential to have a good understanding of the engineering science behind the operation of the key mechanical components in order to address the issues related to friction and wear. Today, tribology is a broad, multidisciplinary area of study and research is being carried out at larger scale (Gupta et al., 2022; Kichloo et al., 2022; Rouf et al., 2021; Shafi et al., 2018). Therefore, research in numerous nations have measured and outlined the advantages of tribology, particularly sustainable tribology (Raina et al., 2021). The Chinese study may be regarded as the most notable of these investigations (Jost & Schofield, 1981). Si-Wei Zhang described in a paper from the Chinese Academy of Engineering that asserted that the UK GNP (Gross National Product) would decrease by 1.1%–1.5% if tribology was extensively utilized in the United Kingdom (Sw, 2009). In addition, the significance of tribology on a global scale is demonstrated after a careful analysis of the findings of a 2-year investigation in China that involved six industries and conservatively estimated savings of $414 billion annually, at 2006 rates, equivalent to 1.55% GDP savings in China (gross domestic product; Zhang & Xie, 2009; Haq et al., 2021a; Slathia et al., 2020).

Agro-industrial waste is being used increasingly frequently in MMCs because it may be used to strengthen metal matrix particles, improving the strength qualities of the composites. Utilizing waste materials could cut down the pollution and disposal space. Therefore, one of the main concerns of today's researchers is recycling waste material for use in the building and automotive industries. Some examples of waste products, which can be used in the building and automotive industries are red mud, fly ash (FA), palm oil fuel ash (POFA), coconut husks, palm oil clinker (POC), and sugarcane bagasse. Despite the extensive research that has been carried out, partial reinforcing of composites with waste materials is still being developed. To address the needs of people in both rural and urban areas, there is a strong market potential for alternative materials made from solid waste that are energy and environmentally friendly and economical (Md & Mohd, 2010).

Various agro-industrial wastes have been used as additives (Figure 2.5). One study revealed that despite having little to no commercial value, coconut shells can be used to make carbon black because of their superior natural structure and low ash content. Coconut shell ash has a density of 2.05 g/cm³ and can tolerate temperatures up to 1,500°C, making it

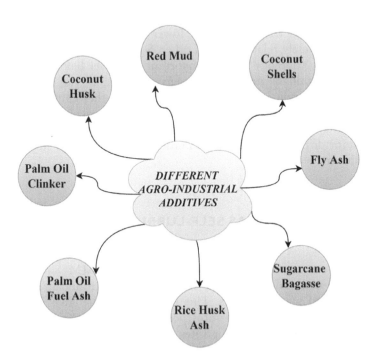

Figure 2.5 Different agro-industrial wastes as additives.

appropriate for use in the development of lightweight metal matrix composite parts with strong thermal resistance (Madakson et al., 2012). FA is a type of particle reinforcement that is used to make composite metals out of potentially discontinuous dispersoids. FA is relatively inexpensive and also less dense reinforcement. It can be found in high amounts in thermal power plants as a waste by-product. Its use in MMCs is therefore very common (Rajan et al., 2007). FA's main chemical components are Al_2O_3, SiO_2, Fe_2O_3, and CaO (Tavman, 1996). Bagasse, a fibrous by-product left over from crushing and extracting the juice from sugarcane, is one of the major agricultural waste products in the world. The primary components of Sugarcane Bagasse (SCB), according to one investigation, are cellulose, hemicellulose, lignin, ash, and wax (Walford, 2008). For developing the novel materials with distinctive physical and chemical properties, SCB's composition makes it an excellent component to be used as reinforcing fibre in composite materials. An agricultural waste and by-product of rice husk is called rice husk ash (RHA). It is a significant seasonal environmental and health problem and was expected to create 21,000,000 tonnes globally in 2002. Among the family of other agricultural wastes, RHA is one of the most silica-rich raw materials, holding between 90% and 98% silica after combustion. In Malaysia, POFA, which is a common agricultural solid waste, is very siliceous. The most important component of POFA, which can comprise up to 40% of it, is silica, also known as silicon dioxide (SiO_2; Zainudin et al., 2005).

The reduction in frictional losses, lower wear, and enhanced sustainability are intimately related to each other. So, it can be easily understood that sustainable tribology is more important than ever before, in order to promote the sustainable growth and to ensure the stability of our planet. In this aspect, agro-industrial wastes are beneficial as they cause lower environmental contamination. With the availability of agricultural waste and their potential to be used in the composites, an attempt can be made to use them for different applications in order to put forward a step towards sustainable tribology.

2.5 NANOADDITIVES AS SELF-LUBRICATING MATERIALS

Due to excellent tribological qualities, self-lubricating MMCs are a significant class of materials, which are steadily gaining ground in aerospace, automotive, and marine sectors. This will improve the environment and will decrease the usage of harmful petroleum-based lubricants. In Self Lubricating Metal Matrix composites (SLMMCs), solid lubricant components such as hexagonal boron nitride (h-BN), molybdenum disulphide (MoS_2), and carbonous materials are used in various MMCs as reinforcements to create a unique composite with enticing self-lubricating capabilities. Solid lubricant compounds are drawing researchers to synthesis because of their lubricious characteristics (Omrani et al., 2016; Gupta et al., 2018).

2.5.1 Aluminium alloys

Aluminium alloys that are lightweight and have high strength are extremely desirable to designers due to the rising demand for efficient materials for usage in low-maintenance, energy-efficient engineering systems (Slathia et al., 2018). Due to their superior qualities, aluminium alloys and aluminium matrix composites (AMC) are increasingly used in a variety of industries (Skeldon et al., 1997; Rawal, 2001). These qualities include excellent corrosion resistance, high specific strength, and high damping capacity. In general, compared to aluminium composite reinforced with other ceramic particles, including Al_2O_3 and SiC, graphite-reinforced aluminium matrix has better tribological properties. When compared to unreinforced matrix alloys, the wear rate as well as friction are dramatically decreased by the use of solid lubricants as reinforcements due to the inclusion of graphite particles (Rohatgi et al., 1992). Ames and Alpas (1995) examined how graphite affected the composite materials' wear regime. According to the findings, graphite-reinforced composites remained in the light wear zone even under high loads and didn't show any signs of extreme wear, whereas its alloy showed signs of severe wear under high loads.

2.5.2 Copper composites

A low thermal expansion coefficient and excellent thermal and electrical conductivities are benefits of copper composites enhanced by solid lubricants because they can keep the attributes of copper matrix materials (Omrani et al., 2016). Moustafa et al. (2002) looked into how varied typical loads (50–500 N) and graphite content (8, 15, and 20 wt. %) affected the tribological behaviour of composites made using powder metallurgy. Either the graphite powder with copper coating or a combination of graphite-copper powders was employed. The copper/graphite composite demonstrated less wear rate as compared to sintered copper compacts of up to 200 N normal loads. This is due to the formation of a graphite layer on the surface, which acted as a solid lubricant.

2.5.3 Magnesium alloys

Magnesium alloys, which offer great features like high specific strength, low density and stiffness, excellent machinability, acceptable damping characteristics, and castability, are frequently used in a variety of industries, including the aerospace and automotive sectors. However, their application is not as widespread as that of aluminium alloys due to their low corrosion resistance and wear resistance (Chen & Alpas, 2000). Qi (2006) looked into how the friction and wear characteristics of the AZ91 magnesium alloy matrix composite were affected by the implantation of graphite particles into the magnesium matrix. In terms of wear rate and COF variation, under comparable testing conditions, the wear rate and COF of the composite improved over that of magnesium alloy. In addition, each composite's wear mass loss tends to be decreased with the addition of more graphite. Mindivan et al. (2014) looked into how magnesium self-lubricating materials' tribological characteristics were affected by nano-solid lubricants. The coefficient of friction and wear rate were improved when carbon nanotubes (CNTs) at 0.5 wt.% were inserted into the magnesium matrix.

2.5.4 Nickel-graphite composites

Nickel-graphite composites have excellent high-temperature performance, which make them suitable for use in high-efficiency engines (Etaati et al., 2010). There have been efforts to investigate how testing conditions and material choices affect the tribological behaviour of nickel-graphite composites. Chen et al. (2006) and Scharf et al. (2009) looked into how CNTs affected the tribological behaviour of nickel composites that self-lubricate. The outcomes demonstrated that inserting CNT nanoparticles decreased COF. In addition, they found that incorporating CNTs into the nickel matrix is much more effective than integrating graphite micro-particles into the nickel matrix because of good mechanical properties of CNTs.

Moreover, the demand for environment-friendly materials has increased vastly because of worries about environmental contamination in recent years. Liquid or grease-based lubricants, which are used to reduce friction and wear, might harm the environment. As a result, solid lubricants including boric acid, hexagonal boron nitride, MoS_2, and allotropes of carbon are effective alternatives to traditional lubricants. Self-lubricating composites thus provide a number of benefits, and further research must go on towards the development of these novel materials (Table 2.1).

2.6 NANOMATERIALS USED IN DIFFERENT COMPOSITES

A number of nanomaterials have been used in different matrix composites with the aim to achieve improved material properties. In this regard, carbon nanotubes and multi-wall CNTs have been well explored by researchers. Furthermore, the studies on other nanomaterials and hybrid reinforced alloys have been discussed in this section.

Nanocomposites and tribology – Overview 37

Table 2.1 Effect of micro/nanofiller on tribological properties of different metal matrix composites (MMCs)

Metal Matrix	Micro/nanofillers	Preparation technique	Testing conditions	Influence on wear	Influence on friction	Ref.
Aluminium	Graphene (3% by weight)	Powder metallurgy	POD, AL:50N	−50% compared to composite with 0% graphene	–	Ghazaly et al. (2016)
Copper	Nano-graphite NG (15% by weight)	Powder metallurgy	POD, EN30 steel, SS:0.77 m/s, AL:36N	−50% compared to composite with 5% NG	−40% compared to composite with 5% NG	Rajkumar and Aravindan (2013)
Aluminium	Multiwalled carbon nanotubes (MWCNT) 5% by weight	Ball milling, cold compaction, hot extrusion	POD, EN31 Steel, SS:1.1 m/s, AL:20N	−78.8% compared to pure Aluminium	Significant decrease in COF	Bastwros et al. (2013)
Chromium	MWCNT	Electrodeposition, ultrasonication	AL:20N	−38% compared to simple Chromium	–	Zeng and Lin (2009)
Aluminium (AA 6061)	SiC, graphite nanoparticles (GNP)	Friction stir process	–	−90% compared to base material	Reduction in COF	Moustafa et al. (2021)
Aluminium (AA 6061)	SiC, Al$_2$O$_3$	Friction stir process	–	−70% compared to base material	Reduction in COF	Moustafa et al. (2021)
Aluminium (AA 6061)	SiC	Friction stir process	–	−60% compared to base material	Reduction in COF	Moustafa et al. (2021)
Aluminium (Al7010)	B$_4$C and BN	Ultrasonic assisted stir-casting	POD, EN31 Steel, AL:20N, SS:1 m/s	−65% reduction compared to matrix without reinforcement	−6% reduction compared to matrix without reinforcement	Dirisenapu et al. (2021)

2.6.1 Studies related to carbon-related materials

Recently, ceramic nanoparticles and carbon-based nanotubes have become much more affordable than they were in the late 1980s or early 1990s, which has resulted in a remarkable growth of the study and creation of light metals reinforced with nanoparticles. Owing to their remarkable strength, electrical conductivity, and stiffness, CNTs have a special importance. Zhang et al. (2020) fabricated AA 7075/CNTs composite by powder metallurgy and the influence of aging behaviour on the mechanical properties and microstructure was studied. It was observed that hardness and tensile strength of the composite was more than the base alloy.

Ogawa et al. (2019) fabricated carbon nanofibre-reinforced AMC by ball milling and spark plasma sintering (SPS). Thermal conductivity of composite fabricated by SPS was more and the tensile properties of the composite fabricated by ball milling were higher. Hanizam et al. (2019a) fabricated A356/multiwalled carbon nanotubes (MWCNTs) composite by mechanical stirring and followed by thixoforming and T6 heat treatment. Hardness and tensile strength of the composite were analysed by using Taguchi's design of experiment. Turan (2019) used semi-powder metallurgy technique and hot extrusion process for fabrication of AMC reinforced with MWCNT, graphene nanoplatelets (GNPs), and fullerene. Yield strength and tensile strength of the composites improved with even 0.25% reinforcement. Yuan et al. (2019) fabricated AMC reinforced with CNT by the flake powder metallurgy technique. Hardness, yield strength, and ultimate tensile strength of the composite increased with increasing wt. % of reinforcement. Hanizam et al. (2019b) studied the effect of thixoforming and short T6 heat treatment on the mechanical properties and microstructure of A356/MWCNT composite reinforced at 0.5 wt. %. The yield strength, ultimate tensile strength, and elongation to failure of composite first increased with reinforcement and then increased by thixoforming and later further increased by short T6 heat treatment.

Ghasali et al. (2018) fabricated AMC reinforced with graphene and CNT by SPS, microwave, and conventional method. Maximum bending strength and relative hardness were obtained by SPS technique, and maximum microhardness was obtained by the microwave technique. Fan et al. (2018) fabricated AMC reinforced with CNT by flake powder metallurgy. High-energy milling for 2 hours was introduced, which improved the strength, ductility, and interfacial bonding. Cavaliere et al. (2017) fabricated AMC reinforced with CNTs at 0.5 and 1 wt. % by SPS and analysed the microstructure and the mechanical properties. Increase in strength and porosity was observed. Najimi and Shahverdi (2017) studied the microstructure and mechanical properties of AA 6061/CNT composite fabricated by different milling methods. Different modes of planetary and horizontal attrition or milling were used. Salama et al. (2017) fabricated single matrix and dual matrix structures of AMC reinforced with CNT at 1 and 2.5 wt. %. In dual matrix, ductility was found to be 14.8% more than single matrix while the hardness and the modulus of the matrix were nearly similar.

Huang et al. (2017) studied the superplastic deformation of AA 6061/CNT composite by uniaxial tensile tests at different temperatures and strain rates fabricated by flake powder metallurgy. Shin et al. (2016) studied the mechanical and thermal properties of AA 2024/MWCNTs composite fabricated by powder metallurgy. It was noticed that yield stress was up to two times and compressive strength was up to three times higher than unreinforced composite. Zhou et al. (2016) investigated the interface and the interfacial reactions of the AMC reinforced with MWCNTs. Li et al. (2016) studied the effect of different sintering time and temperature on the AMC reinforced with CNTs to obtain optimal enhancement in mechanical properties. Mansoor and Shahid (2016) fabricated AMC reinforced with CNT by induction melting and observed 77% increase in yield strength, 52% increase in tensile strength, 44% increase in ductility, and 45% increase in hardness.

2.6.2 Studies related to other nanomaterials

Almost every facet of modern life is being transformed by applications of nanotechnology. The number of research publications (informative and scientific) on engineered nanomaterials (ENMs) has expanded as a result of the increased usage of ENMs in a variety of fields, including energy, information technology, chemical and medical equipment, and others (Raina & Anand, 2018; Anand et al., 2021; Shafi et al., 2019; Haq et al., 2021b; Subramanian et al., 2021). Nanomaterials have also been used as reinforcements to enhance the properties of different metals. Mohanavel et al. (2020) reinforced AA 6351 with nanoparticles of Si_3N_4 at 0, 1, 2, and 3 wt. % with the particle size of 30 nm by the stir-casting technique. Yield strength, ultimate tensile strength, impact strength, compressive strength, and hardness increased, while the % Elongation and wear rate decreased with the increasing wt. % of reinforcement. Shayan et al. (2020) fabricated AA 2024/TiO_2 nanocomposite at 0.5 and 1 vol. % by the stir-casting technique. Hardness increased by 25%, ultimate tensile strength increased by 28%, yield strength increased by 4%, and elongation increased by 163%.

Khoshghadam-Pireyousefan et al. (2020) fabricated nanocomposite of Al-Graphene by SPS. Yield strength increased by 79%, ultimate tensile strength increased by 49%, and hardness increased by 44% for 1 wt. % graphene. Barati et al. (2019) studied the microstructure, mechanical properties, wear behaviour, and corrosion behaviour of AA 6061 reinforced with nanoparticles of SiO_2 fabricated by friction stir processing (FSP) and friction stir vibration processing (FSVP). Results showed that FSVP was more advantageous for improving the microstructure and mechanical properties of the composite. Al-Salihi et al. (2019) studied the mechanical properties and wear behaviour of AA 7075 reinforced with nanoparticles of Al_2O_3 fabricated by stir-casting technique at 0, 1, 3, and 5 wt. % reinforcement. Ultimate tensile strength increased by 14.3%, yield tensile strength increased by 34.3%, and hardness increased by 26.3%.

Kumar et al. (2019) fabricated nanocomposite of AA 6061 reinforced with nano-sized ZrO_2 by stir-casting technique at 0, 2, 4, and 6 wt. % reinforcement. Hardness and tensile strength increased while % Elongation decreased with increasing wt. % of reinforcement. Nourbakhsh et al. (2018) fabricated AA 2024/TiO_2 nanocomposite by stir-casting technique at 0.5, 1, and 1.5 vol. % reinforcement. Maximum strength was observed at 1% reinforcement and hardness increased by 31.75% at 1.5% reinforcement. In another study researchers fabricated nanocomposite of AA 7068 reinforced with TiO_2 at 3, 6, and 9 wt. % by powder metallurgy technique. Maximum microhardness was observed at 9 wt. % reinforcement. Qasim et al. (2017) fabricated nanocomposite of A356 reinforced with nanorod structures of ZnO and studied the tensile strength, hardness and ductility of the composites. Manivannan et al. (2017) fabricated AA 6061/SiC nanocomposite by liquid metallurgy technique and studied the wear behaviour.

From the studies discussed above, it can be deduced that addition of nanomaterials in MMCs have enhanced various mechanical properties like tensile strength, compressive strength, impact strength, fatigue strength, etc. Tribological properties, particularly the wear resistance, also improved.

2.6.3 Studies related to hybrid nanomaterials

Hybrid metals and alloys are extensively being manufactured due to their superior mechanical properties like high specific strength, good ductility while being lightweight. They are used to produce different components in aerospace, automotive, and electronic industries (Han et al., 2020). Ahamad et al. (2020) worked on Al-Al_2O_3-Carbon hybrid composite at 0, 2.5, 5, 7.5, and 10 wt. % reinforcement and studied the hardness and wear behaviour of the hybrid composite. Hardness value increased and the wear rate decreased with the increasing wt. % of reinforcement. Boppana et al. (2020) fabricated AA 6061-ZrO_2-graphene hybrid composite using 1 wt. % ZrO_2 and 0.5 and 0.75 wt. % nano graphene. Ultimate tensile strength and yield strength increased while the % Elongation decreased with the increasing wt. % of graphene reinforcement. Shuvho et al. (2020) fabricated AA 6063/TiO_2/Al_2O_3/SiC hybrid composite using 1 wt. % TiO_2, 1 wt. % Al_2O_3, and varying wt. % (2.5, 5, 7.5 and 10) of SiC reinforcement. Hardness, ultimate tensile strength, and yield strength of the hybrid composite increased with increasing wt. % of SiC in hybrid composite.

Ahamad et al. (2020) studied the microstructure and the mechanical properties of Al-Al_2O_3-TiO_2 hybrid composite fabricated by stir-casting technique at 0, 2.5, 5, 7.5, and 10 wt. % of Al_2O_3 and TiO_2 reinforcement. Maximum ultimate tensile strength was observed at 5 wt. % of reinforcement. Impact strength decreased while density and hardness decreased with increasing wt. % of reinforcement. Singh et al. (2020) fabricated AA 2024/Al_2O_3/ZrO_2/graphite-reinforced hybrid composite at 3, 4, 6, and 10 wt. % of Al_2O_3, ZrO_2 and graphite reinforcement by the stir-casting technique. Hardness and tensile strength increased with increasing wt. % of reinforcement.

Hossain et al. (2020) fabricated AA 6103-Al_2O_3-SiC hybrid composite by the stir-casting technique using 1 wt. % Al_2O_3 reinforcement and varying wt. % (0, 2, 4, 6 and 8) of SiC. Results showed increase in hardness and decrease in wear rate with increasing wt. % of SiC reinforcement. Jamwal et al. (2019) fabricated AA 1100-Al_2O_3-TiC hybrid composite by the stir-casting technique using 5, 10, 15, and 20 wt. % of Al_2O_3 and TiC reinforcement. Results showed that hardness and tensile strength enhanced while a decrease in wear resistance at higher wt. % of reinforcements was observed. Fenghong et al. (2019) fabricated AA 6061-SiC-WC hybrid composite by the stir-casting technique at 0, 5, 7.5, and 10 wt. % of reinforcements and observed the increase in yield strength, ultimate tensile strength, compressive strength, and hardness with increasing wt. % of reinforcement. Dhandapani et al. (2016) fabricated Al-MWCNT-B_4C hybrid composite by powder metallurgy technique and studied the microstructure and mechanical properties of the composite. Improvement in mechanical properties was observed with increasing wt. % of MWCNT reinforcement.

It can be concluded that hybrid nanomaterials offer a variety of features that are not possible with the use of base materials. Due to their functional features, hybrid nanomaterials and nanocomposites have attracted significant interest for a variety of applications. For the development of different components, it is important to create macroscopic assemblies of hybrid nanomaterials and nanomaterials in nanocomposites with precise control over orientation and placement.

2.7 CHALLENGES AND FUTURE SCOPE

Despite a number of advantages offered by MMnCs, their practical use is still hindered due to some challenges. For the development of MMnCs with exceptional qualities, the choice of synthesis processes and process parameters is essential. Particular focus has been placed on the liquid-based production techniques, where it is still very difficult to get a uniform homogeneous dispersion of nanoparticles in the melt due to the poor wettability of the nanoparticles with the metal matrix. There aren't enough experimental or simulation investigations to determine the ideal dispersing conditions, despite the benefits of dispersing techniques (such as melt ultrasonication) during casting and solidification to prevent agglomeration and improve wetting conditions.

A precise understanding of the processing, methods, and characteristics of metal matrix nanocomposites will be given by *in situ* imaging of deagglomeration of nanoparticle aggregates in molten metals. In addition to large-scale atomistic modelling, which can provide the same knowledge, such studies can offer accurate understanding of the de-agglomeration dynamics in certain treatment processes. Quantum-based chemical potentials for different MMnCs must be created in future in order to do large-scale atomistic modelling and analyse the dynamics of deagglomeration.

In addition, from an industrial standpoint, one of the most difficult problems that need to be researched in upcoming projects is the upscaling of dependable processes for the industrial use of MMnCs. Last but not least, MMnCs are typically processed at a lab level, whereas a continual industrial scale, where the melt runs into the mould and nanoparticles are always uniformly dispersed to strengthen the metal matrix, has not yet been established.

Including nanometric molecules in the molten matrix, dispersing the molecules consistently in the matrix, and forming a strong interfacial bond among the reinforcement and the matrix are the major problems encountered while manufacturing MMnCs via stir mixing and casting. The challenge is mostly brought on by the fact that full wetting becomes more challenging to achieve as particle size reduces. This is because as the nanometric particle starts to pierce through molten metal, the surface energy needed by it to bend to a small radius increases. It is also worth mentioning that as the particles are not moistened by the molten matrix, the possibility of agglomeration and clustering rises. The Gibb's energy of the system must be minimized in order to minimize agglomeration and provide appropriate dispersion.

The distribution of the reinforcements is significantly influenced by the method chosen for adding the particles and blending them with the molten matrix. While there are numerous methods for adding and blending particles into a melt, most of these techniques have been not been effective at dispersing nanometric particles. For instance, gas injection of particles incorporates a significant amount of gas in the melt without greatly incorporating the particles. Mechanical mixing, which is frequently utilized, brings floating pollutants and oxide layers into the melt. Excess mixing can elevate the melt's gas concentration, which raises the porosity of the cast sample. For the successful inclusion of nanometric particles in the melt, stirring speed and time, melt temperature, the atmosphere over the melt, the number and type of the particles being incorporated in the matrix, and other essential elements must be carefully monitored and regulated. These factors are shown in Figure 2.6.

Along with the thermodynamic influence, the larger surface area of nanometric particulates encourages gas absorption and leads to the formation of a gas layer on the surface. This gas layer keeps the surfaces of the particles from coming into contact with the molten matrix. It can be inferred from this that there are two different sorts of hurdles to particle integration in the matrix:

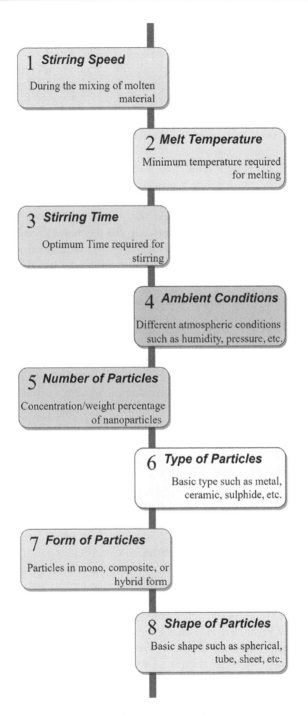

Figure 2.6 Factors effecting nanoparticle inclusion in the melt.

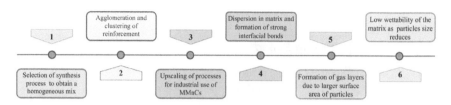

Figure 2.7 Challenges in Fabrication and use of MMnCs.

thermodynamic challenges related to wettability and mechanical challenges brought on by the presence of a gas layer on nanometric particle surfaces (Chatterjee & Mallick, 2013). By actively mixing the melt with a stirrer, electromagnetic stirring while adding the particles in the melt, or employing ultrasonic particle dispersion approach, the mechanical challenges can be overcome (Murthy et al., 2012). Furthermore, by implementing measures to increase the wettability of the particles in the molten matrix, thermodynamic barriers can be minimized. These challenges are summarized in Figure 2.7.

2.8 CONCLUSION

MMnCs are characterized by high strength and less weight and are widely used in a number of industrial and biomedical facets. A range of nanoparticles, such as nanoceramics, nanooxide/non-oxides, carbon-based nano-allotropes, which are high strength are used to reinforce a matrix, which typically consists of a lightweight metal like aluminium or magnesium. The utilization of metal matrix nanocomposites in a wide range of engineering applications makes them very intriguing materials. Recent studies have demonstrated the potential to create composite materials with remarkable mechanical characteristics, which can further be improved by enhancing particle dispersion. The hardness, mechanical strength, creep behaviour, wear resistance, and damping qualities in particular showed amazing results. This category of MMCs could replace the costly heat treatment that is now applied to traditional monolithic alloys, thus expanding the variety of alloys that are suitable for structural and functional purposes. Despite these qualities, there are some areas that still need improvement. The manufacturing of MMnCs is substantially more difficult than the manufacturing of micro-MMCs. As particles downsize from the micro to nanolevel, a number of new problems and challenges are encountered, which must be resolved. It is still currently unknown how CNTs or ceramic nanoparticles interact and react with the matrix. The failure of the composites may be caused by the improper bonding contact. Another critical issue that must be resolved is particle clustering.

REFERENCES

Ahamad, N., Mohammad, A., Sadasivuni, K. K., & Gupta, P. (2020). Structural and mechanical characterization of stir cast Al-Al$_2$O$_3$-TiO$_2$ hybrid metal matrix composites. *Journal of Composite Materials, 54*(21), 2985–2997.

Akbari, M. K., Mirzaee, O., & Baharvandi, H. R. (2013). Fabrication and study on mechanical properties and fracture behavior of nanometric Al$_2$O$_3$ particle-reinforced A356 composites focusing on the parameters of vortex method. *Materials & Design, 46*, 199–205.

Al-Salihi, H. A., Mahmood, A. A., & Alalkawi, H. J. (2019). Mechanical and wear behavior of AA7075 aluminum matrix composites reinforced by Al$_2$O$_3$ nanoparticles. *Nanocomposites, 5*(3), 67–73.

Ames, W., & Alpas, A. T. (1995). Wear mechanisms in hybrid composites of graphite-20 PctSiC in A356 aluminum alloy (Al-7 Pct Si-0.3 Pct Mg). *Metallurgical and Materials Transactions A, 26*(1), 85–98.

Amirkhanlou, S., Jamaati, R., Niroumand, B., & Toroghinejad, M. R. (2011). Fabrication and characterization of Al/SiCp composites by CAR process. *Materials Science and Engineering: A, 528*(13–14), 4462–4467.

Amirkhanlou, S., Ketabchi, M., Parvin, N., Khorsand, S., & Bahrami, R. (2013). Accumulative press bonding; a novel manufacturing process of nanostructured metal matrix composites. *Materials & Design, 51*, 367–374.

Anand, R., Raina, A., Irfan Ul Haq, M., Mir, M. J., Gulzar, O., & Wani, M. F. (2021). Synergism of TiO$_2$ and graphene as nano-additives in bio-based cutting fluid—An experimental investigation. *Tribology Transactions, 64*(2), 350–366.

Bakshi, S. R., Lahiri, D., & Agarwal, A. (2010). Carbon nanotube reinforced metal matrix composites—A review. *International Materials Reviews, 55*(1), 41–64.

Barati, M., Abbasi, M., & Abedini, M. (2019). The effects of friction stir processing and friction stir vibration processing on mechanical, wear and corrosion characteristics of Al6061/SiO$_2$ surface composite. *Journal of Manufacturing Processes, 45*, 491–497.

Bastwros, M. M., Esawi, A. M., & Wifi, A. (2013). Friction and wear behavior of Al–CNT composites. *Wear, 307*(1–2), 164–173.

Bauri, R., Yadav, D., & Suhas, G. (2011). Effect of friction stir processing (FSP) on microstructure and properties of Al–TiC in situ composite. *Materials Science and Engineering: A, 528*(13–14), 4732–4739.

Boppana, S. B., Dayanand, S., Kumar, M. A., Kumar, V., & Aravinda, T. (2020). Synthesis and characterization of nanographene and ZrO$_2$ reinforced Al 6061 metal matrix composites. *Journal of Materials Research and Technology, 9*(4), 7354–7362.

Casati, R., & Vedani, M. (2014). Metal matrix composites reinforced by nano-particles—A review. *Metals, 4*(1), 65–83.

Cavaliere, P., Sadeghi, B., & Shabani, A. (2017). Carbon nanotube reinforced aluminum matrix composites produced by spark plasma sintering. *Journal of Materials Science, 52*(14), 8618–8629.

Ceschini, L., Dahle, A., Gupta, M., Jarfors, A. E. W., Jayalakshmi, S., Morri, A., Rotundo, F., Toschi, S., & Arvind Singh, A. (2017). *Aluminum and Magnesium Metal Matrix Nanocomposites*. Springer, Singapore.

Chatterjee, S., & Mallick, A. B. (2013). Challenges in manufacturing aluminium based metal matrix nanocomposites via stir casting route. In *Materials Science Forum* (Vol. 736, pp. 72–80). Trans Tech Publications Ltd, Stafa-Zurich. Editor: B.S.S. Daniel.

Chen, H., & Alpas, A. T. (2000). Sliding wear map for the magnesium alloy Mg-9Al-0.9 Zn (AZ91). *Wear, 246*(1–2), 106–116.

Chen, X. H., Chen, C. S., Xiao, H. N., Liu, H. B., Zhou, L. P., Li, S. L., & Zhang, G. (2006). Dry friction and wear characteristics of nickel/carbon nanotube electroless composite deposits. *Tribology International, 39*(1), 22–28.

De Cicco, M. P. (2009). Solidification phenomena in metal matrix nanocomposites.

Deng, C. F., Wang, D. Z., Zhang, X. X., & Ma, Y. X. (2007). Damping characteristics of carbon nanotube reinforced aluminum composite. *Materials Letters, 61*(14–15), 3229–3231.

Dhandapani, S., Rajmohan, T., Palanikumar, K., & Charan, M. (2016). Synthesis and characterization of dual particle (MWCT+ B4C) reinforced sintered hybrid aluminum matrix composites. *Particulate Science and Technology, 34*(3), 255–262.

Dirisenapu, G., Dumpala, L., & Reddy, S. P. (2021). Dry sliding tribological behavior of Al7010/B4C/BN hybrid metal matrix nanocomposites prepared by ultrasonic-assisted stir casting. *Transactions of the Indian Institute of Metals, 74*(1), 149–158.

Etaati, A., Shokuhfar, A., Omrani, E., Movahed, P., Bolvardi, H., & Tavakoli, H. (2010). Study on homogenization time and cooling rate on microstructure and hardness of Ni-42.5wt% Ti-3wt% Cu alloy. In *Defect and Diffusion Forum* (Vol. 297, pp. 489–494). Trans Tech Publications Ltd, Stafa-Zurich. Editors: Andreas Öchsner, Graeme E. Murch, Ali Shokuhfar and João M.P.Q. Delgado.

Fan, G., Jiang, Y., Tan, Z., Guo, Q., Xiong, D. B., Su, Y., Lin, R., Hu, L., Li, Z., & Zhang, D. (2018). Enhanced interfacial bonding and mechanical properties in CNT/Al composites fabricated by flake powder metallurgy. *Carbon, 130*, 333–339.

Farooq, S. A., Raina, A., Mohan, S., Arvind Singh, R., Jayalakshmi, S., & Irfan Ul Haq, M. (2022). Nanostructured coatings: Review on processing techniques, corrosion behaviour and tribological performance. *Nanomaterials, 12*(8), 1323.

Fenghong, C., Chang, C., Zhenyu, W., Muthuramalingam, T., & Anbuchezhiyan, G. (2019). Effects of silicon carbide and tungsten carbide in aluminium metal matrix composites. *Silicon, 11*(6), 2625–2632.

Ferkel, H., & Mordike, B. L. (2001). Magnesium strengthened by SiC nanoparticles. *Materials Science and Engineering: A, 298*(1–2), 193–199.

Ghasali, E., Alizadeh, M., & Ebadzadeh, T. (2018). TiO$_2$ ceramic particles-reinforced aluminum matrix composite prepared by conventional, microwave, and spark plasma sintering. *Journal of Composite Materials, 52*(19), 2609–2619.

Ghazaly, A., Seif, B., & Salem, H. G. (2016). Mechanical and tribological properties of AA2124-graphene self lubricatingnanocomposite. In *Light Metals 2013* (pp. 411–415). Springer, Cham.

Gupta, G., Haq, M. I. U., Raina, A., & Shafi, W. K. (2022). Effect of epoxidation and nanoparticle addition on the rheological and tribological properties of canola oil. *Proceedings of the Institution of Mechanical Engineers, Part J: Journal of Engineering Tribology, 236*(9), 1837–1845.

Gupta, G., Kumar, P., Raina, A., & Haq, M. I. U. (2018, August). Effect of SiC reinforcement on mechanical behavior of aluminum alloys–a review. *AIP Conference Proceedings, 2006*(1), 030051.

Hamedan, A. D., & Shahmiri, M. (2012). Production of A356–1 wt% SiCnanocomposite by the modified stir casting method. *Materials Science and Engineering: A*, *556*, 921–926.

Han, J. K., Herndon, T., Jang, J. I., Langdon, T. G., & Kawasaki, M. (2020). Synthesis of hybrid nanocrystalline alloys by mechanical bonding through high-pressure torsion. *Advanced Engineering Materials*, *22*(4), 1901289.

Hanizam, H., Salleh, M. S., Omar, M. Z., & Sulong, A. B. (2019a). Effects of mechanical stirring and short heat treatment on thixoformed of carbon nanotube aluminium alloy composite. *Journal of Alloys and Compounds*, *788*, 83–90.

Hanizam, H., Salleh, M. S., Omar, M. Z., & Sulong, A. B. (2019b). Optimisation of mechanical stir casting parameters for fabrication of carbon nanotubes–aluminium alloy composite through Taguchi method. *Journal of Materials Research and Technology*, *8*(2), 2223–2231.

Haq, M. I. U., Mohan, S., Raina, A., Jayalakshmi, S., Singh, R. A., Chen, X., Konovalov, S., & Gupta, M. (2021a). Mechanical and tribological properties of aluminum based metal matrix nanocomposites. *Encyclopedia of Materials: Composites*, *1*, 402–414.

Haq, M. I. U., Raina, A., Ghazali, M. J., Javaid, M., & Haleem, A. (2021b). Potential of 3D printing technologies in developing applications of polymeric nanocomposites. In *Tribology of Polymer and Polymer Composites for Industry 4.0* (pp. 193–210). Springer, Singapore. Editors: Hemalata Jena, Jitendra Kumar Katiyar, and Amar Patnaik.

Hashemi, M., Jamaati, R., & Toroghinejad, M. R. (2012). Microstructure and mechanical properties of Al/SiO$_2$ composite produced by CAR process. *Materials Science and Engineering: A*, *532*, 275–281.

Hossain, S., Rahman, M. M., Chawla, D., Kumar, A., Seth, P. P., Gupta, P., Kumar, D., Agrawal, R., & Jamwal, A. (2020). Fabrication, microstructural and mechanical behavior of Al-Al$_2$O$_3$-SiC hybrid metal matrix composites. *Materials Today: Proceedings*, *21*, 1458–1461.

Huang, H., Fan, G., Tan, Z., Xiong, D. B., Guo, Q., Guo, C., Li, Z., & Zhang, D. (2017). Superplastic behavior of carbon nanotube reinforced aluminum composites fabricated by flake powder metallurgy. *Materials Science and Engineering: A*, *699*, 55–61.

Jamaati, R., & Toroghinejad, M. R. (2010). High-strength and highly-uniform composite produced by anodizing and accumulative roll bonding processes. *Materials & Design*, *31*(10), 4816–4822.

Jammoria, N. S., Haq, M. I. U., Singh, R. P., & Raina, A. (2022a). Soft computing techniques and aluminum metal matrix composites. In *Optimization of Industrial Systems* (Chapter 29, pp. 367–90).

Jammoria, N. S., Ul Haq, M. I., & Raina, A. (2022b). Carbon-related materials for tribological application. In *Proceedings of Fourth International Conference on Inventive Material Science Applications* (pp. 469–483). Springer, Singapore. Editors: V. Bindh, João Manuel R. S. Tavares, and Ştefan Ţălu.

Jamwal, A., Vates, U. K., Gupta, P., Aggarwal, A., & Sharma, B. P. (2019). Fabrication and characterization of Al$_2$O$_3$–TiC-reinforced aluminum matrix composites. In *Advances in Industrial and Production Engineering* (pp. 349–356). Springer, Singapore. Editors: Kripa Shanker, Ravi Shankar, and Rahul Sindhwani.

Jost, H. P., & Schofield, J. (1981). Energy saving through tribology: A techno-economic study. *Proceedings of the Institution of Mechanical Engineers*, *195*(1), 151–173.

Karimi, M., Toroghinejad, M. R., & Dutkiewicz, J. (2016). Nanostructure formation during accumulative roll bonding of commercial purity titanium. *Materials Characterization*, *122*, 98–103.

Khoshghadam-Pireyousefan, M., Rahmanifard, R., Orovcik, L., Švec, P., & Klemm, V. (2020). Application of a novel method for fabrication of graphene reinforced aluminum matrix nanocomposites: Synthesis, microstructure, and mechanical properties. *Materials Science and Engineering: A*, *772*, 138820.

Kichloo, A. F., Raina, A., Haq, M. I. U., & Wani, M. S. (2022). Impact of carbon fiber reinforcement on mechanical and tribological behavior of 3D-Printed polyethylene terephthalate glycol polymer composites—An experimental investigation. *Journal of Materials Engineering and Performance*, *31*(2), 1021–1038.

Kumar, G. V., Pramod, R., Sekhar, C. G., Kumar, G. P., & Bhanumurthy, T. (2019). Investigation of physical, mechanical and tribological properties of Al6061–ZrO$_2$ nano-composites. *Heliyon*, *5*(11), e02858.

Li, C., Liu, X., Yi, J., Teng, L., Bao, R., Tan, J., Yang, C., & Zou, Z. (2016). Effects of sintering parameters on the microstructure and mechanical properties of carbon nanotubes reinforced aluminum matrix composites. *Journal of Materials Research*, *31*(23), 3757.

Liu, C. Y., Wang, Q., Jia, Y. Z., Zhang, B., Jing, R., Ma, M. Z., Jing, Q., & Liu, R. P. (2012). Effect of W particles on the properties of accumulatively roll-bonded Al/W composites. *Materials Science and Engineering: A*, *547*, 120–124.

Lv, Z., Ren, X., & Hou, H. (2017). Application of accumulative roll bonding process for manufacturing Mg/2 wt.% CNTs nanocomposite with superior mechanical properties. *Journal of Nanoscience and Nanotechnology*, *17*(6), 4022–4031.

Madakson, P. B., Yawas, D. S., & Apasi, A. (2012). Characterization of coconut shell ash for potential utilization in metal matrix composites for automotive applications. *International Journal of Engineering Science and Technology*, *4*(3), 1190–1198.

Malaki, M., Xu, W., Kasar, A. K., Menezes, P. L., Dieringa, H., Varma, R. S., & Gupta, M. (2019). Advanced metal matrix nanocomposites. *Metals*, *9*(3), 330.

Malik, A., Haq, M. I. U., Raina, A., & Gupta, K. (2022). 3D Printing towards implementing Industry 4.0: Sustainability aspects, barriers and challenges. *Industrial Robot: The International Journal of Robotics Research and Application*, *49*(3), 491–511.

Manivannan, I., Ranganathan, S., Gopalakannan, S., Suresh, S., Nagakarthigan, K., & Jubendradass, R. (2017). Tribological and surface behavior of silicon carbide reinforced aluminum matrix nanocomposite. *Surfaces and Interfaces*, *8*, 127–136.

Mansoor, M., & Shahid, M. (2016). Carbon nanotube-reinforced aluminum composite produced by induction melting. *Journal of Applied Research and Technology*, *14*(4), 215–224.

Md, S., & Mohd, Z. J. (2010). Utilization of solid wastes in construction materials. *International Journal of Physical Sciences*, *5*(13), 1952–1963.

Mindivan, H., Efe, A., Kosatepe, A. H., & Kayali, E. S. (2014). Fabrication and characterization of carbon nanotube reinforced magnesium matrix composites. *Applied Surface Science*, *318*, 234–243.

Mohanavel, V., Ali, K. A., Prasath, S., & Sathish, T. (2020). Microstructural and tribological characteristics of AA6351/Si3N4 composites manufactured by stir casting. *Journal of Materials Research and Technology*, *9*(6), 14662–14672.

Moustafa, E. B., Abushanab, W. S., Melaibari, A., Yakovtseva, O., & Mosleh, A. O. (2021). The effectiveness of incorporating hybrid reinforcement nanoparticles in the enhancement of the tribological behavior of aluminum metal matrix composites. *JOM, 73*(12), 4338–4348.

Moustafa, S. F., El-Badry, S. A., Sanad, A. M., & Kieback, B. (2002). Friction and wear of copper–graphite composites made with Cu-coated and uncoated graphite powders. *Wear, 253*(7–8), 699–710.

Murthy, I. N., Rao, D. V., & Rao, J. B. (2012). Microstructure and mechanical properties of aluminum–fly ash nano composites made by ultrasonic method. *Materials & Design, 35*, 55–65.

Najimi, A. A., & Shahverdi, H. R. (2017). Effect of milling methods on microstructures and mechanical properties of Al6061-CNT composite fabricated by spark plasma sintering. *Materials Science and Engineering: A, 702*, 87–95.

Nourbakhsh, S. H., Tavakoli, M., & Shahrokhian, M. A. (2018). Investigations of mechanical, microstructural and tribological properties of Al2024 nanocomposite reinforced by TiO_2 nanoparticles. *Materials Research Express, 5*(11), 116531.

Ogawa, F., Yamamoto, S., & Masuda, C. (2019). Thermal conductivity and tensile properties of carbon nanofiber-reinforced aluminum-matrix composites fabricated via powder metallurgy: Effects of ball milling and extrusion conditions on microstructures and resultant composite properties. *Acta Metallurgica Sinica (English Letters), 32*(5), 573–584.

Omrani, E., DorriMoghadam, A., Menezes, P. L., & Rohatgi, P. K. (2016). New emerging self-lubricating metal matrix composites for tribological applications. In *Ecotribology* (pp. 63–103). Springer, Cham. Editor: J. Paulo Davim.

Qasim, Z. S., Jabbar, M. A., & Hassan, J. J. (2017). Enhancement the mechanical properties of aluminum casting alloys (A356) by adding nanorods structures from zinc oxide. *Journal of Material Sciences & Engineering, 6*(2), 2–5.

Qi, Q. J. (2006). Evaluation of sliding wear behavior of graphite particle-containing magnesium alloy composites. *Transactions of Nonferrous Metals Society of China, 16*(5), 1135–1140.

Raina, A., & Anand, A. (2018). Effect of nanodiamond on friction and wear behavior of metal dichalcogenides in synthetic oil. *Applied Nanoscience, 8*(4), 581–591.

Raina, A., Haq, M. I. U., Mohan, S., Anand, A., & Graf, M. (2021). Materials for tribological applications: An overview. In *Tribology and Sustainability* (pp. 3–22). CRC Press, Boca Raton, FL. Editors: Jitendra Kumar Katiyar, Mir Irfan Ul Haq, Ankush Raina, S. Jayalakshmi, and R. Arvind Singh.

Rajan, T. P. D., Pillai, R. M., Pai, B. C., Satyanarayana, K. G., & Rohatgi, P. K. (2007). Fabrication and characterisation of Al–7Si–0.35 Mg/fly ash metal matrix composites processed by different stir casting routes. *Composites Science and Technology, 67*(15–16), 3369–3377.

Rajkumar, K., & Aravindan, S. (2013). Tribological behavior of microwave processed copper–nanographite composites. *Tribology International, 57*, 282–296.

Rawal, S. P. (2001). Metal-matrix composites for space applications. *JOM, 53*(4), 14–17.

Rohatgi, P. K., Ray, S., & Liu, Y. (1992). Tribological properties of metal matrix-graphite particle composites. *International Materials Reviews, 37*(1), 129–152.

Rouf, S., Raina, A., Haq, M. I. U., & Naveed, N. (2021). Sensors and tribological systems: Applications for industry 4.0. *Industrial Robot: The International Journal of Robotics Research and Application, 49*(3), 442–460.

Saito, Y., Tsuji, N., Utsunomiya, H., Sakai, T., & Hong, R. G. (1998). Ultra-fine grained bulk aluminum produced by accumulative roll-bonding (ARB) process. *Scripta Materialia, 39*(9), 1221–1227.

Salama, E. I., Abbas, A., & Esawi, A. M. (2017). Preparation and properties of dual-matrix carbon nanotube-reinforced aluminum composites. *Composites Part A: Applied Science and Manufacturing, 99*, 84–93.

Sanaty-Zadeh, A. (2012). Comparison between current models for the strength of particulate-reinforced metal matrix nanocomposites with emphasis on consideration of Hall–Petch effect. *Materials Science and Engineering: A, 531*, 112–118.

Scharf, T. W., Neira, A., Hwang, J. Y., Tiley, J., & Banerjee, R. (2009). Self-lubricating carbon nanotube reinforced nickel matrix composites. *Journal of Applied Physics, 106*(1), 013508.

Shafi, W. K., Raina, A., & Haq, M. I. U. (2019). Performance evaluation of hazelnut oil with copper nanoparticles—A new entrant for sustainable lubrication. *Industrial Lubrication and Tribology, 71*(6), 749–757.

Shafi, W. K., Raina, A., Haq, M. I. U., & Khajuria, A. (2018). Interdisciplinary aspects of tribology. *International Research Journal of Engineering and Technology, 5*(2), 5–8.

Shayan, M., Eghbali, B., & Niroumand, B. (2020). Fabrication of AA2024– TiO_2 nanocomposites through stir casting process. *Transactions of Nonferrous Metals Society of China, 30*(11), 2891–2903.

Shehata, F., Fathy, A., Abdelhameed, M., & Moustafa, S. F. (2009). Preparation and properties of Al_2O_3 nanoparticle reinforced copper matrix composites by in situ processing. *Materials & Design, 30*(7), 2756–2762.

Shin, S. E., Ko, Y. J., & Bae, D. (2016). Mechanical and thermal properties of nano-carbon-reinforced aluminum matrix composites at elevated temperatures. *Composites Part B: Engineering, 106*, 66–73.

Shuvho, M. B. A., Chowdhury, M. A., Kchaou, M., Roy, B. K., Rahman, A., & Islam, M. A. (2020). Surface characterization and mechanical behavior of aluminum based metal matrix composite reinforced with nano Al_2O_3, SiC, TiO_2 particles. *Chemical Data Collections, 28*, 100442.

Singh, H., Haq, M. I. U., & Raina, A. (2020). Dry sliding friction and wear behaviour of AA6082-TiB2 in situ composites. *Silicon, 12*(6), 1469–1479.

Singh, N., Mir, I. U. H., Raina, A., Anand, A., Kumar, V., & Sharma, S. M. (2018). Synthesis and tribological investigation of Al-SiC based nano hybrid composite. *Alexandria Engineering Journal, 57*(3), 1323–1330.

Skeldon, P., Wang, H. W., & Thompson, G. E. (1997). Formation and characterization of self-lubricating MoS_2 precursor films on anodized aluminium. *Wear, 206*(1–2), 187–196.

Slathia, S., Anand, R., Irfan Ul Haq, M., Raina, A., Mohan, S., Kumar, R., & Anand, A. (2020). Friction and Wear behaviour of AA2024/ZrO_2 composites: Effect of graphite. In *Recent Advances in Mechanical Engineering* (pp. 597–601). Springer, Singapore. Editors: Harish Kumar and Prashant K. Jain.

Slathia, S., Haq, M. I. U., & Raina, A. (2018, August). Fabrication and mechanical characterization of AA2024-ZrO_2-Gr hybrid composite. *AIP Conference Proceedings, 2006*(1), 030047.

Soltani, M. A., Jamaati, R., & Toroghinejad, M. R. (2012). The influence of TiO_2 nano-particles on bond strength of cold roll bonded aluminum strips. *Materials Science and Engineering: A, 550*, 367–374.

Subramanian, J., Ramachandra, A. S., Raina, A., Ul Haq, M. I., Sharma, S. M., & Chen, X. (2021). Polymeric nanostructures for prospective tribological application in miniaturized devices: A review. *Current Nanomaterials, 6*(2), 85–89.

Suryanarayana, C., & Al-Aqeeli, N. (2013). Mechanically alloyed nanocomposites. *Progress in Materials Science, 58*(4), 383–502.

Sw, Z. (2009). *Current Industrial Activities in China and Green Tribology*. Institution of Engineering & Technology, London, UK.

Tavman, I. H. (1996). Thermal and mechanical properties of aluminum powder-filled high-density polyethylene composites. *Journal of Applied Polymer Science, 62*(12), 2161–2167.

Turan, M. E. (2019). Investigation of mechanical properties of carbonaceous (MWCNT, GNPs and C60) reinforced hot-extruded aluminum matrix composites. *Journal of Alloys and Compounds, 788,* 352–360.

Tzanakis, I., Hadfield, M., Thomas, B., Noya, S. M., Henshaw, I., & Austen, S. (2012). Future perspectives on sustainable tribology. *Renewable and Sustainable Energy Reviews, 16*(6), 4126–4140.

Ul Haq, M. I., & Anand, A. (2018). Dry sliding friction and wear behavior of AA7075-Si3N4 composite. *Silicon, 10*(5), 1819–1829.

Walford, S. N. (2008). Sugarcane bagasse: How easy is it to measure its constituents? *Proceedings of the Annual Congress-South African Sugar Technologists' Association, 81,* 266–273.

Yazdani, A., & Salahinejad, E. (2011). Evolution of reinforcement distribution in Al–B4C composites during accumulative roll bonding. *Materials & Design, 32*(6), 3137–3142.

Ye, J., He, J., & Schoenung, J. M. (2006). Cryomilling for the fabrication of a particulate B4C reinforced Al nanocomposite: Part I. Effects of process conditions on structure. *Metallurgical and Materials Transactions A, 37*(10), 3099–3109.

Yuan, C., Tan, Z., Fan, G., Chen, M., Zheng, Q., & Li, Z. (2019). Fabrication and mechanical properties of CNT/Al composites via shift-speed ball milling and hot-rolling. *Journal of Materials Research, 34*(15), 2609–2619.

Zainudin, N. F., Lee, K. T., Kamaruddin, A. H., Bhatia, S., & Mohamed, A. R. (2005). Study of adsorbent prepared from oil palm ash (OPA) for flue gas desulfurization. *Separation and Purification Technology, 45*(1), 50–60.

Zeng, Z., & Lin, Y. (2009). Mechanical properties of hard Cr–MWNT composite coatings. *Surface and Coatings Technology, 203*(23), 3610–3613.

Zhang, H. B., Wang, B., Zhang, Y. T., Li, Y., He, J. L., & Zhang, Y. F. (2020). Influence of aging treatment on the microstructure and mechanical properties of CNTs/7075 Al composites. *Journal of Alloys and Compounds, 814,* 152357.

Zhang, S. W., & Xie, Y. B. (2009). Tribology Science Industrial Application Status and Development Strategy. The Investigation on Position and Function of Tribology in Industrial Energy Conservation, Consumption and Emission Reduction (Report of 2 year Chinese Investigation); Zhang, S.-W., Xie, Y.-B., Eds.

Zhou, W., Bang, S., Kurita, H., Miyazaki, T., Fan, Y., & Kawasaki, A. (2016). Interface and interfacial reactions in multi-walled carbon nanotube-reinforced aluminum matrix composites. *Carbon, 96,* 919–928.

Chapter 3

Tribological behaviour of aluminium metal composites reinforced with nanoparticles

Pawandeep Singh and Vivudh Gupta
Shri Mata Vaishno Devi University

Md Irfan ul Haque Siddiqui
King Saud University

CONTENTS

3.1　Introduction　53
3.2　Processing methods　54
　　3.2.1　Liquid processing　54
　　3.2.2　Semi-solid processing　54
　　3.2.3　Solid processing　55
　　3.2.4　*In situ* processing　55
3.3　Tribological behaviour of Al-NMMCs　56
3.4　Conclusions　56
References　61

3.1 INTRODUCTION

Composite materials are a blend of two or more materials having distinct physical and chemical attributes [1]. Metal matrix composites (MMCs) satisfy the requirements of modern engineering applications and moreover act as a replacement to conventional materials [2]. Aluminium MMCs are used in automotive and aerospace sectors, but the size of reinforcement particles used in such composites is large ranging from one to several micrometres. During mechanical loading, ceramic particles of large size are prone to cracks, which leads to early failure and a drop in composites' ductility [3]. To overcome these limitations, nano-sized reinforcement particles have been introduced in past years to develop nano-MMCs to enhance nanocomposites' mechanical and tribological characteristics [4]. The properties of the composites depend upon the selection of reinforcements and their weight percentage (Wt. %). The reinforcements that are to be selected must offer higher strength-to-weight ratio to the composites [5]. Various reinforcements offer higher hardness, compressive strength and resistance to wear in Al-based NMMCs, such as Al_2O_3, SiC, B_4C, TiB_2, and ZrO_2 [6–10]. Moreover, current research is focused on improving the hardness, wear resistance, fatigue and service life of the aluminium nano metal matrix comosites (Al-NMMCs) in a quest to develop advanced materials with better properties [11]. The materials subjected to tribological applications are influenced by friction and wear. Wear is associated with progressive material loss due to interaction between surfaces of the materials in relative motion under the applied load. Al-Si system-based alloys are used in industrial wear applications and are primarily employed in sliding and abrasive wear applications. Friction and wear type of tribological properties are influenced by extrinsic factors like sliding speed, sliding distance, applied load, tribolayer, etc. and intrinsic factors such as type of reinforcement, microstructure of the composites, reinforcement shape and size, etc. [12]. Also, to analyse the impact of diverse nano-sized reinforcements on the tribological behaviour of Al-NMMCs, various studies have been conducted over the past years by varying extrinsic and intrinsic factors. Therefore, this review presents the experimental results related to the tribological behaviour of Al-NMMCs.

3.2 PROCESSING METHODS

3.2.1 Liquid processing

Nanocomposites fabricated using a conventional casting route cause extreme agglomeration of nanoparticles in the composites due to mechanical stirring, inferior wettability and elevated viscosity generated in the molten matrix led by a large surface-to-volume ratio of the nanoparticles [13]. To overcome such limitations of conventional casting method, high-intensity ultrasonic waves were used, which produces acoustic transient cavitation effects and leads to the micro-bubbles collapsing. This cavitation effect produces an impulse, which breaks the cluster of nanoparticles and promotes the uniform dispersal of nanoparticles. Li et al. [14] by the application of the experimental set-up presented in Figure 3.1 fabricated 2 vol. % SiC nanoparticles-reinforced A356 composite and recorded an 20% improvement in hardness analogized to the base alloy. Moreover, when the same set-up was used by Lan et al. [15] to produce 5 wt. % nano-SiC/AZ91D composites, microhardness of the composites improved to 75% compared to the unreinforced alloy.

3.2.2 Semi-solid processing

Limited studies were carried out using semi-solid processing for the fabrication of Al-NMMCs, including a mixture of rheocasting and squeeze casting. Al_2O_3/A356 nanocomposites were produced by combining rheocasting and squeeze casting [16]. Rheocasting was performed at a lower

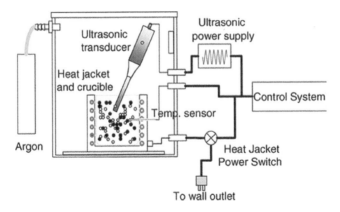

Figure 3.1 Experimental set-up.

temperature than the conventional casting method, causing decreased thermochemical degradation of the reinforcement surface. During rheocasting, nanoparticles were preheated and added to a semi-solid slurry, then stirred vigorously to reach a uniform dispersal of reinforcement particles. Afterwards, the slurry is pressed with the help of a hydraulic press.

3.2.3 Solid processing

Powder metallurgy (PM) is another diverse process used to fabricate nano metal matrix composites (NMMCS). It includes mechanical alloying (MA), where metal powder in pure form and alloying material are added and milled with the help of high-energy ball mill, resulting in continuous fracture and cold welding of the component particles [17]. Furthermore, in the MA process, nanoparticles are distributed evenly with the presence of agglomeration, as in the case of conventional casting methods.

3.2.4 *In situ* processing

There is also an alternative technique employed for the fabrication of nanocomposites, namely, friction stir processing (FSP) [18]. FSP is an *in situ* processing technology, which has been displayed as an efficacious mechanism for refining the grain and has the ability to produce surface composites with the help of a rotating tool that has specially devised pin and shoulder, which are inserted into the plate edges that are to be joined and then traversed along the line of joint as presented in Figure 3.2. The workpiece is heated by the rotating action of the tool and the tool movement produces the joint [19].

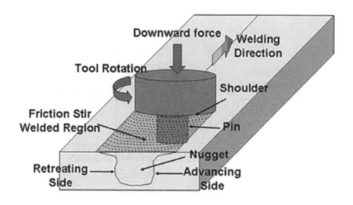

Figure 3.2 Friction stir processing.

3.3 TRIBOLOGICAL BEHAVIOUR OF AL-NMMCs

To enhance the tribological properties of Al-NMMCs, different nanoreinforcements have been used by various researchers and their observations are presented in Table 3.1.

It is evident from the literature review presented in Table 3.1 that different combinations of reinforcements were incorporated into the base alloy to fabricate mono and hybrid composites. Variation in physical and chemical characteristics of nanoparticle reinforcements was beneficial in improving the tribological performance of the composites and also led to the significant improvement in mechanical properties. Reinforcements are effective in transferring the load and increasing the load-bearing capacity of composites, which further enhance the hardness and wear resistance.

3.4 CONCLUSIONS

Based on the literature review, the following conclusions can be drawn from the current study:

1. Ultrasonic-assisted stir casting and powder metallurgy were the most common methods used for fabrication Al-based nanocomposites.
2. Adding higher amount of reinforcement particles leads to particle agglomeration and reduction in the properties, which lead to the restriction of reinforcement content up to 5 wt. %.
3. Hybrid nanocomposites provide better mechanical and tribological properties than single reinforced nanocomposites and base alloys.

Table 3.1 Tribological behaviour of Al-NMMCs

Matrix	Reinforcement	Process	Observations	Ref.
LM9	Multi-walled carbon nanotubes (MWCNTs)	Ultrasonic-assisted stir casting	Nanocomposites' hardness increased, whereas wear rate and friction coefficient decreased with the addition of MWCNTs.	[20]
Al	B_4C	Powder metallurgy	Higher wt. % of B_4C increased the hardness and compressive strength, and decreased the wear rate of nanocomposites. Mechanically mixed layer formation was also observed on the surface of worn-out nanocomposites.	[21]
Al	Graphene	Powder metallurgy	The incorporation of graphene nanoplatelets showed a substantial gain in the mechanical and wear properties of nanocomposites.	[22]
Al	$TiB_2/Al2O_3$	Ultrasonic-assisted stir casting	Developed hybrid nanocomposites showed improved mechanical and tribological properties. Presence of Al_2O_3 acts as obstacles while oxidation of TiB_2 during wear tests acts as a solid lubricant.	[23]
AA2219	SiC	Stir casting	Addition of nano-SiC diminished the subsurface fatigue and decreased the wear rate. Also, there was an improvement in mechanical properties.	[24]
Al6061	SiC/Gr	Ultrasonic-assisted stir casting	With the increase in Gr content in hybrid nanocomposites, wear rate and COF decreased.	[25]
Al	WS_2	Spark plasma sintering	Hardness and compressive strength of nanocomposites increased due to reduced interparticle space. Wear loss and friction coefficient were also reduced with the adhesive type of wear mechanism in pure Al, while abrasive wear mechanism was observed in nanocomposites.	[26]
Al alloy	SiO_2	Stir casting	Compressive strength of nanocomposites decreases, whereas yield strength, elastic modulus and wear resistance of the nanocomposites improved with the incorporation of SiO_2 nanoparticles. Also, the dominant mechanism of wear was abrasive wear.	[27]

(Continued)

Table 3.1 (Continued) Tribological behaviour of Al-NMMCs

Matrix	Reinforcement	Process	Observations	Ref.
Al	TiO_2	Powder metallurgy	Addition of TiO_2 nanoparticles significantly improved ultimate tensile strength and yield strength. The presence of hard nanoparticles acts as barriers to the dislocation movement and increased the hardness and wear resistance.	[28]
Al6061	Ag	Stir casting	The nanocomposites showed improvement in hardness, compressive strength and wear resistance. While the friction coefficient increased with the addition of Ag nanoparticles.	[29]
Al6061	Al_2O_3	Mechanical milling and hot pressing	Longer milling time resulted in significant gain in hardness and wear resistance. Abrasion and delamination were the dominant wear mechanisms.	[30]
Al2024	Graphene/TiC	Powder metallurgy	Uniform dispersal of nanoparticles in the matrix improved the wear resistance of nanocomposites. Wear mechanism changed from adhesion to abrasion in the case of hybrid nanocomposites.	[31]
Al2024	SiC/h-BN	Ultrasonic-assisted stir casting	Hybrid nanocomposites exhibited enhanced wear resistance at all temperatures and applied loads. Formation of thin lubricating film decreased the wear rate, whereas abrasive, adhesive and fracture mode of wear mechanism was responsible for wear.	[32]
Al7075	Graphene	Stir casting	Mechanical and tribological properties of graphene nanoparticles-reinforced Al7075 nanocomposites were improved significantly. Friction coefficient and wear rate were also reported to have decreased by the addition of self-lubricating graphene in the base matrix.	[33]
Al	MWCNTs/Al_2O_3	Powder metallurgy	Tribological and mechanical properties of hybrid nanocomposites improved by substitution of Al_2O_3. Moreover, when MWCNT was added, there was further increase in properties due to its self-lubrication effect and carbon film formation, which covered the wear surface.	[34]

(Continued)

Table 3.1 (Continued) Tribological behaviour of Al-NMMCs

Matrix	Reinforcement	Process	Observations	Ref.
Al6061	Al_2O_3	Ultrasonic-assisted stir casting	Specific wear rate and COF were influenced significantly by the wt. % of Al_2O_3 followed by load and sliding distance. Improved performance of the nanocomposites was due to the formation of mechanically mixed layer (MML).	[35]
Al6061	SiC/MWCNTs	Powder metallurgy	Microhardness of the hybrid nanocomposites was three times higher than the composites reinforced with single reinforcement. Reduced wear rate was also observed for nanocomposites owing to homogeneous distribution of nanoparticles.	[36]
Al	Al_2O_3/GO	Powder metallurgy	Hardness of the nanocomposites wear increased by 48%, and specific wear rate and COF were reduced by 55% and 5%, respectively. Prominent wear mechanism was observed to be abrasive wear because of plastic deformation.	[37]
AA6061	MWCNTs	Friction stir processing	Uniform dispersion of CNTs in the fabricated composites formed a layer of solid lubricant, which decreased the delamination and stabilized the COF, thereby reducing the specific wear rate.	[38]
Al7075	Al_2O_3/BN	Ultrasonic-assisted stir casting	COF and specific wear rate of the nanocomposites were reduced by 18.5% and 63%, respectively, with the addition of nanoparticles. Applied load and sliding velocity were found to be the dominant parameter influencing the COF and specific wear rate (SWR).	[39]
AA7050	TiC/graphite	Powder metallurgy	High wt. % of graphite decreased the hardness and hybrid nanocomposites exhibited significantly lower wear rate and COF than single reinforced nanocomposite.	[40]

(Continued)

Table 3.1 (Continued) Tribological behaviour of Al-NMMCs

Matrix	Reinforcement	Process	Observations	Ref.
AA7075	Gr	Powder metallurgy	Grain refinement increased the hardness of the nanocomposites. Furthermore, wear rate and COF also decreased due to the addition of reinforcement particles. Abrasive type of wear mechanism was observed for nanocomposites.	[41]
AA7150	hBN	Double stir casting	Influence of the various test parameters were analysed using the Taguchi technique. Applied load has the higher significance on the physical and statistical properties of nanocomposites, whereas sliding velocity greatly influences the COF.	[42]
Al7075	SiC	Hot dynamic compaction	Microhardness and compressive strength of the nanocomposites were improved mainly due to Orowan mechanism and the Hall–Petch effect. Wear rate of the nanocomposites was also reduced with the increase in SiC content.	[43]
Al	SiC/graphite	Powder metallurgy	SiC addition compensates the detrimental effects of graphite on hardness and strength of the hybrid nanocomposites. COF and wear rate of the composites decreased with abrasives and delamination was observed to be a prominent wear mechanism.	[44]
Al	Tungsten carbide (WC)	Ultrasonic-assisted stir casting	Hardness increased with the increase in wt.% of WC. Wear resistance and friction behaviour of the composites were also improved. Mechanism of wear shifted from adhesion to abrasion with the increase in WC content.	[45]
Al7150	hBN	Ultrasonic-assisted stir casting	Hardness and tensile strength of the nanocomposites improved, whereas wear loss decreased with the incorporation of hBN nanoparticles. Abrasive type of wear was the dominant mechanism of wear.	[46]

4. COF of nanocomposites is also reduced with the addition of nanoparticles, while solid lubricant reinforcements are more effective in reducing the COF.
5. Applied load strongly influences the wear rate, followed by sliding speed and distance.
6. The mechanism responsible for wear was adhesion, abrasion, plastic deformation and delamination.

REFERENCES

[1] P. Singh, R. Mishra, and B. Singh, "Microstructural and mechanical characterization of lamb bone ash and boron carbide reinforced ZA-27 hybrid metal matrix composites," *Proceedings of the Institution of Mechanical Engineers, Part L: Journal of Materials: Design and Applications*, vol. 235, no. 11, 2021.

[2] P. Singh, R. Mishra, and B. Singh, "Mechanical characterization of eggshell ash and boron carbide reinforced ZA-27 hybrid metal matrix composites," *Proceedings of the Institution of Mechanical Engineers, Part C: Journal of Mechanical Engineering Science*, vol. 236, no. 3, pp. 1766–1779, 2022.

[3] S. C. Tjong, "Novel nanoparticle-reinforced metal matrix composites with enhanced mechanical properties," *Advanced Engineering Materials*, vol. 9, no. 8, pp. 639–652, 2007.

[4] S. Banerjee, P. Sahoo, and J. P. Davim, "Tribological characterisation of magnesium matrix nanocomposites: A review," *Advances in Mechanical Engineering*, vol. 13, no. 4, 2021.

[5] L. Singh, B. Singh, and K. K. Saxena, "Manufacturing techniques for metal matrix composites (MMC): An overview," *Advances in Materials and Processing Technologies*, vol. 6, no. 2, pp. 441–457, 2020.

[6] H. Wang, Q. Jiang, Y. Wang, B. Ma, and F. Zhao, "Fabrication of TiB2 particulate reinforced magnesium matrix composites by powder metallurgy," *Materials Letters*, vol. 58, no. 27–28, pp. 3509–3513, 2004.

[7] G. Li, Y. Zhao, H. Wang, G. Chen, Q. Dai, and X. Cheng, "Fabrication and properties of in situ (Al3Zr+ Al_2O_3) p/A356 composites cast by permanent mould and squeeze casting," *Journal of Alloys and Compounds*, vol. 471, no. 1–2, pp. 530–535, 2009.

[8] A. A. Yar, M. Montazerian, H. Abdizadeh, and H. Baharvandi, "Microstructure and mechanical properties of aluminum alloy matrix composite reinforced with nano-particle MgO," *Journal of Alloys and Compounds*, vol. 484, no. 1–2, pp. 400–404, 2009.

[9] J. Lai, Z. Zhang, and X.-G. Chen, "Precipitation strengthening of Al-B4C metal matrix composites alloyed with Sc and Zr," *Journal of Alloys and Compounds*, vol. 552, pp. 227–235, 2013.

[10] Q. Zhang, X. Ma, and G. Wu, "Interfacial microstructure of SiCp/Al composite produced by the pressure less infiltration technique," *Ceramics International*, vol. 39, no. 5, pp. 4893–4897, 2013.

[11] T. Dursun and C. Soutis, "Recent developments in advanced aircraft aluminium alloys," *Materials & Design (1980–2015)*, vol. 56, pp. 862–871, 2014.

[12] P. Sharma, D. Khanduja, and S. Sharma, "Tribological and mechanical behavior of particulate aluminum matrix composites," *Journal of Reinforced Plastics and Composites*, vol. 33, no. 23, pp. 2192–2202, 2014.

[13] M. Azouni and P. Casses, "Thermophysical properties effects on segregation during solidification," *Advances in Colloid and Interface Science*, vol. 75, no. 2, pp. 83–106, 1998.

[14] X. Li, Y. Yang, and X. Cheng, "Ultrasonic-assisted fabrication of metal matrix nanocomposites," *Journal of Materials Science*, vol. 39, no. 9, pp. 3211–3212, 2004.

[15] J. Lan, Y. Yang, and X. Li, "Microstructure and microhardness of SiC nanoparticles reinforced magnesium composites fabricated by ultrasonic method," *Materials Science and Engineering: A*, vol. 386, no. 1–2, pp. 284–290, 2004.

[16] E.-S. Y. El-Kady, T. S. Mahmoud, and A. A.-A. Ali, "On the electrical and thermal conductivities of Cast A356/Al$_2$O$_3$ metal matrix nanocomposites," *Materials Sciences and Applications*, vol. 2, no. 09, p. 1180, 2011.

[17] C. Suryanarayana, "Mechanical alloying and milling," *Progress in Materials Science*, vol. 46, no. 1–2, pp. 1–184, 2001.

[18] C. Hsu, C. Chang, P. Kao, N. Ho, and C. Chang, "Al-Al$_3$Ti nanocomposites produced in situ by friction stir processing," *Acta Materialia*, vol. 54, no. 19, pp. 5241–5249, 2006.

[19] Z. Ma, "Friction stir processing technology: A review," *Metallurgical and Materials Transactions A*, vol. 39, no. 3, pp. 642–658, 2008.

[20] V. Srinivas, A. Jayaraj, V. Venkataramana, T. Avinash, and P. Dhanyakanth, "Effect of ultrasonic stir casting technique on mechanical and tribological properties of aluminium-multi-walled carbon nanotube nanocomposites," *Journal of Bio- and Tribo-Corrosion*, vol. 6, no. 2, pp. 1–10, 2020.

[21] E. M. Sharifi, F. Karimzadeh, and M. Enayati, "Fabrication and evaluation of mechanical and tribological properties of boron carbide reinforced aluminum matrix nanocomposites," *Materials & Design*, vol. 32, no. 6, pp. 3263–3271, 2011.

[22] M. C. Şenel, M. Gürbüz, and E. Koç, "Mechanical and tribological behaviours of aluminium matrix composites reinforced by graphene nanoplatelets," *Materials Science and Technology*, vol. 34, no. 16, pp. 1980–1989, 2018.

[23] A. Dorri Moghadam, E. Omrani, P. L. Menezes, and P. K. Rohatgi, "Effect of in-situ processing parameters on the mechanical and tribological properties of self-lubricating hybrid aluminum nanocomposites," *Tribology Letters*, vol. 62, no. 2, pp. 1–10, 2016.

[24] N. Faisal and K. Kumar, "Mechanical and tribological behaviour of nano scaled silicon carbide reinforced aluminium composites," *Journal of Experimental Nanoscience*, vol. 13, no. sup1, pp. S1–S13, 2018.

[25] A. Prasad Reddy, P. Vamsi Krishna, and R. Rao, "Tribological behaviour of Al6061-2SiC-xGr hybrid metal matrix nanocomposites fabricated through ultrasonically assisted stir casting technique," *Silicon*, vol. 11, no. 6, pp. 2853–2871, 2019.

[26] H. S. Vaziri, A. Shokuhfar, and S. S. S. Afghahi, "Investigation of mechanical and tribological properties of aluminum reinforced with Tungsten Disulfide (WS2) nanoparticles," *Materials Research Express*, vol. 6, no. 4, p. 045018, 2019.

[27] M. Azadi, M. Zolfaghari, S. Rezanezhad, and M. Azadi, "Effects of SiO_2 nano-particles on tribological and mechanical properties of aluminum matrix composites by different dispersion methods," *Applied Physics A*, vol. 124, no. 5, pp. 1–13, 2018.

[28] A. E. Nassar and E. E. Nassar, "Properties of aluminum matrix Nano composites prepared by powder metallurgy processing," *Journal of King Saud University-Engineering Sciences*, vol. 29, no. 3, pp. 295–299, 2017.

[29] G. Pitchayyapillai, P. Seenikannan, P. Balasundar, and P. Narayanasamy, "Effect of nano-silver on microstructure, mechanical and tribological properties of cast 6061 aluminum alloy," *Transactions of Nonferrous Metals Society of China*, vol. 27, no. 10, pp. 2137–2145, 2017.

[30] N. Hosseini, F. Karimzadeh, M. Abbasi, and M. H. Enayati, "Tribological properties of Al6061-Al_2O_3 nanocomposite prepared by milling and hot pressing," *Materials & Design*, vol. 31, no. 10, pp. 4777–4785, 2010.

[31] F. Lin, J. Wang, H. Wu, F. Jia, Y. Lu, M. Ren, M. Yang, Z. Chen, and Z. Jiang, "Synergistic effects of TiC and graphene on the microstructure and tribological properties of Al2024 matrix composites," *Advanced Powder Technology*, vol. 32, no. 10, pp. 3635–3649, 2021.

[32] P. Paulraj and R. Harichandran, "The tribological behavior of hybrid aluminum alloy nanocomposites at High temperature: Role of nanoparticles," *Journal of Materials Research and Technology*, vol. 9, no. 5, pp. 11517–11530, 2020.

[33] V. Chak and H. Chattopadhyay, "Synthesis of graphene-aluminium matrix nanocomposites: Mechanical and tribological properties," *Materials Science and Technology*, vol. 37, no. 5, pp. 467–477, 2021.

[34] A. S. Zayed, B. M. Kamel, T. Abdelsadek Osman, O. A. Elkady, and S. Ali, "Experimental study of tribological and mechanical properties of aluminum matrix reinforced by Al_2O_3/CNTs," *Fullerenes, Nanotubes and Carbon Nanostructures*, vol. 27, no. 7, pp. 538–544, 2019.

[35] I. Manivannan, S. Ranganathan, and S. Gopalakannan, "Tribological behavior of aluminum nanocomposites studied by application of response surface methodology," *Advanced Composites and Hybrid Materials*, vol. 2, no. 4, pp. 777–789, 2019.

[36] V. Sivamaran, V. Balasubramanian, M. Gopalakrishnan, V. Viswabaskaran, A. G. Rao, and G. Sivakumar, "Mechanical and tribological properties of Self-Lubricating Al 6061 hybrid nano metal matrix composites reinforced by nSiC and MWCNTs," *Surfaces and Interfaces*, vol. 21, p. 100781, 2020.

[37] A. S. Mohammed, O. S. Aljebreen, A. S. Hakeem, T. Laoui, F. Patel, and M. M. Ali Baig, "Tribological behavior of aluminum hybrid nanocomposites reinforced with alumina and graphene oxide," *Materials*, vol. 15, no. 3, p. 865, 2022.

[38] A. Sharma, H. Fujii, and J. Paul, "Influence of reinforcement incorporation approach on mechanical and tribological properties of AA6061-CNT nanocomposite fabricated via FSP," *Journal of Manufacturing Processes*, vol. 59, pp. 604–620, 2020.

[39] C. Kannan, R. Ramanujam, and A. Balan, "Mathematical modeling and optimization of tribological behaviour of Al 7075 based hybrid nanocomposites," *Proceedings of the Institution of Mechanical Engineers, Part J: Journal of Engineering Tribology*, vol. 235, no. 8, pp. 1561–1574, 2021.

[40] H. Fallahdoost, A. Nouri, and A. Azimi, "Dual functions of TiC nanoparticles on tribological performance of Al/graphite composites," *Journal of Physics and Chemistry of Solids,* vol. 93, pp. 137–144, 2016.

[41] R. R. Raj, J. Yoganandh, M. Saravanan, and S. S. Kumar, "Effect of graphene addition on the mechanical characteristics of AA7075 aluminium nanocomposites," *Carbon Letters,* vol. 31, no. 1, pp. 125–136, 2021.

[42] P. Madhukar, N. Selvaraj, C. Rao, and G. V. Kumar, "Tribological behavior of ultrasonic assisted double stir casted novel nano-composite material (AA7150-hBN) using Taguchi technique," *Composites Part B: Engineering,* vol. 175, p. 107136, 2019.

[43] G. Majzoobi, A. Atrian, and M. Enayati, "Tribological properties of Al7075-SiC nanocomposite prepared by hot dynamic compaction," *Composite Interfaces,* vol. 22, no. 7, pp. 579–593, 2015.

[44] S. Mosleh-Shirazi and F. Akhlaghi, "Tribological behavior of Al/SiC and Al/SiC/2vol% Gr nanocomposites containing different amounts of nano SiC particles," *Materials Research Express,* vol. 6, no. 6, p. 065039, 2019.

[45] A. Pal, S. Poria, G. Sutradhar, and P. Sahoo, "Tribological behavior of Al-WC nano-composites fabricated by ultrasonic cavitation assisted stir-cast method," *Materials Research Express,* vol. 5, no. 3, p. 036521, 2018.

[46] P. Madhukar, N. Selvaraj, C. Rao, and G. V. Kumar, "Fabrication and characterization two step stir casting with ultrasonic assisted novel AA7150-hBN nanocomposites," *Journal of Alloys and Compounds,* vol. 815, p. 152464, 2020.

Chapter 4

Tribological performance of RGO and Al$_2$O$_3$ nanodispersions in synthetic lubricant

Pranav Dev Srivyas
Indian Institute of Technology, Guwahati

M. S. Charoo
National Institute of Technology, Srinagar

Soundhar Arumugam and Tanmoy Medhi
Indian Institute of Technology, Guwahati

CONTENTS

4.1 Introduction	65
4.2 Experimental details	66
4.2.1 Materials and methods	66
4.2.2 Preparation of RGO and Al$_2$O$_3$ nanolubricants	66
4.2.3 Tribological testing	67
4.3 Results and discussion	67
4.3.1 Effect of Al$_2$O$_3$ concentration on friction and wear behavior	67
4.3.2 Effect of reduced graphene oxide concentration on friction and wear	69
4.3.3 Extreme pressure (EP) test	70
4.3.4 Worn surface analysis	71
4.4 Conclusions	73
References	73

4.1 INTRODUCTION

The most significant technological advancement of the 21st century is widely regarded as nanotechnology. Recent research has shown that adding various nanostructures to lubricants can reduce tribological losses in mechanical systems [1,2]. As a result, a plethora of nanomaterials with intriguing anti-friction and anti-wear properties have emerged, particularly graphene nanolubricants [3,4]. Graphitic materials have been extensively studied as lubricant additives. Their tribological properties demonstrate reduced friction and wear when compared to other nanoadditives [5–7]. They are also used to improve lubrication system thermal conductivities.

Carbon nanostructures have incredible overall properties, with good chemical stability [8,9].They are composed of a single layer of carbon packed in a hexagonal lattice with a carbon–carbon distance of 0.142 nm [10–13]. Graphene-based lubricant oil additives are being investigated, including alkylated graphene [14], ultrathin graphene [15], chemically functionalized reduced graphene oxide (RGO) [16], graphene oxide [17], and monolayer RGO [18,19]. Previous studies on the tribological properties of dispersing RGOs in lubricant oils have been published [13–16]. Gupta et al. discovered that RGO dispersions in polyalkylene (PEG) oil have worse friction and wear at low and high RGO concentrations but improve significantly at intermediate concentrations [13]. Li et al. studied the tribological properties of PAO6 dispersions containing various types of RGOs (Oxygen contents ranging from 8.1 to 11.7 wt. %) and concluded that the wear and friction behavior is highly dependent on the synthetic procedure [15]. Schlüter et al. measured Stribeck curves of two nano-dispersions composed of an additivated and fully saturated ester oil and thermally reduced graphite oxide flakes (oxygen content of 14 wt%). It was discovered that the latter had better lubricating properties than the former [16].

In this study, RGO sheets and Al_2O_3 nanoparticles were monodispersed in the PAO-4 lubricant. Pin-on-disc testing in a reciprocating design was used to evaluate the friction and wear characteristics. The lubrication performance and related mechanisms of base lubricant, RGO lubricant, and Al_2O_3 lubricant were discovered through the investigation of wear.

4.2 EXPERIMENTAL DETAILS

4.2.1 Materials and methods

Because mild steel and bronze have a wide range of uses in bushings and bearings, these two materials are the most often used tribo-pair in mechanical systems. The pins are made of bronze. A disc made of AISI 4340 alloy steel is utilized to combat the bronze. The disc material is an alloy with chromium (Cr), nickel (Ni), and molybdenum (Mo) other than iron (Fe) and carbon that has been heat-treated (C). Four-ball EP tests are conducted using 12.7 mm diameter AISI E52100 stainless steel balls. Grade 4 polyalphaolefin is employed as the lubricant. The virgin lubricant is supplemented with Al_2O_3 (20 nm) and RGO (5–10 nm in thickness; multilayered) nanoparticles.

4.2.2 Preparation of RGO and Al_2O_3 nanolubricants

Al_2O_3 nanoparticles with a diameter of 20 nm (provided by Intelligent Material Pvt. Ltd., India) and RGO nanosheets with an average particle size of 20–30 nm (Pt nanocrystal) and a thickness of 5–10 nm were employed as additives. RGO sheets and Al$_2$O$_3$ nanoparticles with variable concentrations (0.05, 0.1, 0.15, 0.2, 0.25 wt. %) were ultrasonically dispersed in PAO-4 at room temperature. After that, the nanolubricants were mechanically stirred and de-agglomerated with a high-intensity ultrasonicator. A chilled water bath was used to keep the suspension temperature stable during ultrasonic agitation.

4.2.3 Tribological testing

Pin-on-disc friction testing on a tribometer was used to analyze the tribological characteristics of the nanolubricants. Before testing, the samples are polished using silicon carbide emery paper on an automatic polisher in accordance with ASTM E3-95 Standard. The surface is polished using diamond paste (8, 6, 3, 1, and 0.25 m) on a velvet cloth with aerosol spray until mirror surface finish is achieved (surface roughness of Ra = 50 ± 5 nm). Each tribotest was performed three times in order to reduce experimental error. Testing for friction was done in line with ASTM G133-05 standards. Tribotests were performed using 2 mm stroke, constant sliding distance of 1,000 m, normal load of 50 N, and a 10 Hz frequency that doesn't change throughout the tribotest. Samples are cleaned with acetone before and after each test for 10 minutes in an ultrasonic bath, and they are then dried in a vacuum oven. A bronze pin with a diameter of 10 mm and a tip diameter of 2 mm is secured in a pin holder that reciprocates over a fixed mild steel disc while being vertically positioned. On a four-ball (Ducom) tester, Extreme Pressure (EP) tests are carried out in accordance with ASTM D2783 standard to assess the extreme pressure characteristics of the virgin and designed lubricants. Starting with a load of 784 N, successive runs are made at progressively greater loads of 981, 1236, 1569, and 1961 N until the balls fuse together, which is referred to as the weld point of the oil.

4.3 RESULTS AND DISCUSSION

4.3.1 Effect of Al$_2$O$_3$ concentration on friction and wear behavior

The effects of particle concentration when using Al$_2$O$_3$ lubricants on the wear, surface roughness (Ra), and coefficient of friction (COF) of the worn surfaces were investigated. The tests were run with a standard load of 50 N and 1,000 m sliding distance. It is important to understand that "0.00 wt%" refers to base oil. The inclusion of Al$_2$O$_3$ nanoparticles resulted in lower values of COF, wear volume (WV), and Ra compared to the baseline, indicating enhanced lubrication of the rubbing contact. According to

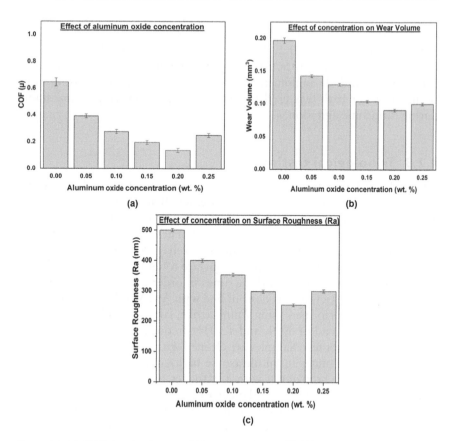

Figure 4.1 (a) COF vs. aluminum oxide concentration. (b) Wear volume vs. aluminum oxide concentration. (c) Surface roughness vs. aluminum oxide concentration.

Figure 4.1a, the COF value of the Al_2O_3 lubricant falls as particle concentration rises from 0.05 to 0.20 wt%. Only a small rise in COF results with further increases in concentration. Similar to WV, Ra also drops with concentration, reaching its lowest point at 0.2 wt%. As indicated in Figure 4.1b and c, WV and Ra also increase with any additional increase. This demonstrates that there is an ideal particle concentration for getting the best performance in terms of friction and wear [17]. If there aren't enough additives, there won't be enough nanoparticles to effectively lubricate the rubbing area. When there are too many additives, nanoparticles have a tendency to clump together at the contact surface, which increases stress under abrasion conditions and causes wear protection to fail [18]. The minimum friction and wear is reported for 0.2 wt. % of Al_2O_3 nanoparticles, which reported 70.41% abasement in the COF and 53.91% reduction in WV with respect to virgin PAO 4 lubricant.

Tribological performance of RGO and Al$_2$O$_3$ nanodispersions 69

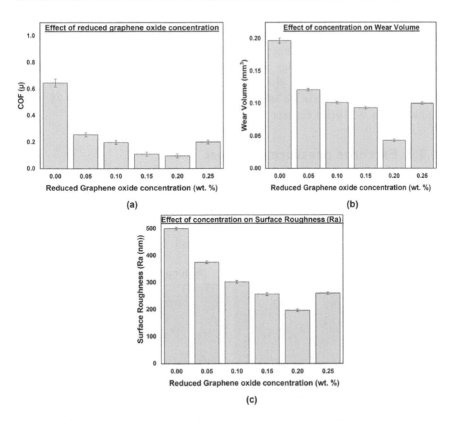

Figure 4.2 (a) COF vs. reduced graphene oxide concentration. (b) Wear volume vs. reduced graphene oxide concentration. (c) Surface roughness vs. reduced graphene oxide concentration.

4.3.2 Effect of reduced graphene oxide concentration on friction and wear

The impact of particle concentration utilizing RGO lubricants on the COF, wear, and surface roughness (Ra) of the worn surfaces were studied. The tests were run with a standard load of 50 N and a 1,000 m sliding distance. It is important to understand that "0.00 wt%" refers to base oil. The inclusion of RGO nanoparticles resulted in lower values of COF, WMW, and Ra compared to the baseline, indicating enhanced lubrication of the rubbing contact. According to Figure 4.2a, the COF value of the RGO lubricant falls as particle concentration rises from 0.05 to 0.20 wt%. Only a small rise in COF results in further increase in concentration. Similar to WV, Ra also drops with concentration, reaching its lowest point at 0.2 wt%. As indicated in Figure 4.2b and c WV and Ra also increases with any additional increase. This demonstrates that there is an ideal particle concentration for getting the best performance in terms of friction and wear [17].

If there aren't enough additives, there won't be enough nanoparticles to effectively lubricate the rubbing area. When there are too many additives, nanoparticles have a tendency to clump together at the contact surface, which increases stress under abrasion conditions and causes wear protection to fail [18]. For RGO, the minimum friction and wear is reported for 0.2 wt. % of nanoparticles with 79.04% reduction in COF and 78.02% reduction in WV with respect to the virgin PAO 4 lubricant.

4.3.3 Extreme pressure (EP) test

EP tests are carried out in accordance with ASTM D2783 standard to assess the extreme pressure characteristics of the virgin and designed lubricants. Starting with a load of 784 N, successive runs are made at progressively greater loads of 981, 1236, 1569, and 1961 N until the balls fuse together, which is referred to as the weld point of the oil. Figure 4.3 displays optical

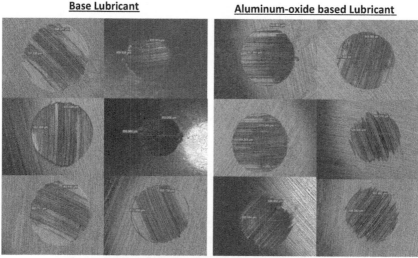

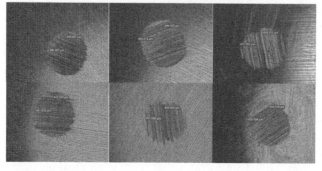

Figure 4.3 Optical micrographs of the balls for various lubricant samples.

micrographs of the balls for various lubricant samples. There is evidence of adhesive wear, which is correlated with the load applied to the contacting surface's asperities. Micro-welded points form as a result of the distortion and adhesion of these asperities. Pits and grooves are produced as a result of the top ball's constant rotation rupturing the micro-welded spots. Al_2O_3, RGO lubricants are found to have lower COF than virgin lubricants and to be more stable under the final seizure load (FSL). The FSL is the stress at which the material welds and the fluid film breaks. Because the lubricant coating grows thinner under greater loading conditions and nanoadditives fill the empty space between the asperities, the tribo-pair is kept in the boundary lubrication regime.

4.3.4 Worn surface analysis

Scanning Electron Microscope (SEM) pictures of lubricated worn surfaces with base lubricant, RGO, and Al_2O_3 lubricants are shown in Figure 4.4. Wide furrows and a few pits can be seen on the surface, which has been worn down while employing the base lubricant. During the sliding, the tribo-oxide layer is generated on parallel scratch locations. When localized oxide layer is separated, the resulting debris heightened wear as well as friction during sliding [19]. The RGO and Al_2O_3 lubricant's worn surface is comparatively smooth with few thin scratches. Raman spectra show distinct D and G bands. This suggests that during sliding, a GO tribofilm developed and left traces on the worn surface. Wide scratches have an ID/IG ratio of 1.067, which is comparable to the 1.080 ratio found for the GO solution. This shows that the GO tribo-film anchored on the GO nanosheets, which in turn decreased the friction-induced damage.

The data presented above unmistakably showed that RGO lubricant's overall tribological performance was better than that of the base lubricant and the Al_2O_3 lubricant. When the base lubricant was utilized, incomplete localized tribo-oxide layer development occurred on the surface at the beginning of sliding. As sliding continued, it was possible for the tribo-oxide layers to locally break, causing severe asperity-asperity contact in the restricted area and consequently extremely high wear and friction. Sliding wear may be further aggravated by the introduction of fragments from fractured layers into the contact zone.

When lubrication is performed using the Al_2O_3 lubricant, the Al_2O_3 nanoparticles form the tribo-layer during subsequent sliding and serve as a load bearer to safeguard the contact pair. In addition, the tribo-layer might aid in removing wear particles from the area of contact, strengthening its anti-wear capability [17]. This localized tribo-oxide protective coating on the furrow surfaces was also anticipated to help guard the contact surface against wear. In order to support load and lessen wear, Al_2O_3 tribo-layers were created in specified locations. The lubricity of the Al_2O_3 tribo-layer might be worsened by the agglomeration of Al_2O_3 in the contact zone, which

72 Nanomaterials for Sustainable Tribology

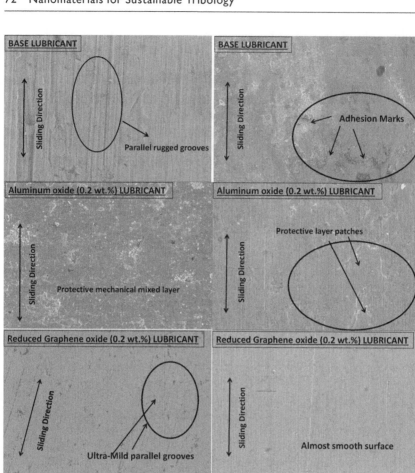

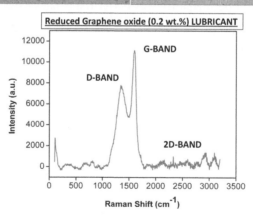

Figure 4.4 SEM and Raman analysis of the wear scar.

disrupts the dynamic equilibrium of the Al_2O_3 tribo-layer distribution. As a result, the Al_2O_3 tribo-layer's performance was significantly diminished.

A GO film is created when RGO sheets adhere to the peaks as well as valleys of a contact zone while sliding. The GO layer greatly minimizes the friction by reducing the resistance to shear [20,21]. However, with quite high contact forces, the GO is quickly damaged. Therefore, the protecting film is damaged when the GO fractures, increasing friction mechanism and wear due to plowing.

4.4 CONCLUSIONS

The development of lubricant with RGO nanosheets and Al_2O_3 nanoparticles as additives was successful. Al_2O_3 nanoparticles were evenly and sparsely distributed. As a result, Al_2O_3 nanoparticles could efficiently access the contact surface. The Al_2O_3 particles also served as load bearers, strengthening the mechanically mixed layer. Nanolubricants were able to dynamically polish asperities and transport wear debris away from the contact zone. RGO nanolubricant performs the best among all, thanks to the low shear resistance provided by protective layer. According to tribological testing, RGO and Al_2O_3 lubricants significantly decreased friction, wear, and surface roughness in comparison to the base lubricant. The minimum friction and wear is reported for 0.2 wt. % of Al_2O_3 nanoparticles, which reported 70.41% abasement in the COF and 53.91% reduction in WV with respect to virgin PAO 4 lubricant. For RGO, the minimum friction and wear is reported for 0.2 wt. % of nanoparticles with 79.04% reduction in COF and 78.02% reduction in WV with respect to virgin PAO 4 lubricant.

REFERENCES

1. J. Padgurskas, R. Rukuiza, I. Prosyčevas, & R. Kreivaitis, Tribological properties of lubricant additives of Fe, Cu and Co nanoparticles, *Tribol. Int.*, 60 (2013), pp. 224–232.
2. A. B. Najan, R. R. Navthar, & M. J. Gitay, Experimental investigation of tribological properties using nanoparticles as modifiers in lubricating oil, *IRJET*, 04 (2017), pp. 1125–1129.
3. S. Shahnazar, S. Bagheri, & S. B. A. Hamid, Enhancing lubricant properties by nanoparticles additives, *Int. J. Hydrog. Energy*, 41 (2016), pp. 3153–3170.
4. D. Berman, A. Erdemir, & A. V. Sumant, Graphene: A new emerging lubricant, *Mater. Today*, 17 (2014), pp. 31–42.
5. A. K. Rasheed, M. Khalid, W. Rashmi, T. C. S. M. Gupta, & A. Chan, Graphene based nanofluids and nanolubricants—Review of recent developments, *Renew. Sust. Energ. Rev.*, 63 (2016), pp. 346–362.
6. A. K. Sharma, A. K. Tiwari, A. R. Dixit, & R. K. Singh, Investigation into performance of SiO2 nanoparticle based cutting fluid in machining process, *Mater. Today*, 4 (2017), pp. 133–141.

7. M. Farsadi, S. Bagheri, & N. A. Ismail, Nanocomposite of functionalized graphene and molybdenum disulfide as friction modifier additive for lubricant, *J. Mol. Liq.*, 244 (2017), pp. 304–308.
8. M. S. Dresselhaus, A. Jorio, & R. Saito, Characterizing graphene, graphite, and carbon nanotubes by Raman spectroscopy, *Annu. Rev. Condens. Matter Phys.*, 1 (2010), pp. 89–108.
9. V. Eswaraiah, V. Sankaranarayanan, & S. Ramaprabhu, Graphene-based engine oil Nanofluids for Tribological applications, *ACS Appl. Mater. Interfaces*, 3 (2011), pp. 4221–4227.
10. D. Berman, A. Erdemir, & A. V. Sumant, Few layer graphene to reduce wear and friction on sliding steel surfaces, *Carbon*, 54 (2013), pp. 454–459.
11. S. Choudhary, H. P. Mungse, & O. P. Khatri, Dispersion of alkylated graphene in organic solvents and its potential for lubrication applications, *J. Mater. Chem.*, 22 (2012), pp. 21032–21039.
12. L. Zhang, Y. He, L. Zhu, C. Yang, Q. Niu, & C. An, In situ alkylated graphene as oil dispersible additive for friction and wear reduction, *Ind. Eng. Chem. Res.*, 56 (2017), pp. 9029–9034.
13. B. Gupta, N. Kumar, K. Panda, S. Dash, & A. K. Tyagi, Energy efficient reduced graphene oxide additives: Mechanism of effective lubrication and antiwear properties, *Sci. Rep.*, 6 (2016), p. 18372.
14. E. Varrla, S. Venkataraman, & R. Sundara, Graphene-based engine oil nanofluids for Tribological applications, *ACS Appl. Mater. Interfaces*, 3 (2011), pp. 4221–4227.
15. Y. Li, J. Zhao, C. Tang, Y. He, Y. Wang, J. Chen, J. Mao, Q. Zhou, B. Wang, F. Wei, J. Luo, & G. Shi, Highly exfoliated reduced graphite oxide powders as efficient lubricant oil additives, *Adv. Mater. Interfaces*, 3 (2016), p. 1600700.
16. B. Schlüter, R. Mülhaupt, & A. Kailer, Synthesis and tribological characterization of stable dispersions of thermally reduced graphite oxide, *Tribol. Lett.*, 53 (2014), pp. 353–363.
17. A. He, S. Huang, J.-H. Yun, H. Wu, Z. Jiang, J. R. Stokes, S. Jiao, L. Wang, & H. Huang, Tribological performance and lubrication mechanism of alumina nanoparticle water-based suspensions in ball-on-three-plate testing, *Tribol. Lett.*, 65 (2017), p. 40.
18. D. Bazrgari, F. Moztarzadeh, A. Sabbagh-Alvani, M. Rasoulianboroujeni, M. Tahriri, & L. Tayebi, Mechanical properties and tribological performance of epoxy/Al_2O_3 nanocomposite, *Ceram. Int.*, 44 (2018), pp. 1220–1224.
19. X. Zheng, Y. Xu, J. Geng, Y. Peng, D. Olson, & X. Hu, Tribological behavior of Fe_3O_4/MoS_2 nano composites additives in aqueous and oil phase media, *Tribol. Int.*, 102 (2016), pp. 79–87.
20. H. Kinoshita, Y. Nishina, A. A. Alias, & M. Fujii, Tribological properties of monolayer graphene oxide sheets as water-based lubricant additives, *Carbon*, 66 (2014), pp. 720–723.
21. H.-J. Song, & N. Li, Frictional behavior of oxide graphene nanosheets as water-base lubricant additive, *Appl. Phys. A*, 105 (2011), pp. 827–832.

Chapter 5

Synergism of the hybrid lubricants to enhanced tribological performance

Pranav Dev Srivyas
Indian Institute of Technology, Guwahati

M. S. Charoo
National Institute of Technology, Srinagar

Soundhar Arumugam
Indian Institute of Technology, Guwahati

Tanmoy Medhi
Indian Institute of Technology, Guwahati

CONTENTS

5.1 Introduction	75
5.2 Experimental details	77
5.2.1 Materials	77
5.2.2 Preparation of RGO-Al$_2$O$_3$ hybrid lubricants	77
5.2.3 Tribological testing	77
5.3 Results and discussion	78
5.3.1 Effect of concentration on friction and wear	78
5.3.2 Extreme pressure (EP) test	78
5.3.3 Worn surface analysis and lubrication mechanism	81
5.4 Conclusions	82
References	83

5.1 INTRODUCTION

In steel forming and other industrial processes, friction and wear have long been a problem because they lead to energy loss and mechanical failure [1–6]. To increase the effectiveness and performance of mechanical systems, lubricants were employed in manufacturing and machining operations [7–11]. Due to the intrinsic toxicity and nonrenewable, nonbiodegradable character of lubricants, their use has, however, given rise to significant environmental problems [4–6]. Recently, there has been a lot of interest in the use of carbon-based nanosheets, as solid lubricants as well as lubricant additives [12–17].

PEGlated graphene was created in a stable aqueous solution by Hu et al. They demonstrated that the PEGlated graphene addition greatly reduced friction and wear [14]. The gains were credited to the development of a protective lubricating layer of carbon coating made up of an adhesive film. Monolayer graphene oxide sheets were studied by Kinoshita et al. for their tribological characteristics as additions [15]. The lubrication of tungsten carbide balls on steel plate was found to be improved by the addition of graphene oxide (GO) sheets. The development of a carbon layer of GO sheets was attributed for the enhanced performance. However, according to recent research [18–22], the van der Waals interactions between the carbon-based sheets make it simple for them to stack again inside the matrix liquid. This resulted in less accessible surface area and restricted generation of the self-lubricating layer. Their ability to reduce friction and resist wear was thereby hampered. Thus, spacers were dispersed across the nanosheets to prevent restacking and preserve their characteristics [23–25]. Al_2O_3 nanoparticles distributed in base fluids with sizes ranging from 1 to 100 nm have demonstrated good friction and wear performance. This is attributed to the effects of rolling and mending mechanism as well as protective layer formation on the contacting surface [26–29]. Due to their high surface activity, Al_2O_3 nanoparticles do, however, have a tendency to aggregate. Agglomeration generates particles' sedimentation, which in turn increases stress under abrasion conditions and eventually results in wear protection failing [30,31]. Although surfactants may reduce agglomeration, they can bring along additional problems such inadequate preparation control, pollution, and greater material costs [32].

Depositing inorganic nanoparticles onto nanosheets without molecular linkers result in dimensionally integrated hybrid nanoadditives, which is one of the most efficient methods for enhancing dispersion. This facilitates nanosheet separation and reduces nanoparticle aggregation in the matrix liquid [32–36]. For instance, by depositing Fe_3O_4 nanoparticles onto MoS_2 sheets, Zheng et al. created Fe_3O_4/MoS_2 nanocomposites. This made the Fe_3O_4 nanoparticles more dispersed and prevented the MoS_2 from stacking back up in aqueous and biodiesel conditions. The evenly dispersed Fe_3O_4 nanoparticles acted as nanoballs throughout the tribological tests, which decreased edge stacking of the MoS_2 sheets. As a result, there was good lubrication in the rubbing area, which decreased wear and friction. In order to create tungsten disulfide (WS2)/graphene nanocomposites, Zheng distributed WS_2 nanoparticles over graphene nanosheets. It was found that WS_2 nanoparticles may be evenly placed over the sheets of graphene to create stable lubricants. The lubricants demonstrated synergistic friction-reduction and anti-wear benefits in addition to maintaining the inherent features of each component. Similar to this, an additive that is dimensionally integrated may develop from the dispersion of Al_2O_3 nanoparticles in an RGO.

However, there haven't been many attempts to date to investigate the additive RGO-Al_2O_3 hybrid nanoparticles' synergistic lubricating performance.

In this study, PAO-4 was combined with RGO sheets and Al_2O_3 nanoparticles to create a dimensionally integrated RGO-Al_2O_3 lubricant. Pin-on-disc testing in a reciprocating design was used to evaluate the friction and wear characteristics. The synergistic lubrication performance and related mechanisms of RGO-Al_2O_3 lubricant were ascurtained through the investigation of wear.

5.2 EXPERIMENTAL DETAILS

5.2.1 Materials

Mild steel and bronze have a wide range of uses in bushings and bearings; these two materials are the most often used tribopair in mechanical systems. The pins are made of bronze. A disc made of AISI 4340 alloy steel is utilized to combat the bronze. The disc material is an alloy with chromium (Cr), nickel (Ni), and molybdenum (Mo) other than iron (Fe) and carbon that has been heat-treated (C). Four-ball EP tests are conducted using 12.7 mm diameter AISI E52100 stainless steel balls. Grade 4 polyalphaolefin is employed as lubricant.

5.2.2 Preparation of RGO-Al_2O_3 hybrid lubricants

Al_2O_3 nanoparticles with a diameter of 20 nm (provided by Intelligent material Pvt. Ltd., India) and reduced graphene oxide (RGO) nanosheets with an average particle size of 20–30 nm (Pt Nano crystal) and a thickness of 5–10 nm were employed as additives. RGO sheets and Al_2O_3 nanoparticles were first ultrasonically dispersed in ethanol at room temperature to prepare dimensionally integrated (RGO-Al_2O_3) hybrid nanoparticles. The RGO-Al_2O_3 mixtures were then mechanically stirred and de-agglomerated using a high-intensity ultrasonic sonifier. A circulating chilled water bath was employed to maintain the suspension temperature during the ultrasonic agitation procedure. After that, the suspension was heated to remove any remaining moisture and ethanol. After that, the base oil was added to hybrid additive. Fixed RGO to Al_2O_3 mass ratio of 1:1 hybrid additive was added in various concentrations of 0.05, 0.1, 0.15, 0.20, and 0.25 wt. % to the base lubricant.

5.2.3 Tribological testing

Pin-on-disc friction testing on a tribometer was used to analyze the tribological characteristics of the hybrid nanolubricants. Before testing, the samples were polished using silicon carbide emery paper on an automatic polisher in accordance with the ASTM E3-95 Standard. The surface was polished using diamond paste (8, 6, 3, 1, and 0.25 m) applied on a velvet cloth with aerosol spray until mirror surface finish is achieved (surface roughness of

Ra=50±5 nm). Each tribotest was performed three times in order to reduce experimental error. Testing for friction was done in line with ASTM G133-05 standards. Tribotests were carried out using 2 mm stroke, constant sliding distance of 1,000 m, normal load of 50 N, and a 10 Hz frequency that doesn't change throughout the tribotest. Samples are cleaned with acetone before and after each test for 10 minutes in an ultrasonic bath, and they are then dried in a vacuum oven. A bronze pin with a diameter of 10 mm and a tip diameter of 2 mm is secured in a pin holder that reciprocates over a fixed mild steel disc while being vertically positioned. On a four-ball (Ducom) tester, extreme pressure (EP) tests are carried out in accordance with ASTM D2783 standard to assess the extreme pressure characteristics of the virgin and designed lubricants. Starting with a load of 784 N, successive runs were made at progressively greater loads of 981, 1236, 1569, and 1961 until the balls fuse together, which is referred to as the weld point of the oil.

5.3 RESULTS AND DISCUSSION

5.3.1 Effect of concentration on friction and wear

The impact of particle concentration while utilizing RGO-Al_2O_3 lubricants with a 1:1 mass ratio of RGO to Al_2O_3 on the COF, wear volume, and surface roughness of the worn surfaces was studied. The tests were run with a standard load of 50 N and a 1,000 m sliding distance. It is important to understand that "0.00 wt%" refers to base oil. The inclusion of RGO-Al_2O_3 hybrid nanoparticles resulted in lower values of COF, wear volume, and Ra compared to the baseline, indicating enhanced lubrication of the rubbing contact. According to Figure 5.1a, the COF value of the RGO-Al_2O_3 lubricant falls as particle concentration rises from 0.05 to 0.20 wt%. Only a small rise in COF results with further increases in concentration. Similar to wear volume, Ra also drops with concentration, reaching its lowest point at 0.2 wt%. As indicated in Figure 5.1b and c wear volume, Ra also increases with any additional increase. This demonstrates that there is an ideal particle concentration for getting the best performance in terms of friction and wear [29]. If there aren't enough additives, there won't be enough nanoparticles to effectively lubricate the rubbing area. When there are too many additives, nanoparticles tend to clump together at the contact surface, which increases stress under abrasion conditions and causes wear protection to fail [31].

5.3.2 Extreme pressure (EP) test

EP tests are carried out in accordance with ASTM D2783 standard to assess the extreme pressure characteristics of the virgin and designed lubricants. Starting with a load of 784 N, successive runs are made at progressively

Synergism of the hybrid lubricants 79

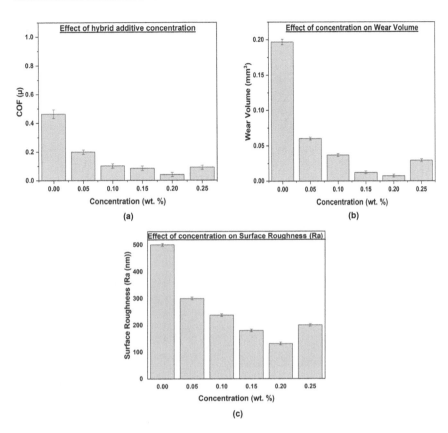

Figure 5.1 (a) Effect of RGO-Al$_2$O$_3$ concentration on COF vs. concentration. (b) Effect of RGO-Al$_2$O$_3$ concentration on wear volume vs. concentration. (c) Effect of RGO-Al$_2$O$_3$ concentration on surface roughness vs. concentration.

greater loads of 981, 1236, 1569, and 1961 N until the balls fuse together, which is referred to as the weld point of the oil. Figure 5.2 displays optical micrographs of the balls for various lubricant samples. There is evidence of adhesive wear, which is correlated with the load applied to the contacting surface's asperities. Micro-welded points form as a result of the distortion and adhesion of these asperities. Pits and grooves are produced as a result of the top ball's constant rotation rupturing the micro-welded spots. There appears to be a total fluid film breakdown with virgin lubricants. RGO/Al$_2$O$_3$ lubricants are found to have lower COF than virgin lubricants and to be more stable under the final seizure load (FSL). The FSL is the stress at which the material welds and the fluid film breaks. Because the lubricant coating grows thinner under greater loading conditions and nanoadditives fill the empty space between the asperities, the tribopair is kept in the boundary lubrication regime.

80 Nanomaterials for Sustainable Tribology

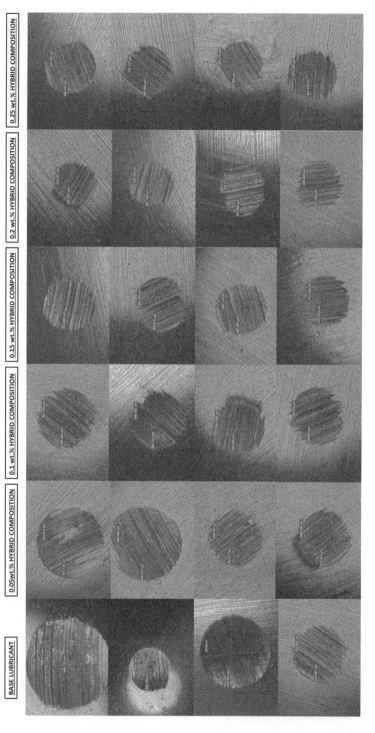

Figure 5.2 Optical micrographs of the balls for various lubricant samples.

5.3.3 Worn surface analysis and lubrication mechanism

Scanning Electron Microscope (SEM) of the lubricated worn surfaces with base lubricant and the RGO-Al$_2$O$_3$ lubricant are shown in Figure 5.3. Wide furrows and a few pits can be seen on the surface that has been worn down while employing virgin lubricants. This indicates that during the sliding, a tribo-oxide layer was generated on the large parallel grooves. When

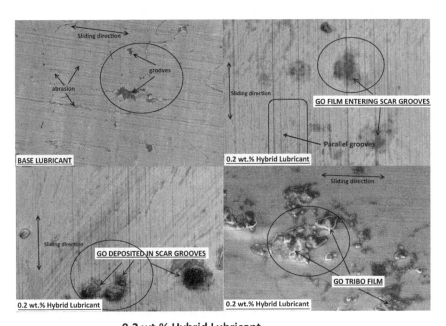

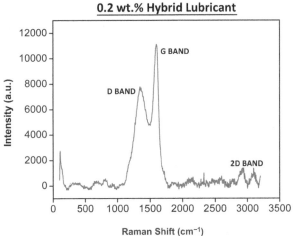

Figure 5.3 SEM and Raman analysis of the wear scar zone.

the oxide layer separated, the resulting debris served as a third body that increases wear and friction [32]. The RGO-Al$_2$O$_3$ lubricant's worn surface is reasonably smooth with thin scratches and has few comparatively wide scratches. Raman spectra show distinct D, G, and 2D bands. This suggests that during sliding, a graphene oxide tribo-film developed and left traces on the worn surface. Wide scratches have an ID/IG ratio of 1.067, which is comparable to the 1.080 ratio found for the RGO lubricant. This shows that the GO tribo-film was reinforced by the nano-Al$_2$O$_3$ anchored on RGO nanosheets, which in turn decreased the damage because of friction.

The RGO-Al$_2$O$_3$ lubricant's overall tribological performance was superior to that of the base lubricant. The synergistic lubrication of RGO and Al$_2$O$_3$ was credited with this outstanding performance. When only the base lubricant was utilized, incomplete localized tribo-oxide layer development occurred on the contact sliding surface at the beginning. As the sliding continued, there is the possibility of the tribo-oxide layers breaking locally. This further leads to severe asperity-asperity contact consequently leading to extremely high friction as well as wear. The sliding wear may be further aggravated by the introduction of fragments from fractured layers into the contact zone.

Al$_2$O$_3$ was dispersed evenly and sparsely across the RGO sheets through mixing after being introduced to RGO suspensions. This resulted in the formation of dimensionally integrated hybrid nanoparticles with Al$_2$O$_3$ as reinforcement. The created hybrid nanoparticles impeded restacking of RGO nanosheets as well as the agglomeration of Al$_2$O$_3$ nanoparticles in the suspension. As a result, the RGO sheets introduced Al$_2$O$_3$ into the contact surface interface to create a composite protective lubricating film. This mechanically mixed layer is made up of a RGO and Al$_2$O$_3$. As the composite protective film covered the rubbing sliding interface, Al$_2$O$_3$ nanoparticles were able to polish asperities and transport debris while also offering low shear resistance. Composite film was strengthened to become much tougher, and it helps to carry a lot of load. Al$_2$O$_3$ nanoparticles could also function more effectively as third-contacting bodies without aggregation, which would lessen friction. As a result of the synergistic interaction between RGO and Al$_2$O$_3$, the tribological behavior improved.

5.4 CONCLUSIONS

The development of a lubricant with RGO-Al$_2$O$_3$ hybrid additives was successful. Al$_2$O$_3$ nanoparticles were evenly and sparsely distributed over the RGO sheets, which prevented agglomeration of the particles and restacking of the RGO nanosheets. As a result, Al$_2$O$_3$ could efficiently access the contact surface and combine with the RGO to form a hybrid composite protective layer. Al$_2$O$_3$ was able to polish asperities and transport wear debris away from the contact zone, thanks to the low shear resistance provided by the RGO in the protective layer. The Al_2O_3 particles also served as load bearers, strengthening the RGO sheets. According to tribological testing, the RGO-Al_2O_3 lubricants significantly decreased friction, wear, and surface roughness. A 0.2 wt.% 1:1 RGO/Al_2O_3 lubricant reduced COF by 91.40%, wear volume by 96%, and surface roughness (Ra) by 72.6% when compared to base lubrication.

REFERENCES

1. Shirizly, A., & Lenard, J. G. (2000). The effect of lubrication on mill loads during hot rolling of low carbon steel strips. *Journal of Materials Processing Technology*, 97, 61–68.
2. Lawal, S. A., Choudhury, I. A., & Nukman, Y. (2014). Evaluation of vegetable and mineral oil-in-water emulsion cutting fluids in turning AISI 4340 steel with coated carbide tools. *Journal of Cleaner Production*, 66, 610–618.
3. Yu, Y., & Lenard, J. G. (2002). Estimating the resistance to deformation of the layer of scale during hot rolling of carbon steel strips. *Journal of Materials Processing Technology*, 121, 60–68.
4. Wu, H., Zhao, J., Xia, W., Cheng, X., He, A., Yun, J. H., Wang, L., Huang, H., Jiao, S., Huang, L., Zhang, S., & Jiang, Z. (2017). A study of the tribological behaviour of TiO_2 nano-additive water-based lubricants. *Tribology International*, 109, 398–408.
5. Alves, S., Barros, B., Trajano, M., Ribeiro, K., & Moura, E. (2013). Tribological behavior of vegetable oil-based lubricants with nanoparticles of oxides in boundary lubrication conditions. *Tribology International*, 65, 28–36.
6. Ji, H., Zhang, X., & Tan, T. (2017). Preparation of a water-based lubricant from lignocellulosic biomass and its tribological properties. *Industrial & Engineering Chemistry Research*, 56, 7858–7864.
7. Srivyas, P. D., Wani, M. F., Sehgal, R., Bisht, C. S. S., Charoo, M. S., Raina, A., & Haq, M. I. U. (2022). Synergetic effect of surface texturing and graphene nanoplatelets on the tribological properties of hybrid self-lubricating composite. *Tribology International*, 168, 107434.
8. Srivyas, P. D., & Charoo, M. S. (2022). Tribological behavior of hybrid aluminum self-lubricating composites under dry sliding conditions at elevated temperature. *Tribology-Materials, Surfaces & Interfaces*, 16(2), 153–167.
9. Srivyas, P. D., & Charoo, M. S. (2021). Nano lubrication behaviour of Graphite, h-BN and Graphene Nano Platelets for reducing friction and wear. *Materials Today: Proceedings*, 44, 7–11.
10. Srivyas, P. D., & Charoo, M. S. (2020). Friction and wear characterization of spark plasma sintered hybrid aluminum composite under different sliding conditions. *Journal of Tribology*, 142(12), 17.
11. Srivyas, P. D., & Charoo, M. S. (2020). Friction and wear reduction properties of GNP nano-particles as nano-additive for Al-Si+ Al_2O_3 composite/ Chromium plated steel tribopair. *Jurnal Tribologi*, 25, 83–101.
12. Tsai, M.-Y., & Jian, S.-X. (2012). Development of a micro-graphite impregnated grinding wheel. *International Journal of Machine Tools and Manufacture*, 56, 94–101.

13. Alberts, M., Kalaitzidou, K., & Melkote, S. (2009). An investigation of graphite nanoplatelets as lubricant in grinding. *International Journal of Machine Tools and Manufacture, 49*, 966–970.
14. Hu, Y., Wang, Y., Zeng, Z., Zhao, H., Ge, X., Wang, K., Wang, L., & Xue, Q. (2018). PEGlated graphene as nanoadditive for enhancing the tribological properties of water-based lubricant. *Carbon, 137*, 41–48.
15. Kinoshita, H., Nishina, Y., Alias, A. A., & Fujii, M. (2014). Tribological properties of monolayer graphene oxide sheets as water-based lubricant additives. *Carbon, 66*, 720–723.
16. Song, H.-J., & Li, N. (2011). Frictional behavior of oxide graphene nanosheets as water-base lubricant additive. *Applied Physics A, 105*, 827–832.
17. Liu, Y., Wang, X., Pan, G., & Luo, J. (2013). A comparative study between graphene oxide and diamond nanoparticles as water-based lubricating additives. *Science China Technological Sciences, 56*, 152–157.
18. Meng, Y., Su, F., & Chen, Y. (2015). A novel nanomaterial of graphene oxide dotted with Ni nanoparticles produced by supercritical CO_2-assisted deposition for reducing friction and wear. ACS *Applied Materials & Interfaces, 7*, 11604–11612.
19. Zhou, Q., Huang, J., Wang, J., Yang, Z., Liu, S., Wang, Z., & Yang, S. (2015). Preparation of a reduced graphene oxide/zirconia nanocomposite and its application as a novel lubricant oil additive. *RSC Advances, 5*, 91802–91812.
20. Meng, Y., Su, F., & Chen, Y. (2015). Synthesis of nano-Cu/graphene oxide composites by supercritical CO_2-assisted deposition as a novel material for reducing friction and wear. *Chemical Engineering Journal, 281*, 11–19.
21. Mensing, J. P., Poochai, C., Kerdpocha, S., Sriprachuabwong, C., Wisitsoraat, A., & Tuantranont, A. (2017). Advances in research on 2D and 3D graphene-based supercapacitors. *Advances in Natural Sciences: Nanoscience and Nanotechnology, 8*, 033001.
22. Xu, L., McGraw, J.-W., Gao, F., Grundy, M., Ye, Z., Gu, Z., & Shepherd, J. L. (2013). Production of high-concentration graphene dispersions in low-boiling-point organic solvents by liquid-phase noncovalent exfoliation of graphite with a hyperbranched polyethylene and formation of graphene/ethylene copolymer composites. *The Journal of Physical Chemistry C, 117*, 10730–10742.
23. Dai, W., Kheireddin, B., Gao, H., & Liang, H. (2016). Roles of nanoparticles in oil lubrication. *Tribology International, 102*, 88–98.
24. Gulzar, M., Masjuki, H., Kalam, M., Varman, M., Zulkifli, N., Mufti, R., & Zahid, R. (2016). Tribological performance of nanoparticles as lubricating oil additives. *Journal of Nanoparticle Research, 18*, 223.
25. Padgurskas, J., Rukuiza, R., Prosyčevas, I., & Kreivaitis, R. (2013). Tribological properties of lubricant additives of Fe, Cu and Co nanoparticles. *Tribology International, 60*, 224–232.
26. Luo, T., Wei, X., Huang, X., Huang, L., & Yang, F. (2014). Tribological properties of Al_2O_3 nanoparticles as lubricating oil additives. *Ceramics International, 40*, 7143–7149.
27. Ali, M. K. A., Xianjun, H., Turkson, R. F., Peng, Z., & Chen, X. (2016). Enhancing the thermophysical properties and tribological behaviour of engine oils using nano-lubricant additives. *RSC Advances, 6*, 77913–77924.

28. Ali, M. K. A., Xianjun, H., Elagouz, A., Essa, F., Abdelkareem, M. A. (2016). Minimizing of the boundary friction coefficient in automotive engines using Al$_2$O$_3$ and TiO$_2$ nanoparticles. *Journal of Nanoparticle Research*, 18, 377.
29. He, A., Huang, S., Yun, J.-H., Wu, H., Jiang, Z., Stokes, J., Jiao, S., Wang, L., & Huang, H. (2017). Tribological performance and lubrication mechanism of alumina nanoparticle water-based suspensions in ball-on-three-plate testing. *Tribology Letters*, 65, 40.
30. Zhang, H. J., Zhang, Z. Z., Guo, F., & Liu, W. M. (2009). Friction and wear behavior of the hybrid PTFE/cotton fabric composites filled with TiO$_2$ nanoparticles and modified TiO$_2$ nanoparticles. *Polymer Engineering & Science*, 49, 115–122.
31. Bazrgari, D., Moztarzadeh, F., Sabbagh, A., Rasoulianboroujeni, M., Tahriri, M., & Tayebi, L. (2018). Mechanical properties and tribological performance of epoxy/Al$_2$O$_3$ nanocomposite. *Ceramics International*, 44, 1220–1224.
32. Zheng, X., Xu, Y., Geng, J., Peng, Y., Olson, D., & Hu, X. (2016). Tribological behavior of Fe3O4/MoS2 nanocomposites additives in aqueous and oil phase media. *Tribology International*, 102, 79–87.
33. Zheng, D., Wu, Y.-P., Li, Z.-Y., & Cai, Z.-B. (2017). Tribological properties of WS2/graphene nanocomposites as lubricating oil additives. *RSC Advances*, 7, 14060–14068.
34. Yang, J., Zhang, H., Chen, B., Tang, H., Li, C., & Zhang, Z. (2015). Fabrication of the gC3N4/Cu nanocomposite and its potential for lubrication applications. *RSC Advances*, 5, 64254–64260.
35. Shao, W., Liu, X., Min, H., Dong, G., Feng, Q., & Zuo, S. (2015). Preparation, characterization, and antibacterial activity of silver nanoparticle-decorated graphene oxide nanocomposite. *ACS Applied Materials & Interfaces*, 7, 6966–6973.
36. Li, X., Zhao, Y., Wu, W., Chen, J., Chu, G., & Zou, H. (2014). Synthesis and characterizations of graphene–copper nanocomposites and their anti-friction application. *Journal of Industrial and Engineering Chemistry*, 20, 2043–2049.

Chapter 6

Recent progress on the application of nano-cutting fluid in turning process

Mohd Bilal Naim Shaikh and Mohammed Ali
Aligarh Muslim University

CONTENTS

6.1	Introduction	87
6.2	Nanofluids: Preparation, mechanisms and as a cutting fluid	88
6.3	Recent studies on the application of nano-cutting fluid in the turning process	88
6.4	Challenges and future outlook	89
6.5	Conclusions	96
References		96

6.1 INTRODUCTION

Metal cutting is the most exploited process among the manufacturing processes, which utilizes enough energy to cut off the excess material from the workpiece by the relative movement between the cutting tool and the workpiece. During metal cutting, an unusual amount of friction is produced along with the shearing of material, which in turn produces very high temperatures at the tool, work and chip interfaces. The employment of cutting fluids improves the tribological conditions at interfaces by virtue of their cooling and lubricating abilities. The employment of cutting fluids improves the machining efficiency and productivity by their cooling effect (which cause reduction in localised heating zone, thermal expansion and distortion of the machined surface) and by their lubrication effect (which cause reduction of cutting forces, extending of tool life, etc.) [1,2]. In addition, it also flushes away the chips from the machining zone, and improves the surface integrity to some extent. An efficient cutting fluid requires the following characteristics: high thermal conductivity, high thermal diffusivity, high specific heat, high lubricity, low viscosity, chemical stability, low friction-ability and high wettability etc. It should be non-toxic, non-corrosive and should not react with the tool or workpiece material [3,4]. However, cutting fluids' usage is accompanied by several drawbacks like the high cost of procurement, disposal and many others. Therefore, other alternatives have been proposed to restrict the extravagant use of cutting

DOI: 10.1201/9781003306276-6

fluids. Minimum quantity lubrication (MQL) is one of the alternatives in which a small amount of cutting fluid is sprayed with high pressure at the cutting zone [5–7]. To cope with the requirements, incorporating metallic, non-metallic, ceramics or carbon-based nanoparticles in conventional cutting fluid has also been employed which shows better thermo-physical and tribological properties along with no negative effect on pressure drop [5,8].

6.2 NANOFLUIDS: PREPARATION, MECHANISMS AND AS A CUTTING FLUID

Nanofluids are suspensions of nanoparticles (at least one of the critical dimensions must be less than about 100 nm), which form a stable colloid and constitute a quasi-single phase media. The overall nature of these fluids have a strong dependence on the properties of base fluids and nanoparticles, which are influenced by the fabrication technique of nanofluids [9,10]. To date, one-step or two-step techniques have been used majorly. However, other techniques are also being used depending on the combination of nanoparticle material and fluids. Table 6.1 and Figure 6.1 represent a detailed information and schematic of the one-step and two-step methods of preparation [9,11,12].

From literature [1,3,5,13,14], it can be concluded that nanoparticles' incorporation in cutting fluids showed remarkable improvement in the thermo-physical properties, tribological properties, wettability and heat-carrying capacity. Rolling, sliding and filming actions of nanoparticles improve the tribological conditions as the machining zone, which in turn reduces power consumption, tool wear, cutting forces, machining zone temperature and friction at interfaces as well as superior dimensional accuracy. The improvement in heat-carrying capacity as well as heat transfer capability due to various phenomena like Brownian motions, fluid layer over nanoparticle, etc. yields better machinability performance. Figure 6.2 enlists the various effects and mechanisms behind the nanofluid performances.

6.3 RECENT STUDIES ON THE APPLICATION OF NANO-CUTTING FLUID IN THE TURNING PROCESS

Several peers [7,15–39] have performed detailed investigations to explain the insights and performance exhibited by nano-cutting fluids during the turning of steels. Table 6.2 lists the recent works related to turning of steel using nano-cutting fluids, exclusively. It reports the type of nanoparticles used, their size and concentration, method of application and workpiece material employed by the researchers to perform machining. It can be noted that a wide range of nanoparticles (type and size), their concentrations (0%–2%) and different base oil-nanoparticle combinations are being utilized. Table 6.3 shows the

Table 6.1 Production methods of nanofluids and their description

Method	Procedure	Advantages and limitations	Remarks
One-step method	Simultaneous production and dispersion of nanoparticles in the base-oil, i.e., both processes occur in parallel	**Pros:** Better control over the size & shape of nanoparticles, superior dispersibility of nanoparticles, higher stability of prepared nanofluids, prevent oxidation of particles, collection, storage & transportation of nanoparticles eliminated. **Cons:** Expensive, limited to batch production, presence of residual reactants in nanofluids due to poor reaction or poor stabilization, low yield.	More often used for the preparation of metallic nanoparticles–based nanofluids. Suitable for low vapour pressure fluid. E.g. Submerged arc nanoparticles synthesis system, microwave-assisted chemical solution method, etc.
Two-step method	**Step 1:** Production and drying of nanoparticles **Step 2:** Dispersion of produced nanoparticles in base-oil through different physical & chemical processes	**Pros:** Simple & economical, large-scale production. **Cons:** Additional steps like drying, storage, etc., more prone for particle's agglomeration tendency, lesser control over size and shape, requires stabiliser and stirring process.	Preferable for ceramic-based nanofluids. E.g. Ultra-sonication, magnetic stirrer, milling, mixing or homogenizing, etc.

effect of nanofluids on machinability measures like cutting forces, surface quality, cutting temperature, tool wear, etc. Table 6.4 represents the properties of nanofluids evaluated by peers after adding nanoparticles. There is remarkable improvement in lubricating properties, thermos-physical properties, and heat transfer and carrying capacity shown by nanofluids on the addition of nanoparticles. In summary, there are numerous advantages observed from the use of nanoparticles having tiny sizes and comparatively large surface area.

6.4 CHALLENGES AND FUTURE OUTLOOK

Although the improvement in machinability performance raises the necessity of using nanofluids in machining technology, there are very wide grey zones, which need to be explored, analysed and assessed for its commercial application. The following are the points which need more attention from the research community.

90 Nanomaterials for Sustainable Tribology

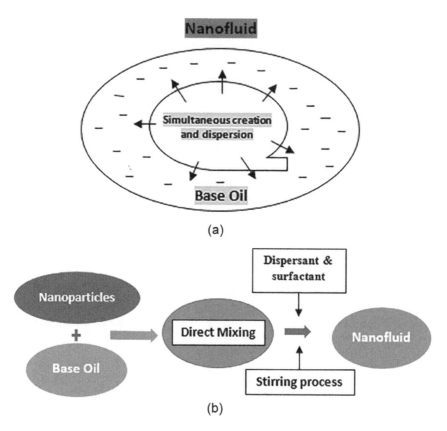

Figure 6.1 Schematic of preparation methods: (a) one-step and (b) two-step.

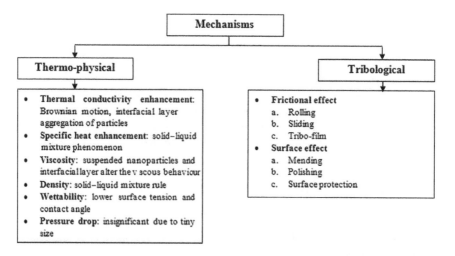

Figure 6.2 Various mechanisms and effects on nanoparticle inclusion.

Table 6.2 Recent works related to application of nano-cutting fluids in turning of various steels

S. No.	Author(s)	Nanoparticle	Conc. (%), size (nm)	Base fluid	Work material	Mode of application
1.	Rajmohan et al. [35]	Multi-walled carbon nanotubes (MWCNT)	0.1 vol.%, 10–20 nm	SAE20W40 emulsion	AISI 316L	Flood cooling
2.	Amrita et al. [17]	Gr	0.1–0.5 wt.%, <80 nm	Water soluble (20:1)	AISI1040	Minimum Quantity Cooling Lubrication (MQCL)
3.	Padmini et al. [31]	MoS$_2$	0.25–1 wt.%	Coconut (CC), sesame (SS) and canola (CAN) oil	AISI1040 steel	Minimum Quantity Lubrication (MQL)
4.	Padmini et al. [32]	Boric acid	0.25–1 wt.%	Coconut (CC), sesame (SS)	AISI 1040 steel	Minimum Quantity Lubrication (MQL)
5.	Saravanakumar et al. [40]	Ag	0.5 vol.%	Emulsion oil	Mild steel	Flood cooling
6.	Tuan et al. [41]	Al$_2$O$_3$ & MoS$_2$	1–3 wt.%, 30 nm	Soybean oil and emulsion oil	90CrSi alloy steel	MQL & MQCL
7.	Elsheikh et al. [23]	Al$_2$O$_3$ & CuO	0.1 % wt. %	Rice bran oil	AISI 4340	MQL
8.	Khan et al. [26]	Cu & Ag	0.1–1 wt.% 10–20 nm(Cu) 20–30 nm (Ag)	Coconut oil	EN 31b	Drop by drop
9.	Das et al. [21]	Al$_2$O$_3$	0.5 vol.%	Conv. soluble oil	AISI 4340	MQL
10.	Gajrani et al. [24]	MoS$_2$	0.3 vol. %	Coconut oil	AISI H-13	MQL
11.	Bhaumik et al. [19]	Reduced grapheneoxide (RGO) & zinc oxide (ZnO)	0.06–0.3 wt. %, 3–6 nm (RGO) 35–46 nm (ZnO)	Glycerol-based lubricants	EN21 steel	Intermittent flow

(Continued)

Table 6.2 (Continued) Recent works related to application of nano-cutting fluids in turning of various steels

S. No.	Author(s)	Nanoparticle	Conc. (%), size (nm)	Base fluid	Work material	Mode of application
12.	Revuru et al. [36]	Boric acid & MoS$_2$	0.25–1 wt. % 80 nm	Coconut oil	AISI 1040	MQL
13.	Khan et al. [42]	Al$_2$O$_3$	0.2–1.2 vol.%	15% solution of Blaser cutting oil	AISI52100	MQL
14.	Duc et al. [22]	Al$_2$O$_3$ & MoS$_2$	1.0–3.0 wt.%	Soybean oil and conv. soluble oil	90CrSi steel	MQL
15.	Öndin et al. [30]	MWCNT	0.6 vol. % 10–20 nm	Vegetable-based cutting oil	PH 13-8 Mo SS	MQL
16.	Abbas et al. [16]	Al$_2$O$_3$	1 wt.% 30-nm	ECO-COOL-MK-3 water solution	AISI 1045 steel	MQL

Table 6.3 Performance of nano-cutting fluid during turning of steel

S. No.	Author(s)	Nanofluid	Performance measures	Remark
1.	Rajmohan et al. [35]	MWCNT in soluble oil.	Surface roughness, tool chip interface temperature.	35% & 26% reduction in temperature & surface roughness, respectively.
2.	Amrita et al. [17]	Gr in soluble oil.	Cutting forces surface roughness, tool wear.	Inclusion of nano-graphite enhanced all the measures.
3.	Padmini et al. [31]	$nMoS_2$ in coconut, sesame & canola oil.	Cutting force, tool wear, tool tip temperatures & surface roughness.	Best performance (reduction of 37%, 21%, 44% & 39% in cutting force, tool wear, tool temperature & surface roughness, respectively) shown by 0.5% $nMoS_2$ in coconut oil.
4.	Padmini et al. [32]	nBoric acid in coconut & sesame oil.	Cutting forces, tool temperatures as well wear & surface finish.	Significant enhancement observed in all the performances assessed.
5.	Saravanakumar et al. [40]	nAg in soluble oil.	Tool tip temperature, cutting force & surface roughness.	nAg inclusion reduces the tool tip temperature, cutting force & surface roughness.
6.	Tuan et al. [41]	nAl_2O_3 & $nMoS_2$ in soyabean & emulsion oil.	Surface finish.	nAl_2O_3 contributes to higher extent compared to $nMoS_2$.
7.	Elsheikh et al. [23]	nAl_2O_3 & nCuO in Rice bran oil	Cutting force, surface roughness & tool wear.	Improved machinability performance exhibited by both nanofluids.
8.	Khan et al. [26]	nCu & nAg in Coconut oil	Cutting force and cutting temperature.	Lower values observed in case of Cu-based nanofluid than others.

(Continued)

Table 6.3 (Continued) Performance of nano-cutting fluid during turning of steel

S. No.	Author(s)	Nanofluid	Performance measures	Remark
9.	Das et al. [21]	nAl$_2$O$_3$ in conv. soluble oil	Cutting forces, tool wear & chip morphology.	Reduction in forces and tool wear as well as least serration in chip produced observed for nanofluids.
10.	Gajrani et al. [24]	nMoS$_2$ in coconut oil	Cutting forces & surface roughness.	Inclusion of nMoS$_2$ results with reduced cutting forces and surface roughness.
11.	Bhaumik et al. [19]	nRGO & nZnO in glycerol-based oil.	Surface roughness.	Graphene oxide–based nanofluid exhibited better surface finish compared to others.
12.	Revuru et al. [36]	nBoric acid & nMoS$_2$ in coconut oil.	Cutting forces, temperatures, surface roughness.	nMoS$_2$ based nanofluid showed better performance in terms of all measures.
13.	Khan et al. [42]	nAl$_2$O$_3$ in Blaser cutting oil.	Surface quality, cutting power, MRV, tool life & production cost.	Nanoparticle inclusion improves all the performances assessed.
14.	Duc et al. [22]	nAl$_2$O$_3$ & nMoS$_2$ in soybean oil and conv. soluble oil.	Cutting forces and surface roughness.	nAl$_2$O$_3$ soybean-based nanofluid exhibited better machinability performance.
15.	Öndin et al. [30]	MWCNTs in Vegetable-based cutting oil.	Surface roughness, cutting zone temperature, tool wear.	Cutting temperature, surface roughness & flank wear reduced significantly with the inclusion of MWCNTs.
16.	Abbas et al. [16]	nAl$_2$O$_3$ in ECO-COOL-MK-3 water solution.	Surface roughness, power consumption & sustainable index.	nAl$_2$O$_3$-based nanofluid improved all the measures.

Table 6.4 Characteristics of nanofluids investigated

S. No.	Nanofluid	Properties of nanofluids investigated	References
1.	0.1% of MWCNTs in SAE20W40 emulsion	Viscosity (40°C & 100°C), Flash point, Fire point, Pour point	Rajmohan et al. [35]
2.	0.05, 0.1, 0.2 & 0.3 vol. % of MWCNTs	Thermal conductivity, Viscosity, Wettability, pH value	Andhare & Raju [18]
3.	0.1, 0.3 & 0.5 wt.% of nano-graphite in water soluble oil	Viscosity	Amrita et al. [15]
4.	0.25, 0.5, 0.75, 1 wt.% of $nMoS_2$ in coconut, sesame & canola oil	Flash & fire points, specific density, stability, thermal conductivity, specific heat & heat transfer coefficients	Padmini et al. [31]
5.	0.25, 0.5, 0.75, 1 wt.% of nBA in coconut & sesame oil	Flash & fire points, kinematic viscosity & thermal conductivity	Padmini et al. [32]
6.	0.1 % wt. % of Al_2O_3 and CuO nanoparticles in rice bran oil	Viscosity, contact angle, thermal conductivity & surface tension	Elsheikh et al. [23]
7.	0.1%, 0.25%, 0.5% and 1% wt. of Cu & Ag nanoparticles in coconut oil	Viscosity, coefficient of friction	Khan et al. [26]
8.	0.06%, 0.08%, 0.1%, and 0.3% wt. % graphene oxide & ZnO	Coefficient of friction	Bhaumik et al. [19]

1. More sustainability-oriented comprehensive investigations are needed for full-fledged adaptation of nanofluid as the cutting fluid with maximum concerns of health and hazard era.
2. Complete life-cycle assessment study for nanofluid based machining process further required to add value with the trustworthy and exhaustive database, which focuses on combinations of different nanoparticles-based fluids-workpiece material.
3. Process hybridization of nanoparticles with other cooling and lubrication systems like electrostatic, ultrasonic, and cryogenic systems is also an era to explore for extracting the best possible sustainable solution with meeting the demand.

Apart from these, many other future prospects which need focus to commercialize the application of nanofluids.

6.5 CONCLUSIONS

This review work conclude the recent research work on the application of nanofluids in turning of steel as workpiece and the induced implications on the machinability performances. The following observations can be made:

1. The inclusion of nanoparticles (metallic, ceramic, carbon based, etc.) enhance machinability by reducing cutting force, surface roughness, tool wear, and cutting temperature.
2. Addition of nanoparticles improved the tribological, thermos-physical and heat transfer capacity of cutting fluids, significantly.
3. It can also be concluded that the ceramic and carbon-based nanoparticle seems to be more effective to alter the tribological condition, which eventually reduces the cutting forces, tool edges and surface roughness.
4. However, the better heat dissipation characteristic of metallic nanoparticle reduces the cutting zone temperature which in turn improved the surface integrity and tool wear.

REFERENCES

[1] A. K. Sharma, A. K. Tiwari, and A. R. Dixit, "Progress of nanofluid application in machining: A review," *Mater. Manuf. Process.*, vol. 30, no. 7, pp. 813–828, 2015, doi: 10.1080/10426914.2014.973583.

[2] J. S. Dureja, V. K. Gupta, V. S. Sharma, M. Dogra, and M. S. Bhatti, "A review of empirical modeling techniques to optimize machining parameters for hard turning applications," *Proc. Inst. Mech. Eng. Part B J. Eng. Manuf.*, vol. 230, no. 3, pp. 389–404, 2016, doi: 10.1177/0954405414558731.

[3] K. Kadirgama, "A comprehensive review on the application of nanofluids in the machining process," *Int. J. Adv. Manuf. Technol.*, vol. 115, no. 9–10, pp. 2669–2681, 2021, doi: 10.1007/s00170-021-07316-8.

[4] R. Sankaranarayanan, H. N. R. Jesudoss, J. Senthil Kumar, and G. M. Krolczyk, "A comprehensive review on research developments of vegetable-oil based cutting fluids for sustainable machining challenges," *J. Manuf. Process.*, vol. 67, pp. 286–313, 2021, doi: 10.1016/j.jmapro.2021.05.002.

[5] M. Hemmat Esfe, M. Bahiraei, and A. Mir, "Application of conventional and hybrid nanofluids in different machining processes: A critical review," *Adv. Colloid Interface Sci.*, vol. 282, p. 102199, 2020, doi: 10.1016/j.cis.2020.102199.

[6] Y. Shokoohi, E. Khosrojerdi, and B. H. Rassolian Shiadhi, "Machining and ecological effects of a new developed cutting fluid in combination with different cooling techniques on turning operation," *J. Clean. Prod.*, vol. 94, pp. 330–339, 2015, doi: 10.1016/j.jclepro.2015.01.055.

[7] K. K. Gajrani, P. S. Suvin, S. V. Kailas, and R. S. Mamilla, "Thermal, rheological, wettability and hard machining performance of MoS2 and CaF2 based minimum quantity hybrid nano-green cutting fluids," *J. Mater. Process. Technol.*, vol. 266, pp. 125–139, 2019, doi: 10.1016/j.jmatprotec.2018.10.036.

[8] A. K. Sharma, A. K. Tiwari, and A. R. Dixit, "Mechanism of nanoparticles functioning and effects in machining processes: A review," *Mater. Today Proc.*, vol. 2, no. 4–5, pp. 3539–3544, 2015, doi: 10.1016/j.matpr.2015.07.331.

[9] A. Ghadimi, R. Saidur, and H. S. C. Metselaar, "A review of nanofluid stability properties and characterization in stationary conditions," *Int. J. Heat Mass Transf.*, vol. 54, no. 17–18, pp. 4051–4068, 2011, doi: 10.1016/j.ijheatmasstransfer.2011.04.014.

[10] S. Chakraborty and P. K. Panigrahi, "Stability of nanofluid: A review," *Appl. Therm. Eng.*, vol. 174, p. 115259, 2020, doi: 10.1016/j.applthermaleng.2020.115259.

[11] D. K. Devendiran and V. A. Amirtham, "A review on preparation, characterization, properties and applications of nanofluids," *Renew. Sustain. Energy Rev.*, vol. 60, pp. 21–40, 2016, doi: 10.1016/j.rser.2016.01.055.

[12] Y. Li, J. Zhou, S. Tung, E. Schneider, and S. Xi, "A review on development of nanofluid preparation and characterization," *Powder Technol.*, vol. 196, no. 2, pp. 89–101, 2009, doi: 10.1016/j.powtec.2009.07.025.

[13] R. B. Ganvir, P. V. Walke, and V. M. Kriplani, "Heat transfer characteristics in nanofluid—A review," *Renew. Sustain. Energy Rev.*, vol. 75, pp. 451–460, 2017, doi: 10.1016/j.rser.2016.11.010.

[14] Z. Said, M. Gupta, H. Hegab, N. Arora, A. M. Khan, M. Jamil, and E. Bellos, "A comprehensive review on minimum quantity lubrication (MQL) in machining processes using nano-cutting fluids," *Int. J. Adv. Manuf. Technol.*, vol. 105, no. 5–6, pp. 2057–2086, 2019, doi: 10.1007/s00170-019-04382-x.

[15] M. Amrita, S. A. Shariq, Manoj, and C. Gopal, "Experimental investigation on application of emulsifier oil based nano cutting fluids in metal cutting process," *Procedia Eng.*, vol. 97, pp. 115–124, 2014, doi: 10.1016/j.proeng.2014.12.231.

[16] A. T. Abbas, M. K. Gupta, M. S. Soliman, M. Mia, H. Hegab, M. Luqman, and D. Y. Pimenov, "Sustainability assessment associated with surface roughness and power consumption characteristics in nanofluid MQL-assisted turning of AISI 1045 steel," *Int. J. Adv. Manuf. Technol.*, vol. 105, no. 1–4, pp. 1311–1327, 2019, doi: 10.1007/s00170-019-04325-6.

[17] M. Amrita, R. R. Srikant, and A. V. Sitaramaraju, "Evaluation of cutting fluid with nanoinclusions," *J. Nanotechnol. Eng. Med.*, vol. 4, no. 3, p. 031007, 2013, doi: 10.1115/1.4026843.

[18] A. B. Andhare and R. A. Raju, "Properties of dispersion of multiwalled carbon nanotubes as cutting fluid," *Tribol. Trans.*, vol. 59, no. 4, pp. 663–670, 2016, doi: 10.1080/10402004.2015.1102369.

[19] S. Bhaumik, V. Paleu, S. Sharma, S. Dwivedi, S. Borkar, and M. Kamaraj, "Nano and micro additivated glycerol as a promising alternative to existing non-biodegradable and skin unfriendly synthetic cutting fluids," *J. Clean. Prod.*, vol. 263, p. 121383, 2020, doi: 10.1016/j.jclepro.2020.121383.

[20] C. Y. Chan, W. B. Lee, and H. Wang, "Enhancement of surface finish using water-miscible nano-cutting fluid in ultra-precision turning," *Int. J. Mach. Tools Manuf.*, vol. 73, pp. 62–70, 2013, doi: 10.1016/j.ijmachtools.2013.06.006.

[21] A. Das, S. K. Patel, B. B. Biswal, N. Sahoo, and A. Pradhan, "Performance evaluation of various cutting fluids using MQL technique in hard turning of AISI 4340 alloy steel," *Measurement*, vol. 150, p. 107079, 2020, doi: 10.1016/j.measurement.2019.107079.

[22] T. M. Duc, T. T. Long, and T. Q. Chien, "Performance evaluation of MQL parameters using Al2O3 and MoS2 nanofluids in hard turning 90CrSi steel," *Lubricants*, vol. 7, no. 5, p. 40, 2019, doi: 10.3390/lubricants7050040.

[23] A. H. Elsheikh, M. A. Elaziz, S. R. Das, T. Muthuramalingam, and S. Lu, "A new optimized predictive model based on political optimizer for eco-friendly MQL-turning of AISI 4340 alloy with nano-lubricants," *J. Manuf. Process.*, vol. 67, pp. 562–578, 2021, doi: 10.1016/j.jmapro.2021.05.014.

[24] K. K. Gajrani, P. S. Suvin, S. V. Kailas, and M. R. Sankar, "Hard machining performance of indigenously developed green cutting fluid using flood cooling and minimum quantity cutting fluid," *J. Clean. Prod.*, vol. 206, pp. 108–123, 2019, doi: 10.1016/j.jclepro.2018.09.178.

[25] M. K. Gupta, P. K. Sood, and V. S. Sharma, "Optimization of machining parameters and cutting fluids during nano-fluid based minimum quantity lubrication turning of titanium alloy by using evolutionary techniques," *J. Clean. Prod.*, vol. 135, pp. 1276–1288, 2016, doi: 10.1016/j.jclepro.2016.06.184.

[26] M. S. Khan, M. S. Sisodia, S. Gupta, M. Feroskhan, S. Kannan, and K. Krishnasamy, "Measurement of tribological properties of Cu and Ag blended coconut oil nanofluids for metal cutting," *Eng. Sci. Technol. Int. J.*, vol. 22, no. 6, pp. 1187–1192, 2019, doi: 10.1016/j.jestch.2019.04.005.

[27] E. Kilincarslan, S. Kabave Kilincarslan, and M. H. Cetin, "Evaluation of the clean nano-cutting fluid by considering the tribological performance and cost parameters," *Tribol. Int.*, vol. 157, p. 106916, 2021, doi: 10.1016/j.triboint.2021.106916.

[28] A. Kumar Sharma, A. Kumar Tiwari, A. Rai Dixit, and R. Kumar Singh, "Measurement of machining forces and surface roughness in turning of AISI 304 steel using alumina-MWCNT hybrid nanoparticles enriched cutting fluid," *Measurement*, vol. 150, p. 107078, 2020, doi: 10.1016/j.measurement.2019.107078.

[29] N. Nagabhooshanam, S. Baskar, T. R. Prabhu, and S. Arumugam, "Evaluation of tribological characteristics of nano zirconia dispersed biodegradable canola oil methyl ester metalworking fluid," *Tribol. Int.*, vol. 151, p. 106510, 2020, doi: 10.1016/j.triboint.2020.106510.

[30] O. Öndin, T. Kıvak, M. Sarıkaya, and Ç. V. Yıldırım, "Investigation of the influence of MWCNTs mixed nanofluid on the machinability characteristics of PH 13-8 Mo stainless steel," *Tribol. Int.*, vol. 148, p. 106323, 2020, doi: 10.1016/j.triboint.2020.106323.

[31] R. Padmini, P. Vamsi Krishna, and G. Krishna Mohana Rao, "Effectiveness of vegetable oil based nanofluids as potential cutting fluids in turning AISI 1040 steel," *Tribol. Int.*, vol. 94, pp. 490–501, 2016, doi: 10.1016/j.triboint.2015.10.006.

[32] R. Padmini, P. V. Krishna, and G. K. Mohana Rao, "Experimental evaluation of nano-molybdenum disulphide and nano-boric acid suspensions in vegetable oils as prospective cutting fluids during turning of AISI 1040 steel," *Proc. Inst. Mech. Eng. Part J J. Eng. Tribol.*, vol. 230, no. 5, pp. 493–505, 2016, doi: 10.1177/1350650115601694.

[33] P. B. Patole and V. V. Kulkarni, "Prediction of surface roughness and cutting force under MQL turning of AISI 4340 with nano fluid by using response surface methodology," *Manuf. Rev.*, vol. 5, p. 5, 2018, doi: 10.1051/mfreview/2018002.

[34] T. Rajmohan, S. D. Sathishkumar, K. Palanikumar, and S. Ranganathan, "Modeling and analysis of cutting force in turning of AISI 316L stainless steel (SS) under nano cutting environment," *Appl. Mech. Mater.*, vol. 766–767, pp. 949–955, 2015, doi: 10.4028/www.scientific.net/AMM.766-767.949.

[35] T. Rajmohan, S. D. Sathishkumar, and K. Palanikumar, "Experimental investigation of machining parameters during turning of AISI 316L stainless steel using nano cutting environment," *Appl. Mech. Mater.*, vol. 787, pp. 361–365, 2015, doi: 10.4028/www.scientific.net/AMM.787.361.

[36] R. S. Revuru, V. K. Pasam, I. Syed, and U. K. Paliwal, "Development of finite element based model for performance evaluation of nano cutting fluids in minimum quantity lubrication," *CIRP J. Manuf. Sci. Technol.*, vol. 21, pp. 75–85, 2018, doi: 10.1016/j.cirpj.2018.02.005.

[37] H. Salimi-Yasar, S. Zeinali Heris, M. Shanbedi, A. Amiri, and A. Kameli, "Experimental investigation of thermal properties of cutting fluid using soluble oil-based TiO2 nanofluid," *Powder Technol.*, vol. 310, pp. 213–220, 2017, doi: 10.1016/j.powtec.2016.12.078.

[38] A. K. Sharma, R. K. Singh, A. R. Dixit, and A. K. Tiwari, "Novel uses of alumina-MoS2 hybrid nanoparticle enriched cutting fluid in hard turning of AISI 304 steel," *J. Manuf. Process.*, vol. 30, pp. 467–482, 2017, doi: 10.1016/j.jmapro.2017.10.016.

[39] V. Singh, A. K. Sharma, R. K. Sahu, and J. K. Katiyar, "Novel application of graphite-talc hybrid nanoparticle enriched cutting fluid in turning operation," *J. Manuf. Process.*, vol. 62, pp. 378–387, 2021, doi: 10.1016/j.jmapro.2020.12.017.

[40] N. Saravanakumar, L. Prabu, M. Karthik, and A. Rajamanickam, "Experimental analysis on cutting fluid dispersed with silver nano particles," *J. Mech. Sci. Technol.*, vol. 28, no. 2, pp. 645–651, 2014, doi: 10.1007/s12206-013-1192-6.

[41] N. M. Tuan, T. M. Duc, T. T. Long, V. L. Hoang, and T. B. Ngoc, "Investigation of machining performance of MQL and MQCL hard turning using nano cutting fluids," *Fluids*, vol. 7, no. 5, p. 143, 2022, doi: 10.3390/fluids7050143.

[42] A. M. Khan, M. K. Gupta, H. Hegab, M. Jamil, M. Mia, N. He, Q. Song, Z. Liu, and C. I. Pruncu, "Energy-based cost integrated modelling and sustainability assessment of Al-GnP hybrid nanofluid assisted turning of AISI52100 steel," *J. Clean. Prod.*, vol. 257, p. 120502, 2020, doi: 10.1016/j.jclepro.2020.120502.

Chapter 7

Dispersion stability of nanoparticles and stability measurement techniques

Pranav Dev Srivyas
Indian Institute of Technology, Guwahati

Sanjay Kumar
National Institute of Technology, Srinagar

Soundhar Arumugam
Indian Institute of Technology, Guwahati

Manoj Kumar
National Institute of Technology, Srinagar

CONTENTS

7.1 Introduction	101
7.2 Stability of nanofluids	102
7.3 Stability valuation techniques	104
7.3.1 Zeta-potential	104
7.3.2 Turbidimetry	106
7.3.3 Ultraviolet–visible (UV-VIS) spectroscopy	107
7.3.4 Sedimentation technique	107
7.3.5 Electron microscopy and light-scattering methods	108
7.4 Stability enhancement procedures	108
7.4.1 Addition of surfactants	108
7.4.2 Surface modification techniques	109
7.4.3 Ultrasonic agitation and stirring	109
7.5 Conclusion	109
References	110

7.1 INTRODUCTION

Our society's continuous growth necessitates the development of new methods and techniques for analyzing. Colloidal suspensions are very common in day-to-day life. A product that isn't a colloid suspension is hard to imagine. Today, almost all cosmetic products, like emulsions, are colloidal suspensions. Furthermore, colloidal suspension systems are used for the manufacture of food, beverages, paints, medicines, and other several products [1,2]. So, the

DOI: 10.1201/9781003306276-7

conception of colloidal suspension system's stability is critical. Colloids are made up of two phases: the dispersed and the dispersant. Because colloidal suspension systems are not genuine solutions, but rather a uniform mixture, they are not stable over time [3]. The scattered may eventually form flocks and separate from the continuous. This affects the characteristics of the system. Therefore, it is crucial to keep colloidal systems stable.

The most important criterion for nanofluids is the suspension stability over time. Aggregation and sedimentation are two critical phenomena for the constancy of nanofluids. The method of synthesis of nanoparticles and nanofluids has a role in the constancy of nanofluids. Stability can be classified as kinetic, steric (depletion), electrostatic, and thermodynamic. The presence of an energy barrier that prevents coagulation is associated with kinetic stability. The term "thermodynamic stability" refers to a situation, wherein the coagulated phase has a greater amount of free energy compared to the dispersed phase. Electrostatic stabilization is brought on by electrostatic repulsion forces. Increased stability occurs when particles floating in a liquid state possess the same electrostatic charge on their surfaces because they do not repel one another [4]. On the other hand, the latter type of stability only occurs in systems with tiny particles and when electrostatic repulsion is stronger than attraction forces. The process of electrostatic stabilization is believed to be short [5]. The steric stability is achieved by adding macromolecules. As was already indicated, the presence of surfactants or polymers modifies the interphase characteristics, which affect the constancy/stability mechanism [6–9].

Depending on the surface charge of the particles, the chemicals mentioned above can absorb on solid surfaces and either repel or attract the particles. When molecules are adsorbed on a solid substrate and resist one another, the stability rises. Aggregation and flocculation occur when the granules protected by the polymer or surfactant come together. The systems proposed can stabilise or flocculate when surface-active substances or macromolecules are present [4,5]. Because science is constantly evolving, there are numerous methods for studying the stability of colloidal systems. Zeta-potential, UV-VIS spectroscopy, turbidimetry, and density measurements are among them [10–13]. Only few review articles [14–17] have discussed nanofluid preparation and stability. In this study, we will cover important developments in the methods for producing stable nanofluids and provide an overview of the stability processes.

7.2 STABILITY OF NANOFLUIDS

The constancy/stability of colloidal suspension systems can be explained by the Derjaguin, Landau, Verwey, and Overbeek (DLVO) theory [18,19]. The interaction energy of granules in solution is used to determine stability in the aforementioned hypothesis [20]. Particle interactions are categorized by attractive van der Waals' forces and repulsive electric double layer forces. If the total of the repelling forces is greater than the total of the attractors, the stated system is stable [21]. Particles adhere to one another and flocculation arises. When the appealing compels by permanent-dipole-permanent-dipole-interactions (Keesom), permanent-dipole-induced-dipole forces (Debye), and transitory-dipole-transitory-dipole forces (London)—outweigh the repellent forces [22]. The total potential energy of contacts can be altered by compounds as well as by concentration, ionic strength, or pH [6]. The mechanism of stabilization in the event of macromolecule inclusion influences the stability of the colloidal dispersion. Steric stabilisation, electrosteric stabilisation, and depletion stabilisation are different types of stabilising processes (Figure 7.1) [23–25]. The steric and electrosteric stabilising processes take place once the polymer is attached to the particle's surface. If steric stabilisation occurs, the adsorbed polymer is electrically neutral. An electrosteric stabilisation system is one in which the polymer is ionic [23,24]. The last mechanism—depletion stabilisation—applies if the polymer is not absorbed on the surface of the particle. The free polymer chains, which can take on a variety of shapes, are positioned between the particles, limiting their attraction to one another and affecting stability [25]. The inclusion of the macromolecule may increase stability, but it also has the potential to increase instability. The flocculation process is the inverse of the stabilisation process, which takes place whenever the total of attraction forces is greater than the total of repulsion forces [4]. Polymer-induced flocculation can be divided into two types: bridging flocculation and depletion flocculation (Figure 7.2). When two or more colloidal particles are joined by polymer chains that have been precipitated on the solid substrate, the result is bridging flocculation, which leads to system aggregation. Certain unbound, unabsorbed macromolecules have an impact on particle aggregation in depletion flocculation [25].

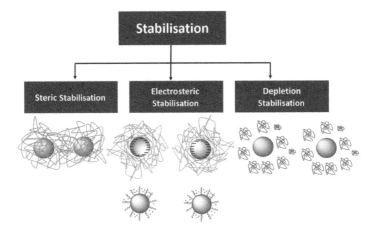

Figure 7.1 Stabilisation mechanisms.

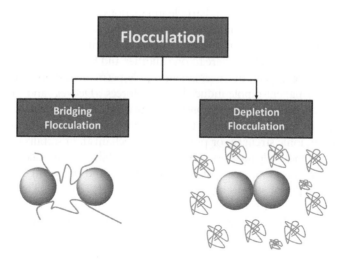

Figure 7.2 Flocculation types.

7.3 STABILITY VALUATION TECHNIQUES

7.3.1 Zeta-potential

Another technique for determining the stability of different colloidal suspension systems is to assess electrophoretic mobility, which is comparable to the zeta-potential. The electrokinetic potential determined in the sliding plane is referred to as the zeta-potential. The zeta-potential decreases with the distance between the sliding plane of the electric double layer and the bulk solution. When new macromolecules or ions are added to the solution, the sliding plane expands and the zeta-potential decreases. It is widely known that a system is considered to be entirely stable [3,26,27] if the zeta-potential is larger than 30 mV. As previously stated, repulsion forces must be greater than attraction forces in order for colloidal suspension to remain stable (Figure 7.3). Brownian movements can occur without solid flocculation if the energy of long-range repulsive electrostatic forces is significant. Utilizing the electrophoretic mobility phenomenon, the zeta-potential is computed by tracking the speed at which molecules move in response to an applied electric field. Granules and molecules will draw toward an electrode if a field is provided. Electrophoretic light scattering (ELS) is the technique employed in the majority of equipment. The intensity of dispersed light is affected by the Doppler Effect as charged particles oscillate in an applied electric field [3]. In fact, three different equations—the Smoluchowski, Henry, and Hückel equations—can be used to get the zeta-potential based on the Ka value.

The zeta-potential (ζ) reflects the repelling force that opposes two particles in electrophoresis theory (Figure 7.4). The pH of the fluid can be adjusted to increase the zeta-potential. Less than 20 mV implies weak stability, less than

Dispersion stability of nanoparticles 105

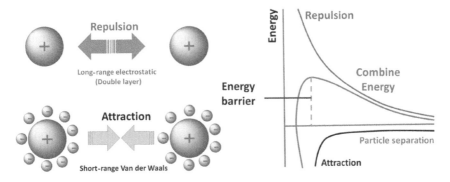

Figure 7.3 Electrostatic repulsion and van der Waals' force.

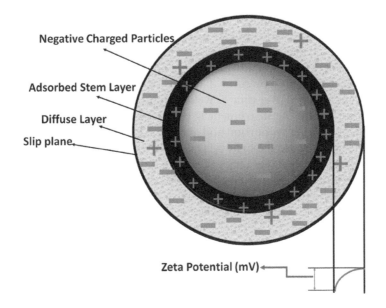

Figure 7.4 Zeta-potential of nanofluids.

5 mV suggests agglomeration, and absolute zeta-potential values more than 60 mV indicate exceptional stability [28]. Many researchers have performed zeta-potential tests on nanofluids for stability analysis. Said [29] determined the zeta-potential value and the typical size of nanoparticles in the base medium using a Zeta-seizer Nano ZS. Nanofluids with a high zeta-potential of 41.8 mV were produced after 30 days. Venkatachalapathy [30] has demonstrated how the electrokinetic properties of nanofluids affect their stability. In order to confirm the stability of nanofluids, Zetasizer performs a zeta-potential test. The test yields a +31.4 mV value, indicating that the nanofluid is stable. Using a DLS analyser, Nikkhah [31] determined the zeta-potential

of CuO in water. The findings demonstrated that for the nanofluids to be stable, the pH must be less than or more than 9.45. Zeta-potential measurements have been shown to be a useful technique for determining the stability of nanofluids by Sadeghinezhad [32]. The consistency of GNP nanofluids was suitable for applications requiring heat transfer, according to the 37.8 mV natural pH zeta-potential. Using the Zeta PALS 190 Plus instrument, Xiaoke Li measured the zeta-potential of SiC nanofluids [33]. Zeta-potentials for the test samples were found to be between 53 and 54 mV. The stability of Al_2O_3-water nanofluids was evaluated using a zeta-potential investigation by Khaleduzzama [34]. After 30 days, the nanofluids were found to be stable.

7.3.2 Turbidimetry

The turbidimetric approach establishes a relationship between the amount of light emitted by the source and the quantity of light that, after passing through the investigated colloidal suspension system, reaches the detector. To assess the change in particle size caused by flocculation, a vertically scanning concentrated liquid dispersion analyser can be utilised. These changes are represented by curve diagrams that show how the average particle size has changed over time as well as thorough sample analysis [4]. The mechanism behind the instrument is based on various scattering of light, in which the volume proportion and average APS of the dispersion medium determine how intense the light is. The instrument has a wide range of particle sizes and can evaluate very concentrated dispersions (up to 95%; from 10 nm to 1 mm). It comprises of a head with two synchronous detectors and an 880 nm light emitting diode to analyse the transparent (transmittance) and focused (backscattering) dispersions (Figure 7.5).

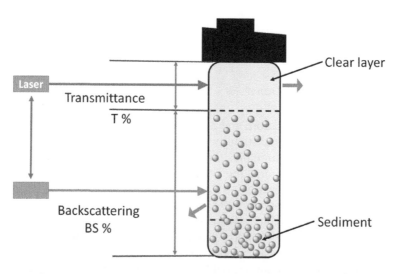

Figure 7.5 Schematic of vertically scanning concentrated liquid dispersion analyser.

7.3.3 Ultraviolet–visible (UV-VIS) spectroscopy

The UV-VIS spectroscopy technique relies on the interaction of light beams with matter. The light beam may experience absorption, refraction, and reflection as it travels through the liquid medium. The concept behind this method is to assess the brightness of light both before and after it enters the system. The ratio of the intensity of the light that is entering through the system (It) to the intensity of the light before it does so is known as the transmittance (I0). The use of a UV-VIS spectrophotometer for spectral analysis is a very valuable technique for assessing the consistency of nanofluids because it offers numerical findings correlating to nanofluid concentration. Mansoor Farbod [35] examined the consistency of refluxed and undisturbed carbon nanotubes (CNTs) using UV-VIS absorption spectra; the nanofluids were discovered to remain steady for 80 days. By regularly assessing the reflecting index (absorption) via a UV-VIS spectrophotometer, Sadeghinezhad [32] discovered that the graphene nanoplatelets (GNP) nanofluids were steady for up to 30 days. Yang [36] assessed the stability of nano refrigerants utilising the transmittance method using a visible spectrophotometer. As the mass fraction increased, the transmittance decreased, and the stability was improved by the dispersant's addition. As transmittance rises with time, stability declines.

7.3.4 Sedimentation technique

It is the simplest basic technique for identifying whether nanofluids are stable [37]. The amount or weight of debris reveals a nanofluid's stability. If the concentration of the supernatant remains largely constant, a nanofluid is said to be stable. In order to identify the iso-electric point, Soner Gumus [38] examined the sedimentation consistency of various nanodiesels while maintaining a constant pH. It was also noted that the interface scattering volume decreased over time. After 1month, the EG-based SiC nanofluids remained homogeneous and stable, according to Xiaoke Li's [33] sedimentation investigation. After observing sedimentation in the GNP nanofluids, Sadeghinezhad [32] concluded that the GNP nanofluids appear to be steady even after the heat transfer. By analysing sedimentation images of nanofluids obtained 30 days after creation, Khaleduzzama [34] examined the durability of nanofluids. By observing TiO_2-H_2O nanofluids with different ratios of Titanium nanotubes and nanosheets, Shao [39] categorised the nanofluids as stable or unstable titanium nanotubes and titanium nanosheets (TiNTs and TiNSs). By observing changes in the sedimentation height over time, it was possible to calculate the sedimentation velocities of nanoparticles in various concentrations of nanofluids (Figure 7.6). It was found that when the concentration proportions of TiNTs to TiNSs are in a reasonable range compared to TiNTs or TiNSs alone, binary nanofluid stabilities may be effectively improved.

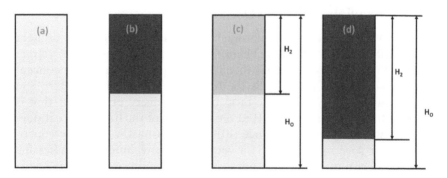

Figure 7.6 Sedimentation of nanoparticles in different concentrations of nanofluids.

7.3.5 Electron microscopy and light-scattering methods

Two fundamental tools for keeping an eye on particle aggregation are microscopy and light-scattering techniques. Very high-resolution microscopes like transmission electron microscopy (TEM) and scanning electron microscope (SEM) are used to take digital images of nanoparticles and electron micrographs. After 24 hours, Xiaoke Li [33] used a SEM to examine the morphology of SiC nanofluids. TEM was used by Khaleduzzama et al. [34] to capture the morphology and element dispersal of an Al_2O_3-water nanofluid. Within the range of 100 nm scales, good particle dispersion was observed. Nonetheless, some nanoparticle overlapping has been observed. Using a laser-scattering method, Hyun Jin Kim [40] quantified the suspended stability of Al_2O_3 nanofluids. The laser sprinkling technique determines the amount of light emitted by each nanofluid solution. Electromagnetic waves, such as lasers, are both scattered and absorbed by nanoparticles suspended in nanofluids. Therefore, nanofluid suspension stability can be assessed by measuring the amplitude of light transmission as a sole function of time.

7.4 STABILITY ENHANCEMENT PROCEDURES

7.4.1 Addition of surfactants

The base fluids are stabilised by the inclusion of surfactants, which lower surface tension, encourage particle immersion, and delay sedimentation. Different types of surfactants have been used in studies for various types of nanofluids. Sodium dodecyl sulphate (SDS) was utilised by Zhai [41] as a surfactant to improve the steadiness of $Al_2O_3+H_2O$ nanofluids. Tiwari [42] improved the stability and dispersion of nanofluids without altering their thermo-physical properties by using a surfactant named CetylTrimethyl Ammonium Bromide (CTAB). Saarinen's [43] studies suggest that emulsions can only hold up over time if Ostwald ripening and coalescence are prevented. Nonionic surfactant mixtures of sorbitantrioleate and polysorbate have been proven to be efficient stabilisers of n-decane/emulsions because of the steric impacts of their large, polar head groups. The surfactant sodium dodecyl benzene sulphate (SDBS) is most suited for the long-term durability of nanofluids, according to Prasad et al. [44]. Surfactants can improve the thermal resistance between fluids and nanoparticles, which can boost thermal conductivity [45].

7.4.2 Surface modification techniques

Long-term stability for the nanofluids can be achieved by directly injecting functional nanoparticles into the basic fluids. The need for surface modification to make nanoparticles steady in a base fluid was shown by Gumus [38]. In order to induce strong repulsive forces, in addition, it is possible to create a suspension with a substantial charge density. The pH of the nanodiesel dispersion, which was controlled using reagent grade hydrochloric acid and sodium hydroxide, affects surface charge density. Farbod [35] discovered that surface modification and relatively short CNT length improved nanofluid stability, with no substantial sedimentation witnessed after 80 days.

7.4.3 Ultrasonic agitation and stirring

Multi-walled carbon nanotubes (MWCNTs) nanoparticles were dispersed in the base oil using an ultrasonic blender for 6 hours without the use of a surfactant by Saarinen [43]. This disperses big nanoparticle aggregations in the fluid and produces a stable suspension. Prasad et al. [44] agitated the nanofluids with an ultrasonic vibrator and discovered that vibration for 8 hours effectively avoided nanopowder sediment. Sajedi [46] used a 240W ultrasonic cleaner and a magnetic stirrer to create stable SiO_2/water nanofluids. Lee et al. [47] proved that stirring the nanofluids for 1 hour results in dispersion stability. They show how SiO_2 and Al_2O_3 nanoparticle cluster sizes change over time, both with and without stirring.

7.5 CONCLUSION

This chapter provides a brief outline of current research on the preparation and stability of nanofluids. It is difficult to produce consistent, long-lasting nanofluids with minimal agglomeration and without altering thermophysical characteristics. Some scientists created their nanofluids without the use of surfactants or pH adjustments because they wanted to alter the thermo-physical characteristics. As a result, particle sedimentation cannot be stopped without the use of dispersants or pH regulation. An ideal pH value and surfactant concentration can be determined in order to maintain the physical qualities. Furthermore, the time of the ultrasonic process cannot be successfully regulated with respect to various nanoparticles and base fluids. Nanofluid stability over time is essential for both theoretical and real-world applications. It should be emphasized that various nanoparticles require different stability mechanisms. Most researched nanofluids have not yet had their long-term stability verified, hence more fundamental theoretical and experimental work is needed to increase nanofluid stability.

REFERENCES

1. Dickinson, E. (2015). Colloids in food: Ingredients, structure, and stability. *Annual Review of Food Science and Technology*, 6, 211–233.
2. Kralchevsky, P. A., Danov, K. D., Denkov, N. D., & Birdi, K. S. (1997). *Handbook of surface and colloid chemistry*. Boca Raton, FL: CRC Press.
3. Uskoković, V. (2012). Dynamic light scattering based microelectrophoresis: Main prospects and limitations. *Journal of Dispersion Science and Technology*, 33(12), 1762–1786.
4. Grządka, E. (2014). Stability of manganese dioxide by guar gum in the absence or presence of surfactants. *Cellulose*, 21(3), 1641–1654.
5. Wiśniewska, M., Terpiłowski, K., Chibowski, S., Urban, T., Zarko, V. I., & Gun'ko, V. M. (2013). Stability of colloidal silica modified by macromolecular polyacrylic acid (PAA)—Application of turbidymetry method. *Journal of Macromolecular Science, Part A*, 50(6), 639–643.
6. Grządka, E. (2015). Factors influencing the stability of the polysucrose/alumina system. *Colloid and Polymer Science*, 293(10), 2845–2853.
7. Napper, D. H. (1983). *Polymeric stabilization of colloidal dispersions* (Vol. 3). London, UK: Academic Press.
8. Somasundaran, P. (Ed.). (2006). *Encyclopedia of surface and colloid science* (Vol. 5). Boca Raton, FL: CRC Press.
9. Grządka, E. (2011). Competitive adsorption in the system: Carboxymethylcellulose/surfactant/electrolyte/Al2O3. *Cellulose*, 18(2), 291–308.
10. Wiśniewska, M., Chibowski, S., & Urban, T. (2012). Effect of the type of polymer functional groups on the structure of its film formed on the alumina surface–suspension stability. *Reactive and Functional Polymers*, 72(11), 791–798.
11. Wiśniewska, M., Chibowski, S., & Urban, T. (2012). Investigation of the stability of an alumina suspension in the presence of ionic polyacrylamide. *Thin Solid Films*, 520(19), 6158–6164.
12. Chibowski, S., Wiśniewska, M., & Urban, T. (2010). Influence of solution pH on stability of aluminum oxide suspension in presence of polyacrylic acid. *Adsorption*, 16(4), 321–332.
13. Hackley, V., Somasundaran, P., & Lewis, J. (2001). *Polymers in particulate systems: Properties and applications*. Boca Raton, FL: CRC Press.

14. Sharma, A. K., Tiwari, A. K., & Dixit, A. R. (2016). Rheological behaviour of nanofluids: A review. *Renewable and Sustainable Energy Reviews, 53*, 779–791.
15. Haddad, Z., Abid, C., Oztop, H. F., & Mataoui, A. (2014). A review on how the researchers prepare their nanofluids. *International Journal of Thermal Sciences, 76*, 168–189.
16. Sidik, N. A. C., Mohammed, H. A., Alawi, O. A., & Samion, S. (2014). A review on preparation methods and challenges of nanofluids. *International Communications in Heat and Mass Transfer, 54*, 115–125.
17. Alawi, O. A., Sidik, N. A. C., & Mohammed, H. A. (2014). A comprehensive review of fundamentals, preparation and applications of nanorefrigerants. *International Communications in Heat and Mass Transfer, 54*, 81–95.
18. Derjaguin, B. V. (1941). Acta physicochim. *USSR, 1*, 14.
19. Verwey, E. J. W., & Overbeek, J. T. G. (1955). Theory of the stability of lyophobic colloids. *Journal of Colloid Science, 10*(2), 224–225.
20. Matusiak, J., & Grządka, E. (2017). Stability of colloidal systems-a review of the stability measurements methods. *Annales Universitatis Mariae Curie-Sklodowska, sectio AA–Chemia, 72*(1), 33.
21. Grządka, E., Wiśniewska, M., Gun'ko, V. M., & Zarko, V. I. (2015). Adsorption, electrokinetic and stabilizing properties of the guar gum/surfactant/alumina system. *Journal of Surfactants and Detergents, 18*(3), 445–453.
22. Sato, T. (1980). Stabilization of colloidal dispersions by polymer adsorption. Acta Polymerica, 32(9), 582.
23. Sung, A. M., & Piirma, I. (1994). Electrosteric stabilization of polymer colloids. *Langmuir, 10*(5), 1393–1398.
24. Fischer, E. W., Sterzel, H. J., & Wegner, G. K. Z. Z. (1973). Investigation of the structure of solution grown crystals of lactide copolymers by means of chemical reactions. *Colloid and Polymer Science, 251*, 980–990.
25. Semenov, A. N., & Shvets, A. A. (2015). Theory of colloid depletion stabilization by unattached and adsorbed polymers. *Soft Matter, 11*(45), 8863–8878.
26. Rumiantcev, B. M., Zhukov, A. D., Bobrova, E. Y., Romanova, I. P., Zelenshikov, D. B., & Smirnova, T. V. (2016). The systems of insulation and a methodology for assessing the durability. In *MATEC Web of Conferences* (Vol. 86, p. 04036). Les Ulis, France: EDP Sciences.
27. Tucker, I. M., Corbett, J. C. W., Fatkin, J., Jack, R. O., Kaszuba, M., MacCreath, B., & McNeil-Watson, F. (2015). Laser Doppler Electrophoresis applied to colloids and surfaces. *Current Opinion in Colloid & Interface Science, 20*(4), 215–226.
28. Ajitha, B., Divya, A., & Reddy, P. S. (2013). Impact of pH on the properties of spherical silver nanoparticles capped by PVA. *Advanced Materials Manufacturing & Characterization, 3*(1), 403–406.
29. Said, Z., Sabiha, M. A., Saidur, R., Hepbasli, A., Rahim, N. A., Mekhilef, S., & Ward, T. A. (2015). Performance enhancement of a flat plate solar collector using titanium dioxide nanofluid and polyethylene glycol dispersant. *Journal of Cleaner Production, 92*, 343–353.
30. Venkatachalapathy, S., Kumaresan, G., & Suresh, S. (2015). Performance analysis of cylindrical heat pipe using nanofluids—An experimental study. *International Journal of Multiphase Flow, 72*, 188–197.

31. Nikkhah, V., Sarafraz, M. M., Hormozi, F., & Peyghambarzadeh, S. M. (2015). Particulate fouling of CuO-water nanofluid at isothermal diffusive condition inside the conventional heat exchanger-experimental and modeling. *Experimental Thermal and Fluid Science, 60*, 83–95.
32. Sadeghinezhad, E., Togun, H., Mehrali, M., Nejad, P. S., Latibari, S. T., Abdulrazzaq, T., Kazi, S. N., & Metselaar, H. S. C. (2015). An experimental and numerical investigation of heat transfer enhancement for graphenenano-platelets nanofluids in turbulent flow conditions. *International Journal of Heat and Mass Transfer, 81*, 41–51.
33. Li, X., Zou, C., Lei, X., & Li, W. (2015). Stability and enhanced thermal conductivity of ethylene glycol-based SiC nanofluids. *International Journal of Heat and Mass Transfer, 89*, 613–619.
34. Khaleduzzaman, S. S., Sohel, M. R., Saidur, R., & Selvaraj, J. (2015). Stability of Al2O3-water nanofluid for electronics cooling system. *Procedia Engineering, 105*, 406–411.
35. Farbod, M., Ahangarpour, A., & Etemad, S. G. (2015). Stability and thermal conductivity of water-based carbon nanotube nanofluids. *Particuology, 22*, 59–65.
36. Yang, D., Sun, B., Li, H., & Fan, X. (2015). Experimental study on the heat transfer and flow characteristics of nanorefrigerants inside a corrugated tube. *International Journal of Refrigeration, 56*, 213–223.
37. Mitra, I., Manna, N., Manna, J. S., & Mitra, M. K. (2014). Synthesis of chlorophyll entrapped red luminescent silica nanoparticles for bioimaging application. *Procedia Materials Science, 6*, 770–774.
38. Gumus, S., Ozcan, H., Ozbey, M., & Topaloglu, B. (2016). Aluminum oxide and copper oxide nanodiesel fuel properties and usage in a compression ignition engine. *Fuel, 163*, 80–87.
39. Shao, X., Chen, Y., Mo, S., Cheng, Z., & Yin, T. (2015). Dispersion stability of TiO2-H2O nanofluids containing mixed nanotubes and nanosheets. *Energy Procedia, 75*, 2049–2054.
40. Kim, H. J., Lee, S. H., Lee, J. H., & Jang, S. P. (2015). Effect of particle shape on suspension stability and thermal conductivities of water-based bohemite alumina nanofluids. *Energy, 90*, 1290–1297.
41. Zhai, Y. L., Xia, G. D., Liu, X. F., & Li, Y. F. (2015). Heat transfer enhancement of Al2O3-H2O nanofluids flowing through a micro heat sink with complex structure. *International Communications in Heat and Mass Transfer, 66*, 158–166.
42. Tiwari, A. K., Ghosh, P., & Sarkar, J. (2015). Particle concentration levels of various nanofluids in plate heat exchanger for best performance. *International Journal of Heat and Mass Transfer, 89*, 1110–1118.
43. Saarinen, S., Puupponen, S., Meriläinen, A., Joneidi, A., Seppälä, A., Saari, K., & Ala-Nissila, T. (2015). Turbulent heat transfer characteristics in a circular tube and thermal properties of n-decane-in-water nanoemulsion fluids and micelles-in-water fluids. *International Journal of Heat and Mass Transfer, 81*, 246–251.
44. Prasad, P. D., Gupta, A. V. S. S. K. S., & Deepak, K. (2015). Investigation of trapezoidal-cut twisted tape insert in a double pipe u-tube heat exchanger using Al2O3/water nanofluid. *Procedia Materials Science, 10*, 50–63.

45. Yu, W., & Xie, H. (2012). A review on nanofluids: Preparation, stability mechanisms, and applications. *Journal of Nanomaterials, 2012*, 435873.
46. Sajedi, R., Jafari, M., & Taghilou, M. (2016). An experimental study on the effect of conflict measurement criteria for heat transfer enhancement in nanofluidics. *Powder Technology, 297*, 448–456.
47. Lee, J. S., Lee, J. W., & Kang, Y. T. (2015). CO2 absorption/regeneration enhancement in DI water with suspended nanoparticles for energy conversion application. *Applied Energy, 143*, 119–129.

Chapter 8

Natural fiber–reinforced polymer nanocomposites

Asrar Rafiq Bhat and Prateek Saxena
Indian Institute of Technology, Mandi

CONTENTS

8.1	Introduction	115
8.2	Different types of nanofillers used in the composites	116
8.3	Tribological properties of natural fiber–reinforced polymer nanocomposites	118
8.4	Mechanical properties of natural fiber–reinforced polymer nanocomposites	121
8.5	Applications of natural fiber–reinforced polymer nanocomposites	129
	8.5.1 Applications of nanocellulose fiber–reinforced composites in biomedical sector	131
	8.5.2 Applications of nano-chitosan-based composites for biomedical composites	132
8.6	Conclusion and future scope	133
References		134

8.1 INTRODUCTION

In the current times, polymer-based materials are widely used in a variety of industries and applications, domestic goods, including biomedicine, aerospace, military hardware, and the production of spacecraft [1]. A polymer's molecular weight, monomer unit, and composition determine the best uses for it. By incorporating filler particles into the polymer matrix of polymeric composites, the material is strengthened and gives better qualities than the original polymer. Using less expensive materials in place of expensive ones has the added benefit of lowering production costs [2]. In the study of composite materials, the matrix is the component that combines with the filler particles to produce the new material. Fillers called reinforcing materials are incorporated into the matrix of the base material to improve its mechanical, tribological, and other properties. They can be divided into primary and secondary reinforcing materials, where the former are fibers that are frequently used to reinforce a particular material and the latter are fibers that improve the qualities of the composite. In order to increase miscibility, mechanical properties like tensile strength, electrical conductivity as well as other special capabilities like anti-corrosion and flame retardancy, researchers have been integrating other components into polymers [3–5].

The interactions between the filler particles and polymer matrix can impart these and other distinct features as well as performance enhancements. As a result, polymeric composite materials are a viable substitute for common metallic and alloy materials like steel and aluminum. Utilizing natural materials has been pushed by the rising demand for sustainable and environment-friendly products. The negative environmental effects of nonrenewable materials can be considerably reduced by using natural fibers to strengthen polymers. With the support of continuous nanotechnology research, scientists will be able to create superior polymeric nanocomposites that can be considerably personalized to the needs of the products, depending on their use, and lessen their negative effects on the environment [6].

In the current research arena, polymer nanocomposites with improved performances are emerging as a trend, overcoming the constraints of bulk polymer and satisfying societal and market expectations in tribological applications. Conventional fillers are frequently employed to improve their tribological, mechanical, and thermal qualities, especially under harsh working conditions [7,8]. These fillers include mineral filler, metallic powdered filler, natural fibers, and carbon fibers (CF). The development of polymer nanocomposites is driven by the ongoing search for more sophisticated materials. Nanofillers have a large surface area-to-volume ratio, which allows them to have a considerable impact on the tuning of characteristics even at extremely low filler loadings. The potential applications of polymer nanocomposites are expanded by multifunctional nanofillers, enabling them to be developed for a particular application, such as tribological components for intense operating temperature, highly corrosive environments, and working environments with unacceptable levels of lubricants. By reducing resource, energy, and waste consumptions and using newly developed polymer nanocomposites for tribological usage, can increase cost effectiveness [9–11].

8.2 DIFFERENT TYPES OF NANOFILLERS USED IN THE COMPOSITES

Based on their shape, nanofillers are divided into three categories viz. fibrous (rod-like), spherical, and sheet. Researchers [12] discovered that the structure (form and size) of the fillers has an impact on the mobility and number of the interfacial beads that make up the polymer-filler network. The network's strength is influenced by the beads' diffusion coefficient as

Natural fiber–reinforced polymer nanocomposites 117

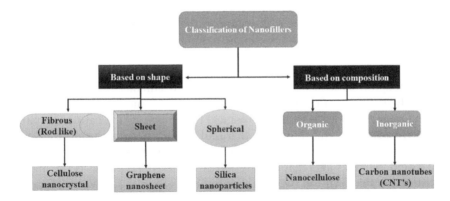

Figure 8.1 Classification of nanofillers based on shape and composition. (Conceptualized from [17].)

well as the stress contribution. A greater filler size increases the voids for all forms while also lowering the number of interfacial beads, mobility, and stress contribution. Rod-shaped fillers show a minor reduction in stress distribution while lowering their diffusion coefficient, thereby improving their network. It was determined that rod-shaped fillers outperform spherical and sheet shapes in terms of matrix reinforcement. In addition, there are two subcategories of nanomaterials: organic, or carbon-based, and inorganic. Figure 8.1 illustrates the classification of nanofiller based on shape and composition. Nanocellulose is an illustration of an organic nanoparticle. Nanocellulose has grown in popularity as a result of the demand for more ecologically friendly engineering materials. It may be manufactured for the purpose of reinforcement in polymer composites, create smart materials, and may be used in a variety of industries and applications. Depending on the extraction methods and starting materials, cellulose, which is largely present in natural fibers as their main component, can be treated to produce several forms of nanomaterials, including bacteria cellulose (BC), cellulose nanofibrils (CNFs), and cellulose nanocrystals (CNCs) [13–15]. Small concentrations of nanoparticles added to polymers can enhance their mechanical and thermal capabilities without having an adverse effect on other qualities like density and toughness. However, greater loading weakens them as a result of the composite's increased void content [16].

In contrast to the high amount of fibers (40%–60%) utilized in typical composites, cellulose nanofibers are used as reinforcement in polymer nanocomposites in medium proportion (about 10%). The nanocomposites were synthesized by researchers by adding 5% weight of various cellulose nanowhiskers to a polylactic acid matrix. The impact of different mass percentages of cellulose nanofiber in the epoxy matrix was assessed [18]. The inclusion of 0.25% and 0.5% cellulose nanofibers resulted in uniformly dispersed polymeric matrix, according to the authors, whereas there was an issue of agglomeration with regard to that of nanofibers in the matrix with a 0.75% concentration. Cellulose nanofibers offer a lot of potential for application as reinforcement in polymer matrices, despite any remaining barriers to their use [19].

Although the use of nanoparticles can greatly enhance polymer composites, there are certain disadvantages as well. The increased surface area of nanomaterials leads to morphological instability, which causes the particles to aggregate and clump together [20]. The creation of clusters that can grow larger than nanoscales is known as agglomeration. Agglomerations are rigid, whereas aggregations are loosely cohesive [21]. This is the major distinction between the two. Although the strong van der Waals contacts among the particles are thought to be the cause of agglomeration, associated particle transport mechanisms and interfacial chemical reactions are actually the main cause for this [22,23]. The degradation in mechanical properties caused by improper dispersion of nanofillers leading to poor morphological stability is one of the main problems in the manufacturing of nanocomposites. According to studies, the shape, loading, and size of the filler can all have an impact on the dispersion, with more uniform shape, smaller size, and less filler inclusion displaying higher homogeneity in the matrix [24–26].

This chapter describes the most recent advancements in polymeric composite materials using nanoparticles and natural fibers. Moreover, tribological and mechanical properties of some of the natural fiber–reinforced (NFR) polymer nanocomposites are also discussed. It has been demonstrated that adding fillers, natural fibers, and nanoparticles to polymers enhances their mechanical and tribological characteristics, and other qualities including flame retardance.

8.3 TRIBOLOGICAL PROPERTIES OF NATURAL FIBER–REINFORCED POLYMER NANOCOMPOSITES

Choosing composites for various industrial applications depends heavily on their tribological qualities. These composites are subjected to a variety of wear under working conditions, including adhesive wear, abrasive wear, and erosive wear, necessitating an evaluation of their tribological properties [27,28]. A significant amount of the energy required to overcome friction can be avoided by employing effective tribological techniques. Another important aspect of the tribological performance of composites is wear resistance. In situations where composites are subject to wear during service, increased wear resistance is crucial. The service life of the nanocomposites is influenced by the wear resistance qualities, which principally

Natural fiber–reinforced polymer nanocomposites 119

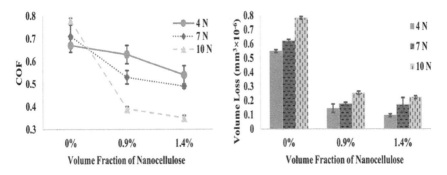

Figure 8.2 Effect of nanocellulose content in bio-epoxy on (a) coefficient of friction and (b) wear volume at different normal loads. (Taken from [30].)

depend on the kind of material. In addition to these, the fiber composition and its matching interfacial strength inside the matrix play a key role in the polymer-based nanocomposites' wear resistance properties [29].

Researchers [30] investigated the tribological behavior of bio-epoxy composites reinforced with nanocellulose fibers (Eucalyptus Kraft pulp). Studying the tribological behavior of the silylated CNF's composites revealed that they had lower coefficients of friction (COF) and wear volumes than the neat bio-epoxy, as illustrated in Figure 8.2. This was because the transfer film that formed on the mating surfaces reduced the composite's "direct contact" with the asperities of the hard metallic counter face. In order to improve the tribological properties of epoxy bionanocomposites, researchers [31] investigated the impact of a unique chemical treatment on sugarcane nanocellulose fibers. The COF of epoxy nanocomposites reinforced with sugarcane nanocellulose fiber was shown to increase with temperature of up to 200°C and to significantly decrease with further increase in temperature. Interestingly, sugarcane fibers treated with both salt and alkaline solution (SAT) sample showed the best wear resistance, followed by those treated with only salt solution (SST). All things considered, it was concluded that SAT-epoxy nanocomposites have better wear and corrosion resistance qualities than neat polymers.

Researchers [32] examined the wear characteristics of polymer nanocomposites reinforced with single-walled carbon nanotubes (SWCNT), and they concluded that the inclusion of SWCNT considerably increased the wear resistance of the composite material when compared to epoxy banana composite. In another study, researchers [33] examined the effects of surface-modified graphene nanoplatelets on the sliding wear characteristics of epoxy composites reinforced with basalt fibers. They discovered that the inclusion of surface-modified graphene nanoparticles (GNPs) significantly enhanced the composites' wear resistance and microhardness. The best microhardness result was obtained with 0.5 wt. % GNPs, which was over

100% better than the sample without GNPs. GNPs produced internal and 3D stresses that strengthened composite materials.

Due to their low price, clay-reinforced polymer nanocomposites appear promising. Polymer chains do, however, only partially penetrate the galleries of clays [34]. As a result, surface-active chemicals typically modify laminated silicates. The integration of nanoclay modified with trimethyl stearyl ammonium and montmorillonite (MMT) produced high stiffness, hardness, and strength epoxy that had enhanced tribological properties. The reinforcing of epoxy resin was improved by upping the nanoclay content to 4 wt. % [35].

Tribological analysis of jute/SiC hybrid epoxy composites has been done. In order to achieve the least wear loss, they employed the Taguchi optimization approach to identify the process parameters that were most ideal. They discovered that the most crucial elements in lowering erosion rate were fiber content, impact velocity, and impact angle. Researchers [36] employed cenosphere, a solid filler made of ceramic that was combined with jute (J) and glass (G) to conduct a study along similar lines. They investigated the erosion wear of the composites in relation to the fabric orientation, layering pattern, and filler content. The analysis showed that Glass-Jute-Jute-Glass (GJJG) fabric orientation produced the lowest wear rate and was the best hybrid composite combination [37]. One of the best hybrid composites with a very low specific wear rate has been found to be cotton fiber and graphite-modified polyester [38]. The frictional force increases if the proportion of cotton fiber is raised while it reduces when the percentage of graphite filler is increased. Due to the conductive nature of graphite particles, the fall in contact temperature provided justification for this behavior.

Researchers [39] have investigated jute/epoxy composites with two distinct ceramic fillers, SiC and Al_2O_3, under dry sliding circumstances. In comparison to composites with SiC filler, they claimed that composites with Al_2O_3 filler exhibited better wear-resistant qualities illustrated in Figure 8.3. In addition, SiC composites had more porosity, cracks, and fiber breakage than Al_2O_3 composites, according to the analysis of worn-out surfaces. Pine bark fibers as a reinforcing medium filled with cement kiln dust (CKD) as tribo-filler is another hybrid composite that has been investigated for tribological qualities under dry sliding conditions [40].

Nanoparticles were utilized by researchers [41] working on sisal/epoxy composites as a supplementary reinforcement. Clay made of MMT was used as reinforcement when the sliding was dry. In comparison to unfilled and higher weight percentages of clay, it was discovered that superior wear performance was obtained at lower weight percentages of nanoclay.

The production of NFR composites utilizing thermoplastics is more complicated. As a result, very few investigations in this area have been conducted. One of these studies [42] involved the tribological analysis of a composite reinforced with nano-ZnO and jute fibers in a polyamide foundation. Finding the ideal process parameters for the least amount of wear

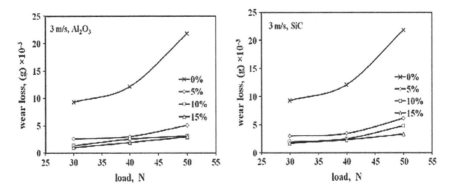

Figure 8.3 Wear loss vs normal load for jute epoxy composites at constant velocity of 3 m/s with fillers: (a) For Al_2O_3 and (b) for SiC. (Taken from [3].)

loss and friction required varying process parameters including sliding velocity, normal force, and reinforcements. It was claimed that jute fiber became more flexible and had higher cohesiveness qualities as a result of the ductile nature of polyamide.

In order to assess the impact of TiO_2 and ZrO_2 filler on the mechanical and wear parameters of the hybrid composite, researchers [43] tested glass/bamboo fiber–reinforced hybrid epoxy composites. They discovered that adding glass to composites made of bamboo fibers improves the qualities of the resulting hybrid composites. They determined that, of the two inorganic fillers, ZrO_2 gives the composite more favorable strength and wear characteristics than TiO_2. The hybrid GBBG (Stacking sequence, Glass Bamboo Bamboo Glass) laminate composite with 9 wt. % ZrO_2 provides the best strength of all the laminate composites. In addition, they discovered that the GBBG laminate composite with 6% ZrO_2 filler had the least amount of wear compared to other hybrid composites and clean polymer composites. Researchers [44] investigated the effects of various parameters on the tribological features of the integration of pineapple, sisal, and TiO_2 filler. Due to the availability of high filler (5 weight %) and hybrid fiber addition (40 wt. %) with the polymer-based composites, low specific wear rate was discovered. The effect of nanoparticles/fibers on tribological properties of various polymer composites is shown in Table 8.1.

8.4 MECHANICAL PROPERTIES OF NATURAL FIBER–REINFORCED POLYMER NANOCOMPOSITES

One of the most crucial factors for researchers to consider is the mechanical properties of fibers, which offer a wealth of potential due to their low cost, low wear, light weight and high specific modulus [45]. NFR polymers

Table 8.1 Effect of nanoparticles/fibers on tribological properties of polymer composites

Polymer matrix	Fiber/nanofibers	Nano/micro-fillers	Test condition	Comparative wear rate	Comparative coefficient of friction	Refs.
Epoxy Resin	Eucalyptus Kraft pulp (Cellulose nanofiber)	—	POD, SS, AL: 10N, SV: 0.15 m/s, SD: 1,000 m	Reduction by 82% compared to neat epoxy	Reduction by 50%	[30]
Epoxy Resin	Sugarcane nanofibers (treated with salt and alkaline solution)	—	POD, EN21 SS, AL: 10N, SV: 0.1 m/s, SD: 1,000 m, TR: 80°C–450°C, Fade and recovery test	Reduction by 21.6% compared to neat epoxy	Increased FR (COF) by 50% and RR (COF) by 35%	[31]
Epoxy Resin	Banana fiber	Single-walled carbon nanotubes (SWCNT)	AL: 30N, SV: 1 m/s, SD: 500 m, gap between fibers: 5 mm	50% reduction	—	[32]
Epoxy Resin	Basalt fiber	Surface-modified graphene nanoplatelets (0.3 wt. % GNP's)	POD, AL: 40N, SV: 0.5 m/s, SD: 1,000 m	26.3% reduction	61% reduction	[33]
Epoxy Resin	—	Nanoclay fillers (4%)	POD, EN31 Steel, AL: 10–30N SV: 100–300 rpm	41% reduction	—	[34]
Epoxy Resin	Jute and glass fibers	Cenosphere fillers (20% by weight)	Erosion test rig (silica particles), impact velocity: 48, 70, 82 m/s, nozzle diameter: 4 mm, time: 5 minutes	1.1%–12.14% improvement in erosion efficiency	—	[36]
Polyester Resin	Cotton fiber	Graphite particles (20 parts per hundred parts of polyester)	POD, SS, AL: 20N, SV: 2.2 m/s, SD: 4,000 m	85% reduction in wear compared to neat polyester	38% reduction	[38]

(Continued)

Table 8.1 (Continued) Effect of nanoparticles/fibers on tribological properties of polymer composites

Polymer matrix	Fiber/nanofibers	Nano/micro-fillers	Test condition	Comparative wear rate	Comparative coefficient of friction	Refs.
Epoxy	Jute fiber	SiC (15%)	POD, AL: 30 N, SV: 4.5 m/s, SD: 1,800 m	82% reduction in wear compared to neat epoxy	10% reduction	[39]
Polyester	Pine bark (0%–12%)	Cement kiln dust (50%)	POD, AL: 10 N, SV: 314 rpm, SD: 3,000–5,000	10% reduction		[40]
Epoxy	Sisal	Nanoclay (5%)	POD (brass disk), contact pressure: 0.15 MPa, SV: 0.35 m/s, SD: 3,600 m	45% reduction	—	[41]
Polyamide	Jute	Nano-ZnO (2%)	POD, AL: 4 N, SV: 0.2 m/s	Significant reduction in wear	Increase in COF	[42]

Remarks: Comparison was done with the neat epoxy and in certain cases with nonhybrid composites.
AL, Applied Load; FR, Fade Ratio; POD, Pin-on-Disc; RR, Recovery Ratio; SD, Sliding Distance; SV, Sliding Velocity; SS, Stainless Steel.

have demonstrated to be biodegradable from the perspective of environmental sustainability. NFR is frequently utilized to improve the mechanical characteristics of thermoset and thermoplastic polymers for a variety of engineering applications [46]. Jute, sisal, banana, jute, ramie, sugarcane, and malva are some of the several NFRs that are frequently employed in composite polymers [47]. In addition to these, other NFRs include pineapple, abaca, and sansevieria leaves [48]. Cellulosic fibers are another name for the aforementioned natural fibers. Their mechanical and physical characteristics, such as specific gravity, diameter, and length, are connected to their primary sources [49]. There are further natural fibers that are less studied but have great mechanical and thermal properties and a chemical makeup that is comparable to that of cellulosic fibers. Sabdariffa, Hibiscus, henequen, pines, esparto, and sabai grass are a few examples of these fibers [50]. Numerous experiments have been done to improve the qualities of composites by adding natural fibers, fillers, and nanomaterials. With the use of untreated Oil Palm Empty Fruit Bunch (OPEFB) fibers, researchers [51] have produced a polycaprolactone composite successfully. A 12.2% filler loading resulted in the maximum elongation and tensile strength. The weak interfacial contacts were thought to be the cause of the composites' decreased tensile strength and elongation as filler loading was increased further.

Researchers [51] have also investigated how well natural fiber composites perform when nanoparticles are used. They proved that altering the loading of nano-OPEFB fibers can enhance the mechanical characteristics of cured epoxy resin. According to the study, a composite with 3% nano-OPEFB filler increased the loss modulus and storage modulus by about 48% and 40%, respectively. Nano-OPEFB filler also prevented the polymeric chain from moving in segments, which led to better dispersion. In comparison to composites with greater loadings, those with a 3% filler loading had better dispersal without voids and an aggregate filler inside the matrix. Researchers [52] investigated the used carbon nanotubes (CNTs) as a filler in order to increase the stiffness of an epoxy composite made of coir fiber and fly ash. It was discovered that adding more CNT to a composite made of fly ash and coir results in a reduction in the shear modulus of the material by 7.0% and 13.6%, respectively, at 0.5 weight % and 1.0 weight % of CNT. The amount of fillers (coir fiber, CNT, and fly ash) has been varied in a few studies between 0 and 2.0 weight %. It was discovered that adding more fly ash and coir raises the shear modulus, whereas adding more CNT only slightly raises it.

Another study found that the interfacial adhesion of natural fibers in a polylactic acid matrix was improved by zinc oxide nanowires. For the growth of nanowires on its surface, researchers [53] submerged sisal fibers in a zinc oxide seed solution. The fibers were coated with multiple dips to connect the nanowires. An increase in the number of dip-coating cycles increased the surface coverage of the nanowires, which were found to be uniform in size at 100nm in diameter and 2μm in length after six cycles. They claimed that the increased number of fibers that underwent dip-coating cycles improved their performance. The debonding energy and interfacial shear strength rose by 403% and 157%, respectively, with six cycles.

The characteristics of nanoclay/epoxy composites were investigated [35]. They claimed that the use of nanoclay improved the tensile and flexural strength of the epoxy resin. At 2 weight % of nanoclay, the greatest flexural strength is attained. The flexural strength has been reduced by the addition of more nanoclay. Addition of 4wt.% of nanoclay resulted in 57.4 MPa tensile strength which was 6% more than pure epoxy. The water barrier capabilities of sisal fiber–reinforced epoxy composites loaded with nanoclay were investigated by researchers [41]. Different studies showed that the qualities of the objects degraded when they were submerged in water; however, the degradation was less pronounced for composite structures filled with nanoclay. According to the findings, nanoclay may be effectively utilized as a filler because it has high water barrier qualities, which enhance the properties that are most likely to be impacted by water or moisture content. Researchers [54] looked at the mechanical characteristics of polyvinyl alchohol (PVA) composites with natural fiber reinforcement and carbon nanofillers in one study. They discovered that the jute component enhanced the PVA/multi-layered graphene (MLG) nanocomposites' storage modulus and toughness.

Recent research has been done on the creation of epoxy with reinforcement of cellulose nanofibers. With increasing reinforcement loading, composites reinforced with cellulose nanofiber exhibit improved thermal stability and mechanical properties [18]. One study [55] evaluated the tension/fracture behavior and swelling of bio-derived epoxy composites. It was also documented as to how to create biocomposites with a high nanocellulose content using an epoxy matrix and evenly disseminated nano-fibrillated cellulose [56]. In a different investigation, epoxy was utilized to modify the surface of cellulose nanofibers that were being reinforced in a PVA matrix. When compared to the composite created using unmodified cellulose nanofibers, PVA reinforced with chemically altered cellulose nanofibers showed improved strength, strain, and elastic modulus [57]. Researchers performed mechanical characterization of bio-epoxy composites reinforced with cellulose nanofiber [30]. The results of the experiments demonstrated that the siloxane groups' reaction with the bio-epoxy rings' curing in the presence of CNFs had catalytic effects, which improved the mechanical properties of the silylated CNF composites relative to the nonsilylated samples.

In order to remove nanocellulose fiber from sugarcane bagasse, researchers [31] developed a unique treatment method. Sugarcane nanocellulose fiber (SNCF) (2.93–17.07 weight %) and sonication duration were used as independent parameters in the statistical experimental design, which was carried out utilizing a central composite design (47.57–132.43 minutes). The mechanical characteristics of SNCF epoxy–based bionanocomposites

Table 8.2 Combinations of matrix, fibers, and fillers for which properties are illustrated in Figure 8.4

S. No.	Combinations (by volume %)
1	Polyester resin+Bamboo fiber (70:30)
2	Polyester resin+E-glass fiber (75:25)
3	Polyester resin+Bamboo fiber+E-glass fiber (70:15:15)
4	Polyester resin+Bamboo fiber+E-glass fiber (70:10:20)
5	Polyester resin+Bamboo fiber+E-glass fiber (70:20:10)
6	Polyester resin+Bamboo fiber+E-glass fiber+Coconut shell powder in microsize (70:15:12:3)
7	Polyester resin+Bamboo fiber+E-glass fiber+Coconut shell powder in nanosize (70:15:12:3)

Source: Modified from [65].

were optimized and assessed with the confirmation experiment in order to verify the efficacy of the established polynomial model. Sugarcane fibers treated with SST and salt solution and alkaline solution (SAT) epoxy nanocomposites had better mechanical and viscoelastic properties than sugarcane raw sample (SRS). In another research [58], researchers investigated the impact of nanoclay on wood fiber and coir fiber–reinforced polypropylene (PP) composites. Results revealed that the tensile strength and modulus improved most for hybrid nanocomposites, which was attributed to good interaction between fiber, polymer, and matrix. Nanoclay strongly improved the interfacial adhesion between PP matrix and fiber, which was evident in Scanning Electron Microscopy (SEM) analysis. Table 8.2 illustrates the effect of nanoparticles/fibers on mechanical characteristics of various polymer composites.

According to one study [59], a hybrid glass/bamboo fiber used as a reinforcement in polyester matrix along with coconut shell powder (nanoparticles) with a 30° orientation generated the highest tensile strength, flexural strength, impact strength, and hardness when compared to other combinations. Hybrid bamboo/glass composites with additional coconut shell powder showed enhanced fatigue life than other composites at a given stress level, according to a fatigue test, which also shows a substantial dispersion in the results.

Researchers [60] looked at the mechanical properties of hybrid natural composites that had aluminum oxide nanoparticles added to them. Results showed by adding 0, 1, 2, and 3 weight % of alumina nanopowder, the initial sisal/coir combination demonstrated improved tensile performance. The tensile characteristics of sisal/banana composites increased similarly. Banana/coir composites improved in characteristics in a similar way (Table 8.3).

Table 8.3 Effect of nanoparticles/fibers on mechanical properties of polymer composites

Polymer matrix	Fiber/nanofibers	Micro/nanofillers	Dispersion technique	Improvement in tensile strength	Improvement in flexural strength	Improvement in impact strength	Refs.
Epoxy Resin	Sisal/Banana 35% by weight (Hybrid)	Al_2O_3 nanoparticles (3% by weight)	Mechanical stirring (Acetone mixture), ultra-sonification	+35.6% compared to composite without nanoparticles	+12.76% compared to composite without nanoparticles	+8.24% compared to composite without nanoparticles	[66]
Epoxy Resin	Sisal/coir 35% by weight (Hybrid)	Al_2O_3 nanoparticles (3% by weight)	Mechanical stirring (Acetone mixture), ultra-sonification	+27.4% compared to composite without nanoparticles	+13.3% compared to composite without nanoparticles	+16.49% compared to composite without nanoparticles	[66]
Epoxy Resin	Banana/coir 35% by weight (Hybrid)	Al_2O_3 nanoparticles (3% by weight)	Mechanical stirring (Acetone mixture), ultra-sonification	+24.4% compared to composite without nanoparticles	+10.3% compared to composite without nanoparticles	+10.42% compared to composite without nanoparticles	[66]
Polypropylene (PP)	Wood/coir (Hybrid each 15% by weight)	Montmorillonite (MMT) nanoclay (2% by weight)	Brabender mixer machine	+20% compared to composite without MMT	—	—	[64]
Polypropylene (PP)	Wood (30% by weight)	Montmorillonite (MMT) nanoclay (2% by weight)	Brabender mixer machine	+33% compared to composite without MMT	—	—	[64]
Polypropylene (PP)	Coir (30% by weight)	Montmorillonite (MMT) nanoclay (2% by weight)	Brabender mixer machine	+18% compared to composite without MMT	—	—	[64]

(Continued)

Table 8.3 (Continued) Effect of nanoparticles/fibers on mechanical properties of polymer composites

Polymer matrix	Fiber/nanofibers	Micro/nanofillers	Dispersion technique	Improvement in tensile strength	Improvement in flexural strength	Improvement in impact strength	Refs.
Polyester	Bamboo/glass fibers (15%:12% by weight)	Nano-coconut shell powder (3% by weight)	—	+45.7% compared to composites without nanopowder	+67% compared to composites without nanopowder	+70% compared to composites without nano powder	[65]
Epoxy Resin	Sugarcane nanocellulose fibers (SCNF) treated with alkaline and salt solution	—	Mechanical stirring, ultra-sonification	+50% compared to pure epoxy	—	+114% compared to pure Epoxy	[63]
Epoxy Resin	Silylated Eucalyptus craft pulp	—	Centrifugation, microfluidizer	Considerable improvement in tensile strength	—	—	[62]
Polyvinyl Alcohol (PVA)	Ball milled jute fiber (5% by weight)	Multi-layered graphene MLG (20% by weight)	Mechanical stirring, ultrasonication	Improvement in storage modulus and hardness	—	—	[57]
Polyvinyl Alcohol (PVA)	Surface-modified Birch pulp nano-fibrillated cellulose+epoxy (0.5% by weight)	—	Scale grinder, microfluidizer	+116% compared to neat PVA	—	—	[61]
Epoxy Resin	—	Nanoclay	Mechanical stirring, ultra-sonification	+6% compared to pure epoxy (at 4% by weight of nanoclay)	+9% compared to pure epoxy (at 2% by weight of nanoclay)	—	[55]
Polycaprolactone (PCL)	—	Oil palm empty fruit bunch (fillers 12.2% by weight)	Thermal Hake blending machine	Increase in tensile strength	—	—	[51]

8.5 APPLICATIONS OF NATURAL FIBER–REINFORCED POLYMER NANOCOMPOSITES

With the rise in the fabrication composite materials that are resin based, the applications of composites also extended to a considerable extent. In that case, the variety of applications is aided by the advanced nanofiber-reinforced polymer composites. As was evident from the previous sections, the primary application area for NFR nanocomposites is the strengthening of the materials' durability and lifespan. As a result, NFR nanocomposites can be used to design structures for the aviation, building, bridge, and rail industries. Their use in the electric and electronic sphere can also be expanded, like lithium-ion batteries, dye-sensitized solar, super capacitors and fuel cells, and cells all appear to benefit from the use of NFR nanocomposites with a possibility to promote efficiency and energy storage [61–65]. Figure 8.2 shows the potential applications of NFR polymer nanocomposites in different sectors.

Similar to this, studies propose a process for creating cellulose nanofiber–reinforced composites using hemp, flax, kraft paper, and rutabaga. The fibers were prepared using both innovative mechanical techniques and chemical treatments. The scientists recommended using nanocellulosic composites for the aerospace and automotive applications due to their lightweight and high-strength characteristics [66]. According to a study on the creation, properties, and potential uses of epoxy clay nanocomposites, numerous specific applications, such as those in the automotive, aerospace, and military industries, may make use of these materials [67] (Figure 8.5).

In addition, NFR nanocomposites are widely utilized membranes for filtration and separation processes in industries including food and water treatment [68]. More work is going on in nanocomposites for discovering potential uses in biomedical engineering, specifically in the fields of scaffolds and drug delivery systems for bone and tissue [69,70]. Recently, there has been a fast increase in interest in using nanofibers as reinforcing agents in dentistry. Currently, fibers are used on a much greater scale in resin composites that are available. For use in dental applications, there are no available commercial nanofiber materials, despite the fact that nanoscale fibers can be utilized to build resin composites that are harder and favorable to be used for this purpose. In addition, the development of new bio-based composites for drug delivery systems for periodontal disorders or remineralization and regeneration in endodontics shows great promise when using natural nanofibers, such as chitin, chitosan, and cellulose [71]. NFR nanocomposites can be utilized as supports for sensors and catalysts [72]. NFR nanocomposites will soon be employed in a variety of applications such as novel materials and customized constructions since the attributes of the composites may be changed by modifying the parameters of the fiber, such as volume percentage, fiber type, and diameter.

130 Nanomaterials for Sustainable Tribology

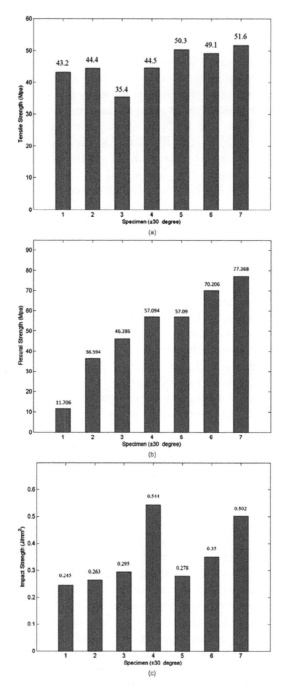

Figure 8.4 Mechanical properties of various composites (see Table 8.2): (a) Tensile Strength. (b) Flexural strength. (c) Impact strength. (Taken from [65].)

Natural fiber–reinforced polymer nanocomposites 131

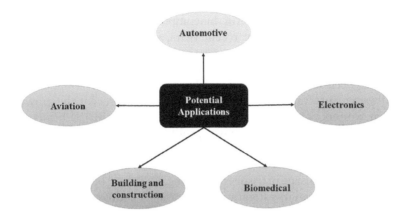

Figure 8.5 Potential applications of natural fiber–reinforced polymer nanocomposites in various sectors.

8.5.1 Applications of nanocellulose fiber–reinforced composites in biomedical sector

Due to its superior mechanical, barrier, and biocompatibility qualities, cellulose materials show considerable potential in a number of biotechnological and biomedical applications [73]. Numerous studies have also been conducted on nano-crystalline cellulose, which is also referred to as CNFs or cellulose nanowhiskers, demonstrating qualities like transparency, nontoxicity, lightness, gas impermeability, stiffness, low thermal expansion, and exceptional mechanical properties. Nanocelluloses are excellent prospects for biomedical applications such tissue engineering, drug delivery, wound healing, and medical implants due to their inherent properties [74]. Due to its renewability, biodegradability, lower cost, availability, sustainability, high surface area, reduced weight, high hydrophilicity, higher mechanical strength, and biocompatibility, nanocellulose has mostly been employed as a filler in nanocomposites. It prevents negative tissue reactions and, unlike proteins, is less immunogenic and nonhemolytic due to its polysaccharide composition. It also fosters cellular connection and tissue growth. It has high wear resistance, can withstand high loads, and is a nondegrading/slow material in vitro and in vivo, making it excellent to be used in scaffolds for longer time. Replacement skin for wounds and burns, blood vessel growth drug releasing systems, gum, nerve, dura mater repair, stent coating, tissue engineering scaffolds, and bone reconstruction are all part of the biomedical sector [75].

Nanocellulose-based biocomposites are intended to function in tissue engineering as matrices or scaffolds to promote essential cellular processes like cell adhesion, proliferation and tissue development. Tissue engineering has identified promising uses for other biocomposites based on nanocellulose, such as CNFs/nanochitin biocomposites, CNFs, CNFs/alginate, and biocomposites [76–78]. In recent years, substantial research has also been done on other nanocelluloses, particularly the cellulose nanofibers for possible wound healing. For applications involving the healing of wounds, CNFs have been shown to be noncytotoxic to boost proliferation, fibroblast cell adhesion, survival, and gene expression, and to prevent the growth of common wound bacteria [79–81]. Bacteria cellulose (BC) is bio-compatible, promotes tissue growth in vivo, proliferates in vitro, and has strong cell attachment according to studies. The mechanical properties and highly porous design of the BC are likewise comparable to the extracellular matrix of human skin. For usage in tissue engineering, BC-based biocomposites have undergone substantial investigation. For example, hydroxyapatite (HA)-BC nanocomposite scaffolds were found to be useful in bone tissue engineering to control proliferation and osteoblastic differentiation of human bone marrow stromal cells, as well as in the bone regeneration of rats with bone defects. For possible usage in diverse tissue engineering applications, BC/collagen and BC/heparin biocomposites have been created. Today, various businesses, like bioprocess and biofill, sell BC for wound healing [82–85].

In order to create biomaterials with micrometer-level precision, nanocellulose dispersions can be used as bio-ink in three-dimensional (3D) printing, according to researchers [86]. To achieve the desired properties, bio-inks can be created using several forms of cellulose, including bacterial cellulose and nanofibrils. Bio-inks must be sufficiently viscous to hold their shape during printing and possess crosslinking capabilities to keep the desired 3D form after printing. In a different study, researchers [78] created multiple bio-inks for 3D printing utilizing formulations based on alginate and CNFs. Alginate solutions and concentrated CNF dispersions were combined to create the bio-inks. The ideal ratio for the best printability was initially determined by measuring the viscosity of the CNF/alginate mixes. The authors then went on to show how the bio-ink made of CNF and alginate could be utilized to create 3D constructions that resemble human nasoseptal chondrocytes (hNCs).

8.5.2 Applications of nano-chitosan-based composites for biomedical composites

Chitosan has a number of benefits, including the capacity to regulate the release of active ingredients and prevent the use of potentially dangerous organic solvents when creating particles. Due to its nontoxicity and capacity to biodegrade without harming the environment, chitosan has several benefits. Chitosan is employed as a contact disinfectant in many biomedical applications as well as in a variety of health care and hygiene fields. Despite significant advancements in the development of antimicrobial agents, a number of diseases that are infectious continue to be challenging to treat for a variety of reasons, including the subpar pharmacological properties of the antimicrobial substances, the absence of antimicrobial structures, and the appearance and spread of certain resistant clones that are currently in use, which can make it challenging for bacterial strains to achieve active concentrations [87].

Contrary to chitosan, it has been discovered that chitosan nanocomposites have a wide range of antibacterial activity against different pathogens (both gram-negative and gram-positive bacteria) [88]. Hybrid nanocomposites based on chitosan have been studied for liver and nerve tissue engineering, cartilage regeneration, and bone tissue engineering [89]. A number of applications related to wound healing have also made use of chitosan-based nanomaterials, which are created in a variety of formulations including chitosan-based sponges, drug-loaded scaffolds, immobilized scaffolds, and composite scaffolds [90].

When used with other nano-sized fillers, chitosan was found to enhance the field of nanotechnology. The properties of each chitosan nanocomposite can be managed, planned, and adjusted in relation to the target tissue depending on the application [91].

The performance of the nanocomposite is intimately tied to the fillers' ability to disperse well inside the polymeric matrix, regardless of the goal. An excellent polymer-filler interface is made possible by successful dispersion, and regardless of the application, this interface always yields high specific interfacial area. Therefore, the key to utilizing nanocomposites to their greatest capacity is a good chitosan/filler interaction. It would also be intriguing to undertake fresh research into different material properties, such as the degradation of nanocomposites. Once the device's toxicity and durability are reduced, these gains may become very significant. Preclinical or in vivo investigations are other crucial areas that need more development. In addition to the abovementioned factors, it is important to address the unresolved problems, such as the emergence of analytical methods for chitosan quality assessment, for both market and scientific reasons [92]. In the near future, it is anticipated that the biomedical area will undergo a revolution due to the immense prospects presented by these materials and their great nanotechnology potential.

8.6 CONCLUSION AND FUTURE SCOPE

In this chapter, the most recent advancements in the development of NFR nanocomposites using natural fibers and fillers of various sizes and forms have been discussed. They have been found to exhibit comparable performances to those of their synthetic counterparts at lower costs, lighter weights, and decreased environmental footprints while reusing currently wasted components; hence, they are the potential alternatives to their synthetic counterparts. The interaction between the filler and matrix is a crucial element in defining how a composite will behave; the stronger the contact, the better the mechanical performance. The proper handling of natural fibers can be used to enhance their interactions and efficiently and economically overcome their drawbacks. Utilizing nanoparticles will enable composites to receive additional advances.

It is clear from the studies provided that there are numerous chances to enhance the performance of composites by using natural fibers and fillers of various sizes. They might be appropriate over diversified spectrum of applications with further development. In recent years, the field of composites reinforced with nanofiber has expanded significantly. Improvements in the composite's mechanical strength, tribological properties, and conductivity can be attributed to both the utilization of electrospun fibers inside the polymer matrix and the insertion of nanomaterials such as tubes, particles, and fibers in the polymer matrix. By utilizing nanoparticles, composites can have additional enhancements. Natural biopolymers and fibers have a lot of inherent qualities that make it possible to utilize them in biomedical applications. More specifically, these qualities consist of development of nanofiber networks, biocompatibility, high controlled porosity, high specific surface area, hydration capacity, lack of toxicity, outstanding barrier properties, mechanical characteristics, and biodegradability.

To create a successful composite material, however, aspects including size, loading, and shape must be taken into account in addition to optimum dispersion. Research opportunities in this field include examining the use of different matrices and their influence with fillers to better understand their underlying interactions. There is still a lot of work to be done, including understanding the connection between structure and property and manufacturing low-cost nanofiber-reinforced composites on a wide scale. Other tasks include successfully inserting nanomaterials into polymer matrix with a high loading content.

REFERENCES

[1] Asyraf, M., Anwar, M., Sheng, L. M., & Danquah, M. K. (2017). Recent development of nanomaterial-doped conductive polymers. *JOM*, 69(12), 2515–2523.

[2] Mostovoy, A., Bekeshev, A., Tastanova, L., Akhmetova, M., Bredihin, P., & Kadykova, Y. (2021). The effect of dispersed filler on mechanical and physicochemical properties of polymer composites. *Polymers and Polymer Composites*, 29(6), 583–590.

[3] Scaffaro, R., & Botta, L. (2013). *Nanostructured Polymer Blends: Chapter 5. Nanofilled Thermoplastic–Thermoplastic Polymer Blends* (Vol. 1). Boston, MA: Elsevier Inc.

[4] He, W., Song, P., Yu, B., Fang, Z., & Wang, H. (2020). Flame retardant polymeric nanocomposites through the combination of nanomaterials and conventional flame retardants. *Progress in Materials Science, 114*, 100687.

[5] Pourhashem, S., Saba, F., Duan, J., Rashidi, A., Guan, F., Nezhad, E. G., & Hou, B. (2020). Polymer/Inorganic nanocomposite coatings with superior corrosion protection performance: A review. *Journal of Industrial and Engineering Chemistry, 88*, 29–57.

[6] Malviya, R. K., Singh, R. K., Purohit, R., & Sinha, R. (2020). Natural fiber reinforced composite materials: Environmentally better life cycle assessment—A case study. *Materials Today: Proceedings, 26*, 3157–3160.

[7] Aldousiri, B., Shalwan, A., & Chin, C. W. (2013). A review on tribological behaviour of polymeric composites and future reinforcements. *Advances in Materials Science and Engineering, 2013*, 645923.

[8] Friedrich, K. (2018). Polymer composites for tribological applications. *Advanced Industrial and Engineering Polymer Research, 1*(1), 3–39.

[9] Oliveira, J. D., Rocha, R. C., & de Sousa Galdino, A. G. (2019). Effect of Al2O3 particles on the adhesion, wear, and corrosion performance of epoxy coatings for protection of umbilical cables accessories for subsea oil and gas production systems. *Journal of Materials Research and Technology, 8*(2), 1729–1736.

[10] Sanjay, M. R., Madhu, P., Jawaid, M., Senthamaraikannan, P., Senthil, S., & Pradeep, S. (2018). Characterization and properties of natural fiber polymer composites: A comprehensive review. *Journal of Cleaner Production, 172*, 566–581.

[11] Sazali, N., & Ngadiman, N. H. A. (2020). Materials for tribology's application: A mini analysis. *Journal of Advanced Research in Fluid Mechanics and Thermal Sciences, 68*(2), 177–185.

[12] Zhang, H., Ma, R., Luo, D., Xu, W., Zhao, Y., Zhao, X., Gao, Y., & Zhang, L. (2020). Understanding the cavitation and crazing behavior in the polymer nanocomposite by tuning shape and size of nanofiller. *Polymer, 188*, 122103.

[13] Kumar, R., Aadil, K. R., Ranjan, S., & Kumar, V. B. (2020). Advances in nanotechnology and nanomaterials based strategies for neural tissue engineering. *Journal of Drug Delivery Science and Technology, 57*, 101617.

[14] Khan, M. N., Rehman, N., Sharif, A., Ahmed, E., Farooqi, Z. H., & Din, M. I. (2020). Environmentally benign extraction of cellulose from dunchifiber for nanocellulose fabrication. *International Journal of Biological Macromolecules, 153*, 72–78.

[15] Mishra, R. K., Sabu, A., & Tiwari, S. K. (2018). Materials chemistry and the futurist eco-friendly applications of nanocellulose: Status and prospect. *Journal of Saudi Chemical Society, 22*(8), 949–978.

[16] Nunes, R. C. R., Fonseca, J. L. C., & Pereira, M. R. (2000). Polymer-filler interactions and mechanical properties of a polyurethane elastomer. *Polymer Testing, 19*(1), 93–103.

[17] Wong, D., Anwar, M., Debnath, S., Hamid, A., & Izman, S. (2021). A review: Recent development of natural fiber-reinforced polymer nanocomposites. *JOM, 73*(8), 2504–2515.

[18] Yusra, A. I., Khalil, H. A., Hossain, M. S., Davoudpour, Y., Astimar, A. A., Zaidon, A., Dungani, R., & Omar, A. M. (2015). Characterization of plant nanofiber-reinforced epoxy composites. *BioResources, 10*(4), 8268–8280.

[19] Petersson, L., Kvien, I., & Oksman, K. (2007). Structure and thermal properties of poly (lactic acid)/cellulose whiskers nanocomposite materials. *Composites Science and Technology, 67*(11–12), 2535–2544.

[20] Zinge, C., & Kandasubramanian, B. (2020). Nanocellulose based biodegradable polymers. *European Polymer Journal, 133*, 109758.

[21] Nichols, G., Byard, S., Bloxham, M. J., Botterill, J., Dawson, N. J., Dennis, A., Diart, V., North, N. C., & Sherwood, J. D. (2002). A review of the terms agglomerate and aggregate with a recommendation for nomenclature used in powder and particle characterization. *Journal of Pharmaceutical Sciences, 91*(10), 2103–2109.

[22] Zhang, W. (2014). Nanoparticle aggregation: Principles and modeling. *Nanomaterial, 811*, 19–43.

[23] Tessema, A., Zhao, D., Moll, J., Xu, S., Yang, R., Li, C., Kumar, S. K., & Kidane, A. (2017). Effect of filler loading, geometry, dispersion and temperature on thermal conductivity of polymer nanocomposites. *Polymer Testing, 57*, 101–106.

[24] Yue, L., Maiorana, A., Khelifa, F., Patel, A., Raquez, J. M., Bonnaud, L., Gross, R., Dubois, P., & Manas-Zloczower, I. (2018). Surface-modified cellulose nanocrystals for biobased epoxy nanocomposites. *Polymer, 134*, 155–162.

[25] Chindaprasirt, P., Sukontasukkul, P., Techaphatthanakon, A., Kongtun, S., Ruttanapun, C., Yoo, D. Y., Tangchirapat, W., Limkatanyu, S., & Banthia, N. (2021). Effect of graphene oxide on single fiber pullout behavior. *Construction and Building Materials, 280*, 122539.

[26] Pang, W. Q., & Xu, Y. (2017). Synthesis and purification at low temperatures. In *Modern Inorganic Synthetic Chemistry*, edited by Ruren Xu and Yan Xu (pp. 45–71). Elsevier.

[27] Krishnan, G. S., Ilayaperumal, K., Babu, L. G., Kumar, S., Sathish, B., & Sanjana, R. (2021). Investigation on the physical and mechanical characteristics of demostachya bipinnata reinforced with polyester composites. *Materials Today: Proceedings, 45*, 1134–1137.

[28] Rouf, S., Raina, A., Haq, M. I. U., & Naveed, N. (2021). Sensors and tribological systems: Applications for industry 4.0. *Industrial Robot, 49*(3), 442–460.

[29] Liu, Y., Xie, J., Wu, N., Wang, L., Ma, Y., & Tong, J. (2019). Influence of silane treatment on the mechanical, tribological and morphological properties of corn stalk fiber reinforced polymer composites. *Tribology International, 131*, 398–405.

[30] Barari, B., Omrani, E., Moghadam, A. D., Menezes, P. L., Pillai, K. M., & Rohatgi, P. K. (2016). Mechanical, physical and tribological characterization of nano-cellulose fibers reinforced bio-epoxy composites: An attempt to fabricate and scale the 'Green' composite. *Carbohydrate Polymers, 147*, 282–293.

[31] Mohit, H., & Selvan, V. (2019). Effect of a novel chemical treatment on nanocellulose fibers for enhancement of mechanical, electrochemical and tribological characteristics of epoxy bio-nanocomposites. *Fibers and Polymers, 20*(9), 1918–1944.

[32] Bellairu, P. K., Bhat, S., & Madhyastha, K. (2019). A study on wear properties of SWCNT reinforced polymer nanocomposite. *AIP Conference Proceedings, 2080*(1), 020013.

[33] Kazemi-Khasragh, E., Bahari-Sambran, F., Siadati, S. M. H., Eslami-Farsani, R., & Arbab Chirani, S. (2019). The effects of surface-modified graphene nanoplatelets on the sliding wear properties of basalt fibers-reinforced epoxy composites. *Journal of Applied Polymer Science, 136*(39), 47986.

[34] Myshkin, N. K., & Kovalev, A. (2018). Polymer mechanics and tribology. *Industrial Lubrication and Tribology, 70*(4), 764–772.

[35] Shettar, M., Kowshik, C. S., Manjunath, M., & Hiremath, P. (2020). Experimental investigation on mechanical and wear properties of nanoclay-epoxy composites. *Journal of Materials Research and Technology, 9*(4), 9108–9116.

[36] Latha, P. S., Rao, M. V., Kumar, V. V. K., Raghavendra, G., Ojha, S., & Inala, R. (2016). Evaluation of mechanical and tribological properties of bamboo—Glass hybrid fiber reinforced polymer composite. *Journal of Industrial Textiles, 46*(1), 3–18.

[37] Dalbehera, S., & Acharya, S. K. (2015). Effect of cenosphere addition on erosive wear behaviour of jute-glass reinforced composite using taguchi experimental design. *Materials Today: Proceedings, 2*(4–5), 2389–2398.

[38] Hashmi, S. A. R., Dwivedi, U. K., & Chand, N. (2007). Graphite modified cotton fiber reinforced polyester composites under sliding wear conditions. *Wear, 262*(11–12), 1426–1432.

[39] Ahmed, K. S., Khalid, S. S., Mallinatha, V., & Kumar, S. J. A. (2012). Dry sliding wear behavior of SiC/Al2O3 filled jute/epoxy composites. *Materials & Design, 36*, 306–315.

[40] Patnaik, A., Satapathy, A., Dwivedy, M., & Biswas, S. (2010). Wear behavior of plant fiber (pine-bark) and cement kiln dust-reinforced polyester composites using Taguchi experimental model. *Journal of Composite Materials, 44*(5), 559–574.

[41] Mohan, T. P., & Kanny, K. (2011). Water barrier properties of nanoclay filled sisal fiber reinforced epoxy composites. *Composites Part A: Applied Science and Manufacturing, 42*(4), 385–393.

[42] Rajasekhar, P., Ganesan, G., & Senthilkumar, C. (2014). Studies on tribological behavior of polyamide filled jute fiber-nano-ZnO hybrid composites. *Procedia Engineering, 97*, 2099–2109.

[43] Latha, P. S., & Rao, M. V. (2018). Investigation into effect of ceramic fillers on mechanical and tribological properties of bamboo-glass hybrid fiber reinforced polymer composites. *Silicon, 10*(4), 1543–1550.

[44] Sumesh, K. R., Saikrishnan, G., Pandiyan, P., Prabhu, L., Gokulkumar, S., Priya, A. K., Spatenka, P., & Krishna, S. (2021). The influence of different parameters in tribological characteristics of pineapple/sisal/TiO$_2$ filler incorporation. *Journal of Industrial Textiles, 51*, 8626S–8644S.

[45] Saba, N., Jawaid, M., Alothman, O. Y., & Paridah, M. T. (2016). A review on dynamic mechanical properties of natural fiber reinforced polymer composites. *Construction and Building Materials, 106*, 149–159.

[46] Nayak, S. K., Mohanty, S., & Samal, S. K. (2009). Influence of short bamboo/glass fiber on the thermal, dynamic mechanical and rheological properties of polypropylene hybrid composites. *Materials Science and Engineering: A, 523*(1–2), 32–38.

[47] Ali, M., Liu, A., Sou, H., & Chouw, N. (2012). Mechanical and dynamic properties of coconut fiber reinforced concrete. *Construction and Building Materials, 30*, 814–825.

[48] Thakur, V. K., & Thakur, M. K. (2014). Processing and characterization of natural cellulose fibers/thermoset polymer composites. *Carbohydrate Polymers, 109*, 102–117.

[49] Khorami, M., & Ganjian, E. (2011). Comparing flexural behaviour of fiber— Cement composites reinforced bagasse: Wheat and eucalyptus. *Construction and Building Materials, 25*(9), 3661–3667.

[50] Pappu, A., Patil, V., Jain, S., Mahindrakar, A., Haque, R., & Thakur, V. K. (2015). Advances in industrial prospective of cellulosic macromolecules enriched banana biofiber resources: A review. *International Journal of Biological Macromolecules, 79*, 449–458.

[51] Ahmad, A. F., Abbas, Z., Obaiys, S. J., & Zainuddin, M. F. (2018). Effect of untreated fiber loading on the thermal, mechanical, dielectric, and microwave absorption properties of polycaprolactone reinforced with oil palm empty fruit bunch biocomposites. *Polymer Composites, 39*(S3), E1778–E1787.

[52] Mahalingam, S., Gopalan, V., Velivela, H., Pragasam, V., Prabhakaran, P., & Suthenthiraveerappa, V. (2020). Studies on shear strength of CNT/coir fiber/fly ash-reinforced epoxy polymer composites. *Emerging Materials Research, 9*(1), 78–88.

[53] Yang, C., Han, R., Nie, M., & Wang, Q. (2020). Interfacial reinforcement mechanism in poly (lactic acid)/natural fiberbiocomposites featuring ZnO nanowires at the interface. *Materials & Design, 186*, 108332.

[54] Joseph, J., Munda, P. R., Kumar, M., Sidpara, A. M., & Paul, J. (2020). Sustainable conducting polymer composites: Study of mechanical and tribological properties of natural fiber reinforced PVA composites with carbon nanofillers. *Polymer-Plastics Technology and Materials, 59*(10), 1088–1099.

[55] Masoodi, R., El-Hajjar, R. F., Pillai, K. M., & Sabo, R. (2012). Mechanical characterization of cellulose nanofiber and bio-based epoxy composite. *Materials & Design (1980–2015), 36*, 570–576.

[56] Ansari, F., Galland, S., Johansson, M., Plummer, C. J., & Berglund, L. A. (2014). Cellulose nanofiber network for moisture stable, strong and ductile biocomposites and increased epoxy curing rate. *Composites Part A: Applied Science and Manufacturing, 63*, 35–44.

[57] Virtanen, S., Vuoti, S., Heikkinen, H., & Lahtinen, P. (2014). High strength modified nanofibrillated cellulose-polyvinyl alcohol films. *Cellulose, 21*(5), 3561–3571.

[58] Islam, M. S., Ahmad, M. B., Hasan, M., Aziz, S. A., Jawaid, M., Haafiz, M. M., & Zakaria, S. A. (2015). Natural fiber-reinforced hybrid polymer nanocomposites: Effect of fiber mixing and nanoclay on physical, mechanical, and biodegradable properties. *BioResources, 10*(1), 1394–1407.

[59] Raja, D. B. P., Prasad, M. S., & Ramanan, G. (2020). Influence of nanoparticles on mechanical and thermal characterization of hybrid fiber reinforced polyester composites. *Materials Today: Proceedings, 24*, 1498–1507.

[60] Sumesh, K. R., & Kanthavel, K. (2019). Green synthesis of aluminium oxide nanoparticles and its applications in mechanical and thermal stability of hybrid natural composites. *Journal of Polymers and the Environment, 27*(10), 2189–2200.

[61] Park, A. M., Wycisk, R. J., Ren, X., Turley, F. E., & Pintauro, P. N. (2016). Crosslinked poly (phenylene oxide)-based nanofiber composite membranes for alkaline fuel cells. *Journal of Materials Chemistry A, 4*(1), 132–141.

[62] Li, Y., Lee, D. K., Kim, J. Y., Kim, B., Park, N. G., Kim, K., Shin, J.-H., Choi, I.-S., & Ko, M. J. (2012). Highly durable and flexible dye-sensitized solar cells fabricated on plastic substrates: PVDF-nanofiber-reinforced TiO2 photoelectrodes. *Energy & Environmental Science, 5*(10), 8950–8957.

[63] Dubal, D. P., Ayyad, O., Ruiz, V., & Gomez-Romero, P. (2015). Hybrid energy storage: The merging of battery and supercapacitor chemistries. *Chemical Society Reviews, 44*(7), 1777–1790.

[64] Xu, J., Wang, L., Guan, J., & Yin, S. (2016). Coupled effect of strain rate and solvent on dynamic mechanical behaviors of separators in lithium ion batteries. *Materials & Design, 95*, 319–328.

[65] Devarayan, K., Lei, D., Kim, H. Y., & Kim, B. S. (2015). Flexible transparent electrode based on PANi nanowire/nylon nanofiber reinforced cellulose acetate thin film as supercapacitor. *Chemical Engineering Journal, 273*, 603–609.

[66] Eloy, F. S., Costa, R. R. C., De Medeiros, R., Ribeiro, M. L., & Tita, V. (2015). Comparison between mechanical properties of bio and synthetic composites for use in aircraft interior structures. Meeting on Aeronautical Composite Materials and Structures.

[67] Bhatnagar, A., & Sain, M. (2005). Processing of cellulose nanofiber-reinforced composites. *Journal of Reinforced Plastics and Composites, 24*(12), 1259–1268.

[68] Feng, Y., Xiong, T., Xu, H., Li, C., & Hou, H. (2016). Polyamide-imide reinforced polytetrafluoroethylene nanofiber membranes with enhanced mechanical properties and thermal stabilities. *Materials Letters, 182*, 59–62.

[69] Sanga Pachuau, L. (2015). A mini review on plant-based nanocellulose: Production, sources, modifications and its potential in drug delivery applications. *Mini Reviews in Medicinal Chemistry, 15*(7), 543–552.

[70] Gao, X., Song, J., Ji, P., Zhang, X., Li, X., Xu, X., Wang, M., Zhang, S., Deng, Y., Deng, F., & Wei, S. (2016). Polydopamine-templated hydroxyapatite reinforced polycaprolactone composite nanofibers with enhanced cytocompatibility and osteogenesis for bone tissue engineering. *ACS Applied Materials & Interfaces, 8*(5), 3499–3515.

[71] Tanimoto, Y. (2015). Dental materials used for metal-free restorations: Recent advances and future challenges. *Journal of Prosthodontic Research, 59*(4), 213–215.

[72] Al-Saleh, M. H., & Sundararaj, U. (2009). A review of vapor grown carbon nanofiber/polymer conductive composites. *Carbon, 47*(1), 2–22.

[73] Halib, N., Perrone, F., Cemazar, M., Dapas, B., Farra, R., Abrami, M., Chiarappa, G., Forte, G., Zanconati, F., Pozzato, G., Murena, L., Fiotti, N., Lapasin, R., Cansolino, L., Grassi, G., & Grassi, M. (2017). Potential applications of nanocellulose-containing materials in the biomedical field. *Materials, 10*(8), 977.

[74] Jorfi, M., & Foster, E. J. (2015). Recent advances in nanocellulose for biomedical applications. *Journal of Applied Polymer Science, 132*(14).

[75] Syverud, K., Pettersen, S. R., Draget, K., & Chinga-Carrasco, G. (2015). Controlling the elastic modulus of cellulose nanofibril hydrogels—Scaffolds with potential in tissue engineering. *Cellulose, 22*(1), 473–481.

[76] Czaja, W. K., Young, D. J., Kawecki, M., & Brown, R. M. (2007). The future prospects of microbial cellulose in biomedical applications. *Biomacromolecules, 8*(1), 1–12.

[77] Markstedt, K., Mantas, A., Tournier, I., Martínez Ávila, H., Hagg, D., & Gatenholm, P. (2015). 3D bioprinting human chondrocytes with nanocellulose-alginate bioink for cartilage tissue engineering applications. *Biomacromolecules, 16*(5), 1489–1496.

[78] Torres-Rendon, J. G., Femmer, T., De Laporte, L., Tigges, T., Rahimi, K., Gremse, F., Zafarnia, S., Lederle, W., Ifuku, S., Wessling, M., Hardy, J. G., & Walther, A. (2015). Bioactive gyroid scaffolds formed by sacrificial templating of nanocellulose and nanochitin hydrogels as instructive platforms for biomimetic tissue engineering. *Advanced Materials, 27*(19), 2989–2995.

[79] Powell, L. C., Khan, S., Chinga-Carrasco, G., Wright, C. J., Hill, K. E., & Thomas, D. W. (2016). An investigation of Pseudomonas aeruginosa biofilm growth on novel nanocellulose fiber dressings. *Carbohydrate Polymers, 137*, 191–197.

[80] Hua, K., Carlsson, D. O., Ålander, E., Lindström, T., Strømme, M., Mihranyan, A., & Ferraz, N. (2014). Translational study between structure and biological response of nanocellulose from wood and green algae. *RSC Advances, 4*(6), 2892–2903.

[81] Mertaniemi, H., Escobedo-Lucea, C., Sanz-Garcia, A., Gandía, C., Mäkitie, A., Partanen, J., Ikkala, O., & Yliperttula, M. (2016). Human stem cell decorated nanocellulose threads for biomedical applications. *Biomaterials, 82*, 208–220.

[82] Bäckdahl, H., Helenius, G., Bodin, A., Nannmark, U., Johansson, B. R., Risberg, B., & Gatenholm, P. (2006). Mechanical properties of bacterial cellulose and interactions with smooth muscle cells. *Biomaterials, 27*(9), 2141–2149.

[83] Saska, S., Barud, H. S., Gaspar, A. M. M., Marchetto, R., Ribeiro, S. J. L., & Messaddeq, Y. (2011). Bacterial cellulose-hydroxyapatite nanocomposites for bone regeneration. *International Journal of Biomaterials, 2011*, 175362.

[84] Saska, S., Teixeira, L. N., de Oliveira, P. T., Gaspar, A. M. M., Ribeiro, S. J. L., Messaddeq, Y., & Marchetto, R. (2012). Bacterial cellulose-collagen nanocomposite for bone tissue engineering. *Journal of Materials Chemistry, 22*(41), 22102–22112.

[85] Petersen, N., & Gatenholm, P. (2011). Bacterial cellulose-based materials and medical devices: Current state and perspectives. *Applied Microbiology and Biotechnology, 91*(5), 1277–1286.

[86] Derby, B. (2012). Printing and prototyping of tissues and scaffolds. *Science, 338*(6109), 921–926.

[87] Zheng, L. Y., & Zhu, J. F. (2003). Study on antimicrobial activity of chitosan with different molecular weights. *Carbohydrate Polymers, 54*(4), 527–530.

[88] Davoodbasha, M., Kim, S. C., Lee, S. Y., & Kim, J. W. (2016). The facile synthesis of chitosan-based silver nano-biocomposites via a solution plasma process and their potential antimicrobial efficacy. *Archives of Biochemistry and Biophysics, 605*, 49–58.

[89] Bhowmick, A., Jana, P., Pramanik, N., Mitra, T., Banerjee, S. L., Gnanamani, A., Das, M., & Kundu, P. P. (2016). Multifunctional zirconium oxide doped chitosan based hybrid nanocomposites as bone tissue engineering materials. *Carbohydrate Polymers, 151*, 879–888.

[90] Ahmed, S., & Ikram, S. (2016). Chitosan based scaffolds and their applications in wound healing. *Achievements in the Life Sciences, 10*(1), 27–37.
[91] Moura, D., Mano, J. F., Paiva, M. C., & Alves, N. M. (2016). Chitosan nanocomposites based on distinct inorganic fillers for biomedical applications. *Science and Technology of Advanced Materials, 17*(1), 626–643.
[92] Luzi, F., Puglia, D., & Torre, L. (2019). Natural fiber biodegradable composites and nanocomposites: A biomedical application. In *Biomass, Biopolymer-Based Materials, and Bioenergy*, edited by Deepak Verma, Elena Fortunati, Siddharth Jain, and Xiaolei Zhan (pp. 179–201). Cambridge, UK: Woodhead Publishing.

Chapter 9

Epoxy-based nanocomposites

Tribological characteristics
and challenges

Vivudh Gupta and Pawandeep Singh
Shri Mata Vaishno Devi University

Nida Naveed
University of Sunderland

CONTENTS

9.1 Introduction 143
9.2 Epoxy nanocomposites: Fabrication aspects and characteristics 144
9.3 Discussion 145
9.4 Challenges and future prospects 150
9.5 Conclusions 150
References 151

9.1 INTRODUCTION

Polymer matrix composites (PMCs) have been widely employed in the manufacturing of various engineering components in different industrial fields including marine, aerospace, and automobile that involve tribological applications. This manifold increase in applications of PMCs is because of their improved properties such as high strength-to-weight ratio, enhanced resistance to chemicals and wear, and improved thermal stability over monolithic materials (Ogbonna et al., 2022). These PMCs possess the potential to replace conventional materials including wood and metals (Parikh and Gohil, 2015). PMCs are constituted of thermosetting or thermoplastic polymers as matrix along with synthetic or natural fibres as the reinforcing phase. Nowadays, continuous demand for the development of sustainable materials has also expedited the use of natural fibres as filler material in PMCs keeping factors such as biodegradability, light weight, economical, eco-friendly, etc. in consideration (Kumar et al., 2019). The tribological properties of PMCs reinforced with either synthetic or natural fibre are dependent on various factors such as fibre orientation, fibre content, volume fraction of fibre, length of fibre, surface treatment undergone by fibre besides different operating parameters such as sliding speed, normal load, sliding distance, temperature, etc. (Parikh and Gohil, 2015;

Kumar et al., 2021). Among the class of thermosetting polymers, epoxy resin is commonly used and has applicability in coatings, laminates, construction materials, adhesives, composites, support material for printed circuit boards, blades for impellers, etc. required in major industrial sectors (Chen et al., 2007; Jia et al., 2006; Fouly and Alkalla, 2020). The major reasons for this increased applicability of epoxy resin is ascribed to the ease in processing, better adhesion properties, higher strength, low shrinkage during curing, low toxicity, chemical stability, electrical insulation, etc. (Kumar and Anand, 2019; Ogbonna et al., 2022). Regardless of all these remarkable characteristics and applications exhibited by epoxy resin, its use is somewhat restricted in tribological applications due to its three-dimensional cross-linked structure after curing, resulting in high brittleness, inferior fatigue, and thermal resistance, along with less resistance to wear. Hence, in order to cope up with poor tribological characteristics of epoxy resins, additional microscale/nanoscale reinforcement particles are added to it for the development of epoxy-based composites. These composites are considered more wear resistant than epoxy resin itself (Wang et al., 2015; Omrani et al., 2015). Nanofiller particles are capable of influencing tribological properties of epoxy polymer even at low concentrations owing to their extensive surface-area-to-volume ratio and better interfacial bonding between composite constituents as compared to traditional microscale reinforcements (Shen et al., 2013; Chan et al., 2021). Work done by various researchers in the field of tribological aspects of epoxy nanocomposites is discussed in subsequent sections.

9.2 EPOXY NANOCOMPOSITES: FABRICATION ASPECTS AND CHARACTERISTICS

Majority of the epoxy nanocomposites are fabricated by a common procedure. Firstly, a mixture of epoxy resin and nanoparticles in a desired ratio are mixed using a mechanical stirrer. Sometimes, the stirring action is also accompanied by ultrasonication process in order to avoid agglomeration of nanoparticles in resin (Kenig et al., 2011). Then, the required quantity of hardener is added to the resin and nanoparticles mixture followed by degassing procedure. Degassing is generally done using a vacuum pump to eradicate air bubbles and other chemicals (such as ethanol) utilized during the ultrasonic process (Shen et al., 2013). This residual mixture is then poured into moulds of desired shape and size and cured at room temperatures and oven at requisite temperatures. Curing aids in inducing cross-linkages in epoxy nanocomposites (Khanam et al., 2015). After curing, the sample is taken out from the mould and machined for a particular size as required for various testing purposes. Moulds used for fabricating such as composites may be made of teflon (Jia et al., 2008), aluminium (Renukappa

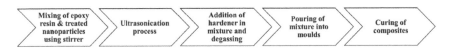

Figure 9.1 Outline depicting the fabrication of epoxy nanocomposites.

et al., 2011), glass (Veena et al., 2011), rubber (Zhang et al., 2013), etc. Brief outline depicting the fabrication of epoxy nanocomposites is shown in Figure 9.1.

Treatment of nanoparticles before embedding them into resin is also carried out. For instance, alumina nanoparticles were washed in acetone and water to remove any contaminants followed by the separation of liquid and nanoparticles using gravity centrifuge and then dried in the oven (Wang et al., 2006). Furthermore, to improve the dispersibility of Carbon nanotubes (CNTs), various functionalization methods like plasma treatment, coating with different coupling agents, and oxidation are utilized. Such processes improve wettability and aids in the creation of more bonding sites between particles and liquid (Chen et al., 2007). Surface treatment of organo-modified montmorillonite (OMMT) particles was carried using octadecylamine surfactant for getting better dispersion properties (Renukappa et al., 2011). Likewise, nanodiamond particles were immersed in the surfactant named tetrahydrofuran and after functionalization, the surfactant was subsequently removed (Ayatollahi et al., 2012). Graphene nanoplatelets were functionalized with aminopropyl trimethoxysilane coupling agent (Amirbeygi et al., 2019). Similarly, (3-glycidoxypropyl) trimethoxysilane was used for the modification of alumina surfaces (Fouly and Alkalla, 2020).

The work of various researchers on the tribological characteristics, i.e., friction and wear aspects of different epoxy-based nanocomposites along with process parameters involved and major findings from their study are compiled in Table 9.1.

9.3 DISCUSSION

Process parameters, viz., sliding speed, normal load, sliding distance, test duration, contact pressure, reinforcement content, etc. play a vital role in affecting friction, wear, and lubrication aspects of any material during sliding. The effect of these variables needs to be carefully studied with a view to understand the changes brought by them in the behaviour of material during tribological test. The various sub-areas in which tribological characteristics of epoxy-based nanocomposites have been studied include the evaluation of tribological performance of epoxy nanocomposites reinforced with different types of nanoparticles, parametric study of tribological behaviour using

Table 9.1 Tribological characteristics of epoxy-based nanocomposites

Composite material	Process parameters involved	Findings from the study	References
Epoxy/carbon-fibre/graphite/PTFE/nano-TiO$_2$ composite	Contact pressure: 0.5–12 MPa Sliding velocity: 0.5–3 m/s	Reduction in the coefficient of friction and contact temperatures with the inclusion of nano-TiO$_2$ particles.	Chang et al. (2005)
Epoxy/MWCNT composites	Sliding speed: 0.431 m/s, Load: 50 N	Best wear and friction characteristics exhibited by composite with 1.5 wt.% of MWCNTs were ascribed to its enhanced mechanical properties along with tubular structure of MWCNTs.	Dong et al. (2005)
Epoxy/SiO$_2$-organo-modified montmorillonite nanocomposites	Sliding time: 30 minutes, Load: 4.20 N, Speed: 300 rpm	Epoxy-based SiO$_2$-organo-modified montmorillonite composites exhibited excellent wear resistance properties due to the synergistic effect of reinforcing agents. Cracks and fatigue wear were predominant.	Jia et al. (2006)
Epoxy/rubber nanocomposites	Load: 30–100 N, Speed: 0.42 m/s, Time: 30–75 minutes	Specific wear rate of nanocomposites increased and hardness decreased as the concentration of rubber nanoparticles exceeded 5 wt.% resulting in their agglomeration.	Yu et al. (2008)
Epoxy/organo-modified montmorillonite nanocomposites	Speed: 0.5–1.5 m/s, Load: 10–30 N, Sliding distance: 1,000–3,000 m	Sliding distance emerged as the most significant parameter affecting wear rate of these composites. Rolling effect of nanoscale reinforcement particles led to improvement in wear resistance of polymer composites.	Renukappa et al. (2011)
Epoxy/silica nanocomposites	Speed: 0.5 m/s, Load: 15–60 N, Sliding distance: 1,000–6,000 m	Mild abrasion wear was observed in silica reinforced epoxy nanocomposites as compared to neat epoxy. Wear loss increased with sliding distance and load.	Veena et al. (2011)
Epoxy/nanodiamond composites	Sliding distance: 0–800 m, Speed: 1 m/s, Pressure: 1 MPa for 1 hour.	Inclusion of 0.1 wt.% nanodiamond decreased the friction and wear rate by 50% & 84%, respectively, owing to the polishing and rolling effect caused by nanodiamond particles.	Ayatollahi et al. (2012)

(Continued)

Table 9.1 (Continued) Tribological characteristics of epoxy-based nanocomposites

Composite material	Process parameters involved	Findings from the study	References
Epoxy/nanoclay composites	Speed: 0.02 m/s, Load: 5 N, Frequency: 3.2 Hz at 11,600 cycles	Composites with 1% of nanoclay particles exhibited slightly higher wear resistance and similar values of coefficient of friction as compared to that of pure epoxy.	Esteves et al. (2013)
Epoxy/wax/ MWCNT composites	Load: 2 N, Sliding speed: 2 cm/s	Reduction in coefficient of friction and wear of composites due to combined lubrication effects shown by wax and CNTs.	Khun et al. (2013)
Epoxy/Ni@NiO nanocomposites	Load: 1 MPa, Sliding speed: 1 m/s, Time: 60 minutes	Composites with 5 wt.% of nanoparticles exhibited 22.2 times increase in wear resistance. Wear mechanism in the case of composites was slight abrasion, whereas it was adhesion and abrasion in the case of neat epoxy.	Wang et al. (2015)
Epoxy/MWCNT/ boron nitride composites	Load: 10 N, Sliding speed: 1.2–1.5 m/s, Distance: 1,800 m	Modification of nanoparticles resulted in wear resistance improvement and friction reduction. Thermal softening is an important aspect for studying friction and wear of polymeric materials.	Düzcükoğlu et al. (2015)
Epoxy/MWCNT and epoxy/silica nanocomposites	Load: 7 N, Sliding speed: 0.93 m/s, Time: 300 seconds	These epoxy-based nanocomposites can be utilized for indoor flooring purposes in different places such as schools, offices, and hospitals.	Ramadan and Reda (2017)
Epoxy/short carbon fibre/ MWCNT composites	Load: 20 N, Sliding speed: 0.7 m/s	Lower wear rate and stable coefficient of friction were obtained during the sliding of hybrid composites. Worn surfaces of hybrid composites revealed the absence of microcracks.	Gbadeyan et al. (2017)
Epoxy/wax/silica nanocomposites	Load: 2–8 N, Sliding speed: 0.08–0.32 m/s	Hybrid reinforcements can aid in yielding better mechanical & tribological properties.	Imani et al. (2018)
Epoxy/ nanocellulose composites	Load: 10 N, Sliding speed: 0.1 m/s, Distance: 1,000 m	Epoxy nanocellulose composites showed improved mechanical, wear, viscoelastic, and corrosion-resistant properties over neat epoxy polymer.	Mohit and Selvan (2019)

(Continued)

Table 9.1 (Continued) Tribological characteristics of epoxy-based nanocomposites

Composite material	Process parameters involved	Findings from the study	References
Epoxy/graphene nanocomposites	Load: 10 N, Sliding speed: 0.09 m/s, Distance: 1,000 m	Decreased coefficient of friction and wear rate was observed in composites having 5 and 8 wt.% of graphene nanoplatelets.	Campo et al. (2020)
Epoxy/graphene/ montmorillonite nanocomposites	Load: 10 N, Sliding speed: 0.5 m/s, Distance: 1,000 m	High hardness of montmorillonite clay and the self-lubricating nature of graphene nanoplatelets decreased the wear rate of epoxy nanocomposites.	Kazemi-Khasragh et al. (2020)
Epoxy/basalt/ nano-alumina composites	Load: 20–40 N, Sliding speed: 3–6 m/s, Distance: 1,800 m	Wear loss in composite with 2.4 wt.% of nano-alumina particles is about 40% less than that in composite without nano-alumina.	Vinay et al. (2021)
Epoxy/Ti$_3$C$_2$ nanocomposites	Sliding Time: 3,600 seconds	Worn surface morphology revealed that as the concentration of Ti$_3$C$_2$ nanosheets increased, wear behaviour changed from fatigue to adhesion.	Xu et al. (2021)
Epoxy/MWCNT composites	Load: 10–20 N	Least abrasion was shown by 0.5 wt.% MWCNT-epoxy composite due to improved interaction between the matrix and reinforcing phase.	Kurien et al. (2021)
Epoxy/MWCNT composites	MWCNT concentration: 0.5–1 wt.%	37% reduction in the coefficient of friction in composite with 1 wt.% of MWCNT. This composite also exhibited improved elastic modulus, wear properties, hardness, and resistance to plastic deformation.	Mirsalehi et al. (2021)

different modelling and optimization techniques, investigation of friction and wear of epoxy nanocomposites after the inclusion of lubricants, tribological performance of hybrid epoxy nanocomposites, etc. These aspects are presented in Figure 9.2.

Due to the inclusion of certain volume % of nano-TiO$_2$ filler particles, steady reduction in coefficient of friction consequently improved the load - bearing capacity of nanocomposites (Chang et al., 2005). Better chemical bonding between the constituents of composites at the interface also results in reducing frictional coefficient and improving resistance to wear during sliding (Luo et al., 2007). Optimal concentration and shape of nanoparticles

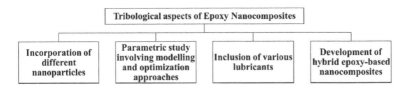

Figure 9.2 Tribological areas studied in the field of epoxy-based nanocomposites.

are also influential in determining the wear rate in nanocomposites (Kenig et al., 2011). Wrinkled surface morphology of graphene nanosheets led to an increase in the coefficient of friction at different normal pressures. However, specific wear rate of such composites decreased when the content of graphene oxide was considerably increased. This was ascribed to an increase in the specific surface area of graphene oxide in composites and the strong adhesion between epoxy and reinforcement causing proper interlocking between the constituents (Shen et al., 2013). Agglomerations in nanocomposite are capable of causing damage to the steel counterbody, which further can cause damage to the nanocomposite by developing deeper grooves in it (Neitzel et al., 2012). The rolling effect produced by nanoparticles during sliding can aid in decreasing contact temperatures and coefficient of friction. Smooth surface morphologies are a characteristic feature of nanocomposites as compared to that of neat polymer caused during sliding (Kurahatti et al., 2014). Formation of tribo-film and high resistance to abrasion in epoxy nanocomposites can result in improving the wear characteristics of polymer (Gao et al., 2016). $NbSe_2$ also provided a constraint to the direct contact between steel balls and epoxy coatings and hence, exhibited improved tribological features (Chen et al., 2016). The wear rate of neat epoxy was found to be exhibiting an increasing trend with an increase in the sliding distance. This happened due to thermal softening of epoxy resin when it was in contact with the counterbody for increased time duration and consequently, heat was also induced from the area of contact during sliding. However, the wear rate of multi-walled carbon nanotubes and short carbon fibre–reinforced epoxy composite was lower than that of neat epoxy with increased sliding distance. This was attributed to the lubrication effect shown by carbon fibre and improved mechanical strength of the nanocomposites (Gbadeyan et al., 2017). Hexagonal boron nitride nanosheets decorated with polydopamine also enhanced the tribological and thermomechanical properties when added in epoxy resin (Song et al., 2018). Improved thermal, tribological, mechanical, electrophysical, and magnetic properties of epoxy nanocomposites reinforced with multi-walled carbon nanotubes make them a potential candidate for various applications in the aerospace field such as development of sandwich and laminated structures, anti-lightning and anti-radar protection devices, heat-resistant paints, etc. (Irzhak et al., 2019). Sugarcane nanocellulose reinforced epoxy bio-nanocomposites also exhibited better dispersion properties in epoxy resin, resulting in improved wear-resistant, mechanical, and anti-corrosion properties (Mohit and Selvan, 2019). Epoxy hybrid nanocomposite coatings reinforced with graphene oxide and titania were successfully developed for various tribological applications (Satheesan and Mohammed, 2021).

9.4 CHALLENGES AND FUTURE PROSPECTS

There are various challenges encountered while developing and studying tribological behaviour of epoxy nanocomposites. Firstly, proper dispersion of nanoparticles after their certain content has been added in the epoxy resin still remains to be improved. Secondly, understanding the development of formation of transfer films in epoxy nanocomposites needs to be properly evaluated. Such challenges can be considered as future prospects in this area. Other future avenues while evaluating the tribological characteristics of epoxy nanocomposite include:

- Development and evaluation of tribological characteristics of hybrid epoxy nanocomposites reinforced with ceramic particles and lubricants.
- Preparation of natural fibre–based epoxy nanocomposites and evaluating their characteristics thereof.
- Comparison of tribological performance of various epoxy nanocomposites developed through different fabrication techniques.
- Investigating the effect of different process parameters on frictional and wear behaviour of epoxy nanocomposites through different modelling and optimization approaches.
- Evaluating the effect of water absorption on tribological features of epoxy nanocomposites.
- Studying the effect of addition of different nanoparticles (with and without chemical modification) on wear characteristics of epoxy-based nanocomposites.

9.5 CONCLUSIONS

This chapter firstly highlighted the importance of polymer composites in various fields. Features and applications of epoxy resin are discussed in detail. Need for the development of epoxy composites is highlighted. Prior to fabrication of epoxy nanocomposites, different surfactants and coupling agents, which are utilized for chemical treatment purpose, are also shown. Various facts, which are responsible for improving the tribological aspects of nanocomposites under different settings of process variables, are included. Other factors including agglomeration of nanoparticles, which deteriorate the tribological performance of such nanocomposites, are also added. Based on the above discussion, it is inferred that inclusion of nanoparticles in epoxy resin aid in improving the friction, wear, and lubrication aspects of epoxy resin.

REFERENCES

Amirbeygi, H., Khosravi, H., & Tohidlou, E. 2019. Reinforcing effects of aminosilane-functionalized graphene on the tribological and mechanical behaviors of epoxy nanocomposites. *Journal of Applied Polymer Science*, 136(18), p. 47410.

Ayatollahi, M., Alishahi, E., Doagou-R. S., & Shadlou, S. 2012. Tribological and mechanical properties of low content nanodiamond/epoxy nanocomposites. *Composites Part B: Engineering*, 43(8), pp. 3425–3430.

Campo, M., Jiménez-Suárez, A., & Ureña, A. 2020. Tribological properties of different types of graphene nanoplatelets as additives for the epoxy resin. *Applied Sciences*, 10(12), p. 4363.

Chan, J. X., Wong, J. F., Petrů, M., Hassan, A., Nirmal, U., Othman, N., & Ilyas, R. A. 2021. Effect of nanofillers on tribological properties of polymer nanocomposites: A review on recent development. *Polymers*, 13(17), p. 2867.

Chang, L., Zhang, Z., Breidt, C., & Friedrich, K. 2005. Tribological properties of epoxy nanocomposites: I. Enhancement of the wear resistance by nano-TiO_2 particles. *Wear*, 258(1–4), pp. 141–148.

Chen, H., Jacobs, O., Wu, W., Rüdiger, G., & Schädel, B. 2007. Effect of dispersion method on tribological properties of carbon nanotube reinforced epoxy resin composites. *Polymer Testing*, 26(3), pp. 351–360.

Chen, J., Yang, J., Chen, B., Liu, S., Dong, J., & Li, C. 2016. Large-scale synthesis of NbSe2 nanosheets and their use as nanofillers for improving the tribological properties of epoxy coatings. *Surface and Coatings Technology*, 305, pp. 23–28.

Dong, B., Yang, Z., Huang, Y., & Li, H.-L. 2005. Study on tribological properties of multi-walled carbon nanotubes/epoxy resin nanocomposites. *Tribology Letters*, 20(3), pp. 251–254.

Düzcükoğlu, H., Ekinci, Ş., Şahin, Ö. S., Avci, A., Ekrem, M., & Ünaldi, M. 2015. Enhancement of wear and friction characteristics of epoxy resin by multiwalled carbon nanotube and boron nitride nanoparticles. *Tribology Transactions*, 58(4), pp. 635–642.

Esteves, M., Ramalho, A., Ferreira, J., & Nobre, J. 2013. Tribological and mechanical behaviour of epoxy/nanoclay composites. *Tribology Letters*, 52(1), pp. 1–10.

Fouly, A., & Alkalla, M. G. 2020. Effect of low nanosized alumina loading fraction on the physicomechanical and tribological behavior of epoxy. *Tribology International*, 152, p. 106550.

Gao, C., Guo, G., Zhao, F., Wang, T., Jim, B., Wetzel, B., Zhang, G., & Wang, Q. 2016. Tribological behaviors of epoxy composites under water lubrication conditions. *Tribology International*, 95, pp. 333–341.

Gbadeyan, O. J., Kanny, K., & Mohan, T. 2017. Influence of the multi-walled carbon nanotube and short carbon fibre composition on tribological properties of epoxy composites. *Tribology-Materials, Surfaces & Interfaces*, 11(2), pp. 59–65.

Imani, A., Zhang, H., Owais, M., Zhao, J., Chu, P., Yang, J., & Zhang, Z. 2018. Wear and friction of epoxy based nanocomposites with silica nanoparticles and wax-containing microcapsules. *Composites Part A: Applied Science and Manufacturing*, 107, pp. 607–615.

Irzhak, V. I., Dzhardimalieva, G. I., & Uflyand, I. E. 2019. Structure and properties of epoxy polymer nanocomposites reinforced with carbon nanotubes. *Journal of Polymer Research*, 26(9), pp. 1–27.

Jia, Q., Shan, S., Wang, Y., Gu, L., & Li, J. 2008. Tribological performance and thermal behavior of epoxy resin nanocomposites containing polyurethane and organoclay. *Polymers for Advanced Technologies*, 19(7), pp. 859–864.

Jia, Q., Zheng, M., Xu, C., & Chen, H. 2006. The mechanical properties and tribological behavior of epoxy resin composites modified by different shape nanofillers. *Polymers for Advanced Technologies*, 17(3), pp. 168–173.

Kazemi-Khasragh, E., Bahari-Sambran, F., Platzer, C., & Eslami-Farsani, R. 2020. The synergistic effect of graphene nanoplatelets-montmorillonite hybrid system on tribological behavior of epoxy-based nanocomposites. *Tribology International*, 151, p. 106472.

Kenig, S., Wagner, H., Zak, A., Moshkovith, A., Rapoport, L., & Tenne, R. 2011. The mechanical and tribological properties of epoxy nanocomposites with WS2 nanotubes. *Sensors and Transducers*, 12, pp. 53–65.

Khanam, A., Mordina, B., & Tiwari, R. 2015. Statistical evaluation of the effect of carbon nanofibre content on tribological properties of epoxy nanocomposites. *Journal of Composite Materials*, 49(20), pp. 2497–2507.

Khun, N. W., Zhang, H., Yang, J., & Liu, E. 2013. Mechanical and tribological properties of epoxy matrix composites modified with microencapsulated mixture of wax lubricant and multi-walled carbon nanotubes. *Friction*, 1(4), pp. 341–349.

Kumar, R., & Anand, A. 2019. Tribological behavior of natural fiber reinforced epoxy based composites: A review. *Materials Today: Proceedings*, 18, pp. 3247–3251.

Kumar, R., Haq, M. I. U., Raina, A., Sharma, S. M., Anand, A., & Abdollah, M. F. B. 2021. Tribological behaviour of natural fibre based polymer composites. In Jena, H., Katiyar, J. K., & Patnaik, A. (eds.), *Tribology of Polymer and Polymer Composites for Industry 4.0*. Singapore: Springer.

Kumar, R., Ul Haq, M. I., Raina, A., & Anand, A. 2019. Industrial applications of natural fibre-reinforced polymer composites-challenges and opportunities. *International Journal of Sustainable Engineering*, 12(3), pp. 212–220.

Kurahatti, R., Surendranathan, A., Ramesh Kumar, A., Auradi, V., Wadageri, C., & Kori, S. 2014. Mechanical and tribological behaviour of epoxy reinforced with nano Al_2O_3 particles. *Applied Mechanics and Materials*, 592–594, pp. 1320–1324.

Kurien, R. A., Selvaraj, D. P., Sekar, M., Koshy, C. P., & Praveen, K. 2021. Comparative mechanical, tribological and morphological properties of epoxy resin composites reinforced with multi-walled carbon nanotubes. *Arabian Journal for Science and Engineering*, 47, pp. 8059–8067.

Luo, Y., Zhi Rong, M., & Qiu Zhang, M. 2007. Tribological behavior of epoxy composites containing reactive SiC nanoparticles. *Journal of Applied Polymer Science*, 104(4), pp. 2608–2619.

Mirsalehi, S., Youzbashi, A., & Sazgar, A. 2021. Nanomechanical and tribological properties of multi-walled carbon nanotubes reinforced epoxy nanocomposite: Biomedical applications. *Polymer Science, Series A*, 63(1), pp. S75–S84.

Mohit, H., & Selvan, V. 2019. Effect of a novel chemical treatment on nanocellulose fibers for enhancement of mechanical, electrochemical and tribological characteristics of epoxy bio-nanocomposites. *Fibers and Polymers*, 20(9), pp. 1918–1944.

Neitzel, I., Mochalin, V., Bares, J. A., Carpick, R. W., Erdemir, A., & Gogotsi, Y. 2012. Tribological properties of nanodiamond-epoxy composites. *Tribology Letters*, 47(2), pp. 195–202.

Ogbonna, V., Popoola, A., Popoola, O., & Adeosun, S. 2022. A review on the recent advances on improving the properties of epoxy nanocomposites for thermal, mechanical, and tribological applications: Challenges and recommendations. *Polymer-Plastics Technology and Materials*, 61(2), pp. 176–195.

Omrani, E., Barari, B., Moghadam, A. D., Rohatgi, P. K., & Pillai, K. M. 2015. Mechanical and tribological properties of self-lubricating bio-based carbon-fabric epoxy composites made using liquid composite molding. *Tribology International*, 92, pp. 222–232.

Parikh, H. H., & Gohil, P. P. 2015. Tribology of fiber reinforced polymer matrix composites—A review. *Journal of Reinforced Plastics and Composites*, 34(16), pp. 1340–1346.

Ramadan, M., & Reda, R. 2017. CNTs, Al_2O_3 and SiO_2 reinforced epoxy: Tribological properties of polymer nanocomposites. *Tribology in Industry*, 39(3), p. 357.

Renukappa, N., Suresha, B., Devarajaiah, R., & Shivakumar, K. 2011. Dry sliding wear behaviour of organo-modified montmorillonite filled epoxy nanocomposites using Taguchi's techniques. *Materials & Design*, 32(8–9), pp. 4528–4536.

Satheesan, B., & Mohammed, A. S. 2021. Tribological characterization of epoxy hybrid nanocomposite coatings reinforced with graphene oxide and titania. *Wear*, 466, p. 203560.

Shen, X.-J., Pei, X.-Q., Fu, S.-Y., & Friedrich, K. 2013. Significantly modified tribological performance of epoxy nanocomposites at very low graphene oxide content. *Polymer*, 54(3), pp. 1234–1242.

Song, J., Dai, Z., Li, J., Tong, X., & Zhao, H. 2018. Polydopamine-decorated boron nitride as nano-reinforcing fillers for epoxy resin with enhanced thermomechanical and tribological properties. *Materials Research Express*, 5(7), p. 075029.

Veena, M., Renukappa, N., Suresha, B., & Shivakumar, K. 2011. Tribological and electrical properties of silica-filled epoxy nanocomposites. *Polymer Composites*, 32(12), pp. 2038–2050.

Vinay, S., Sanjay, M., Siengchin, S., & Venkatesh, C. 2021. Effect of Al_2O_3 nanofillers in basalt/epoxy composites: Mechanical and tribological properties. *Polymer Composites*, 42(4), pp. 1727–1740.

Wang, H., Yan, L., Liu, D., Wang, C., Zhu, Y., & Zhu, J. 2015. Investigation of the tribological properties: Core-shell structured magnetic Ni@ NiO nanoparticles reinforced epoxy nanocomposites. *Tribology International,* 83, pp. 139–145.

Wang, Y., Lim, S., Luo, J., & Xu, Z. 2006. Tribological and corrosion behaviors of Al_2O_3/polymer nanocomposite coatings. *Wear,* 260(9–10), pp. 976–983.

Xu, Z., Shen, X., Wang, T., Yang, Y., Yi, J., Cao, M., Shen, J., Xiao, Y., Guan, J., & Jiang, X. 2021. Investigation on tribological and thermo-mechanical properties of Ti_3C_2 nanosheets/epoxy nanocomposites. *ACS Omega,* 6(43), pp. 29184–29191.

Yu, S., Hu, H., Ma, J., & Yin, J. 2008. Tribological properties of epoxy/rubber nanocomposites. *Tribology International,* 41(12), pp. 1205–1211.

Zhang, J., Chang, L., Deng, S., Ye, L., & Zhang, Z. 2013. Some insights into effects of nanoparticles on sliding wear performance of epoxy nanocomposites. *Wear,* 304(1–2), pp. 138–143.

Chapter 10

Tribo-response of multifunctional polymer nanocomposites

Meghashree Padhan
Indian Institute of Technology, Delhi

CONTENTS

10.1 Introduction to multifunctional polymer nanocomposites (MPNCs) 155
10.2 Multi-component multifunction system–based MPNCs 156
10.3 Single component multifunction system–based MPNCs 157
10.4 Tribo-response of MPNCs 159
    10.4.1 Type of matrix 159
    10.4.2 Type of reinforcement 159
    10.4.3 Fiber/filler-matrix adhesion 160
    10.4.4 Distribution and dispersion of nanofillers 160
    10.4.5 Type of transfer film in the sliding wear mode 160
    10.4.6 Wear mode selected 161
10.5 Application of MPNCs 162
10.6 Future scope of MPNCs 162
References 163

10.1 INTRODUCTION TO MULTIFUNCTIONAL POLYMER NANOCOMPOSITES (MPNCs)

Multifunctional polymer nanocomposites (MPNCs) are the most sought materials in the present era because of their exceptional characteristics of high load-bearing capacity, fatigue resistance, high thermal stability, chemical and corrosion resistance, and efficient tribo-response such as ultra-low friction and wear properties. The MPNCs should incorporate the functions of two or more different components to increase the total composite's efficiency. The extant literature on MPNCs is focused on efforts to design and tailor polymer nanocomposites, which can be offered as a lightweight structural promising substitute for the existing metallic materials (Ul Ain et al. 2021; Ogbonna et al. 2021; Girdhar Shinde and Mangesh Patel 2022; Zhang et al. 2021; Almahri et al. 2022; Chan et al. 2021; Amurin et al. 2022; Novikov et al. 2022; Sengwa and Dhatarwal 2022; Teijido et al. 2022; Zhang et al. 2015; Al-Oqla et al. 2015; Saba et al. 2017; Gul et al. 2016; Delides 2016;

Qi et al. 2022; Andrew et al. 2021; Nisha et al. 2022; Japić et al. 2022; Osman et al. 2022; Aghamohammadi et al. 2021; Rasul et al. 2022; Li et al. 2022; Dorigato and Fredi 2022; Dong et al. 2021). These MPNCs are emerging in applications such as biomedical engineering, electrical capacitors and batteries, remote sensing, space shuttles, environmental and energy storage sectors, automotive, airplanes, etc. Although the primary focus is to develop high-strength lightweight composites, it is equally important to monitor the tribo-response of these composites. Tribo-response of these composites includes the friction and wear behavior of these materials with respect to various operating conditions, such as load, sliding speed, and sliding time.

10.2 MULTI-COMPONENT MULTIFUNCTION SYSTEM–BASED MPNCs

A typical MPNC is also expected to meet the requirements of an efficient tribo-material. An efficient tribo-material requires different levels of friction and wear depending on its applications. The extent of friction and wear requirements, depending on the applications, can be categorized as follows:

- High friction, low wear for brakes and clutches
- High friction, high wear for erasers
- Low friction, low wear for dry bearings, bushes, gears, etc.
- Infinite friction, zero wear for adhesives

The constituents of a typical MPNC and their corresponding functions are shown in Figure 10.1. The research efforts have been diverted from a single reinforcement-matrix system to multiple reinforcements in a single matrix system. Continuous efforts have been focused on the literature to design MPNCs with varying reinforcements such as fibers (long, short, continuous fabric) along with one or multiple fillers with variable shapes, size, and surface functionalization, to achieve synergistic effects (Osman et al. 2022; Aghamohammadi et al. 2021; Podsiadły et al. 2022; Dong et al. 2021; Andrew et al. 2021; Qi et al. 2022).

Panda et al. (2019, 2020, 2021) did a systematic study of various solid lubricants in combination with polyaryletherketone (50 wt.%), short glass fibers (30 wt.%), graphite (10 wt.%), and remaining 10 wt.% with varying combinations of microsolid and nanosolid lubricants (synthetic graphite, molybdenum disulfide, tungsten disulfide, polytetrafluoroethylene (PTFE), hexagonal boron nitride, etc.) to develop composites with high tensile strength and modulus as high as 170 MPa and 18 GPa; specific wear rate $(K_0) \approx 1.83 \times 10^{-16}$ m^3/Nm and coefficient of friction (μ) – 0.040 were tested in the adhesive wear mode under dry sliding conditions.

Qi et al. (2022) studied a multi-component composite system with polyoxymethylene, carbon fiber (10 wt.%), and nanoalumina particles (0.5–8 wt.%).

Tribo-response of multifunctional polymer nanocomposites 157

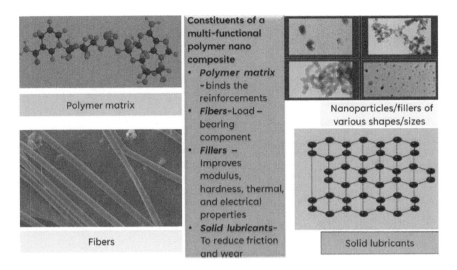

Figure 10.1 Constituents of a multifunctional polymer nanocomposite.

The composites showed the best combination of mechanical and tribological properties at 4 wt.% alumina content because, at higher contents, the nanoparticles tend to agglomerate and act as stress-concentrators leading to failure of the composite. The composites showed tensile strength around 60 MPa and elastic modulus around 6 GPa, along with $K_0 \sim 10^{-15}\,\text{m}^3/\text{Nm}$ and $\mu \sim 0.037$, tested in adhesive wear mode under lubricated conditions.

Dong et al. (2021) systematically reviewed the state-of-the-art multifunctional epoxy nanocomposites reinforced with 2-D nanomaterials. The review summarizes the recent strategies adopted for incorporating the 2-D nanomaterials into epoxy, challenges associated with the fabrication of composites, overall performance, and future prospects.

10.3 SINGLE COMPONENT MULTIFUNCTION SYSTEM–BASED MPNCs

The MPNCs, as the name suggests, perform multiple functions in a single system. The nano-inclusions can be a nanofiber, nanoparticles, nanoflowers, nanorods, nanoplatelets, etc. Based on the type of nanofillers, the MPNCs are broadly classified based on the type of nano-inclusions, as shown in Figure 10.2. The reinforcing nanofiller/nanofiber is chosen in such a way to impart multifunctional attributes simultaneously to result in an MPNC. For example, the addition of carbon nanotubes in a polymer matrix imparts multiple attributes into the system, improves the specific strength along with exceptional modulus of elasticity, imparts electrical and thermal conductivity exceptionally even at ultra-low content, and

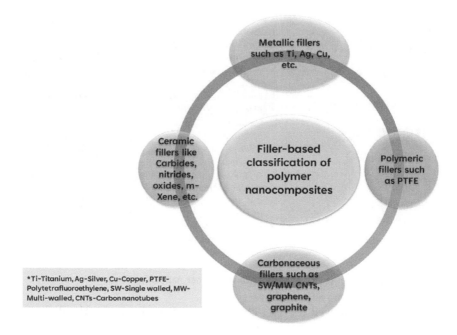

Figure 10.2 Broad classification of MPNCs based on the type of filler.

induces piezo-resistivity in the system. These multifunctional attributes are improved in multi-folds even at ultra-low filler loadings of nanofillers due to their high surface-to-volume ratio, which provides a larger area for interaction with the matrix. Some of the common nanofiller-induced improvements in the polymers, along with enhancement of mechanical properties, are improvements in thermal and electrical properties from the incorporation of CNTs, flame retardant properties in the case of the introduction of nanoclays and metal hydroxides, thermal stability, gas permeability, and so on. Besides these, biocompatible fillers, such as hydroxyapatite, nanocellulose, titanium, and zirconia, are widely explored to develop polymer nanocomposites for biodegradable and biocompatible materials and medical applications such as implants.

Almahri et al. (2022) reported multifunctional 3-D printed auxetic nanocomposite structures with high-density polyethylene and multi-walled carbon nanotubes. The composites showed an exceptional mechanical and tunable piezoresistive response. The composites are targeted for self-sensing smart materials and structures with tunable sensitivity.

Amurin et al. (2022) studied the effect of reduced graphene oxide (r-GO; 0–0.5 wt.%) in ultra-high molecular weight polyethylene to study the multifunctional attributes endowed by r-GO in UHMWPE. r-GO simultaneously improved the hardness, abrasive wear resistance, and crystallinity of UHMWPE.

10.4 TRIBO-RESPONSE OF MPNCs

The MPNCs can also be termed as a specific class of tribo-materials that contain an amalgam of constituents performing various roles, as described in Figure 10.1. The tribo-response of MPNCs is affected by a number of factors as follows:

- Type of matrix
- Type of reinforcement
- Fiber/filler-matrix adhesion
- Distribution and dispersion of nanofillers
- Type of transfer film in the sliding wear mode
- Wear mode in which it is used

10.4.1 Type of matrix

The polymer matrix governs the friction and wear behavior of MPNCs in the following ways:

- Hardness and toughness of the polymer
- Type of film it transfers to the counterface in the case of sliding wear
- How fast the film is transferred to the counterface in the case of sliding wear
- Load-bearing capacity of the polymer
- How effectively the polymers bind the reinforcement

10.4.2 Type of reinforcement

The type of reinforcement in an MPNC can be a filler, fiber, fabric, or a combination of any of these where at least one of the reinforcements is in nano-dimension. In the case of fibers, the aspect ratio of the fibers and the type of fibers play a major role in deciding the tribo-response of the composite. For example, carbon fibers are well known for their high strength, self-lubricity, and thermal and electrical conductivity. This leads to high load-bearing capacity, better frictional heat dissipation, low friction, and low wear of carbon-fiber-based MPNCs in the case of the adhesive wear mode. In contrast to this, aramid fibers are well known for their superior strength and impact resistance. This leads to the superior performance of aramid fiber-based composites in erosive and abrasive wear mode and poor performance in carbon-fiber-based composites. In contrast to this, carbon-fiber-based composites perform the best in the adhesive wear mode and fail in the erosive wear mode. The nanofillers' type, size, and shape govern the mechanisms responsible for the friction and wear behavior at tribo-contacts.

10.4.3 Fiber/filler-matrix adhesion

The fiber/filler-matrix interface is one of the major deciding parameters for the tribo-performance of the composite. The stronger the interface, the lesser the probability of material loss due to wear. Various surface treatments have been employed to functionalize fillers/fiber, which leads to significant improvement of properties of the composite. Padhan et al. (2021) studied the effect of siloxane treatment of alumina nano-spheres in a multifunctional glass fiber–reinforced PolyArylEtherKetones (PAEK) composite lubricated with graphite. It was observed that the inclusion of alumina followed by their functionalization led to improved physical, thermo-physical, and mechanical strength and tribo-properties. The composites showed ultra-low $K_0 - 2 \times 10^{-16}\,m^3/Nm$, ultra-low $\mu - 0.044$, and a very high $PV_{limit} - 112\,MPa\,m/s$. Composites with a strengthened interface of matrix and reinforcement lead to improved mechanical and tribological performance. Surface functionalization of nanofillers also helps in a uniform dispersion of the nanoparticles in the polymer matrix in a deagglomerated form.

10.4.4 Distribution and dispersion of nanofillers

In the case of MPNCs, the most challenging part is handling the nanoparticles, deagglomerating the nanofillers, and dispersing the nanofillers in the matrix effectively. Probe sonication is one of the most preferred approaches to deagglomerate and disperse the nanoparticles in the polymer. However, chemical grafting of the nanofillers in the polymeric chain is an advanced method, where the nanofillers are grafted into the polymeric chain giving the maximum filler-matrix interaction along with the elimination of agglomeration challenge. Although probe sonication effectively addresses the issue of deagglomeration and dispersion of nanoparticles in a polymer matrix, it is equally important to choose the right medium of dispersion, the effective duration of sonication, and validation of deagglomeration by methods like dynamic light scattering.

10.4.5 Type of transfer film in the sliding wear mode

Polymers have the inherent tendency to transfer films onto the counterface in the sliding wear mode, and this, in turn, changes the polymer vs. metal contacts to polymer vs. polymer, leading to reduced friction and wear. In the case of MPNCs, the mechanism of friction and wear reduction is governed by the type of transfer film formed, the rate of transfer film formation, and the chemical composition of the transfer film. When the MPNCs contain a nanofiller, which is a solid lubricant, the transfer film is a combination of polymer matrix and solid lubricant which governs the tribo-mechanisms. Particles like graphite, molybdenum disulfide

(MoS_2), tungsten disulfide (WS_2), and hexaboron nitride (hBN), which possess a layer lattice structure, tend to transfer their uniform film onto the counterface because of weak van der Waals' forces between the layers. In the case of PTFE, the film transferred is very uniform, thin, and homogeneous, and is mainly attributed to the molecular cohesion of carbon and fluorine atoms.

For nanoparticles, the mechanism of friction and wear behavior is also governed by the shape and size of the particles. Nanoparticles act by the following mechanisms to reduce friction and wear in lubricants:

- Rolling effect
- Mending effect
- Protective film formation
- Polishing effect

However, in the case of MPNCs containing nanoparticles, the mechanisms to reduce friction and wear is different. The nano-mechanisms responsible for the improved tribo-performance in case of sliding wear can be the rolling effect, wherein change in sliding contact to point contacts by the embedded nanoparticles serve as load-bearing particles, and protective film formation (Padhan et al. 2021).

Solid lubricants are low surface energy materials, adhesion is minimal, and hence μ is very low. The transfer film can be of various types, depending on the polymer from which it is transferred. Polymers like nylon transfer a thick and patchy film, which leads to high friction. Polymers such as PTFE transfer a thin, coherent, and uniform film leading to low friction (Padhan et al. 2020).

10.4.6 Wear mode selected

The multifunctional polymer nanocomposites (MPNCs) are designed and tailored for some targeted applications involving specific wear modes. Composites containing carbon fibers and solid lubricants are targeted for dry bearing applications and are tested in the adhesive wear mode.

Similarly, the composites containing Kevlar fibers and hard fillers are targeted for abrasion and erosion-resistant applications such as conveyer belts, wind-turbine blades, agricultural bearings, etc.

In another way, it can be said that the extent of improvement of properties of an MPNC designed for the adhesive wear mode can be lower than the deterioration it causes in wear resistance when tested in abrasive and erosive wear modes. However, as an exception to this, a braided fiber composite system of zylon and polyetheretherketone has shown excellent wear resistance in three wear modes simultaneously – adhesive mode, abrasive mode, and erosive wear mode (Padhan et al. 2021).

10.5 APPLICATION OF MPNCs

MPNCs represent an emerging class of materials and have potentially served as an alternative to conventional technologies and are widely used in electromagnetic (EMI) shielding applications, sensors and actuators, aerospace and automotive components, medical implants (hip, knee, dental, cerebral, etc.) and marine applications, smart wearables, energy storage applications, food packaging, and structural batteries. Some of the major application sectors of MPNCs are illustrated in Figure 10.3. The tribo-response of these composites is a vital performance deciding parameter of all the application sectors of MPNCs, since all the materials are used in several operating conditions sliding against another body or impacted with another body. It is important to monitor the friction and wear of the developed MPNCs as every material undergoes wear with time.

10.6 FUTURE SCOPE OF MPNCs

MPNCs have already emerged as a class of materials that have increasingly found potential applications in a lot of sectors discussed in the previous sections. The tribo-response of the existing smart MPNCs such as EMI shielding materials, self-sensing implant, smart wearables, etc. still remains underexplored and needs to be focused along with other functional aspects. The existing MPNCs are aimed to develop materials with multifunctional attributes with advanced technologies, which increase the overall cost.

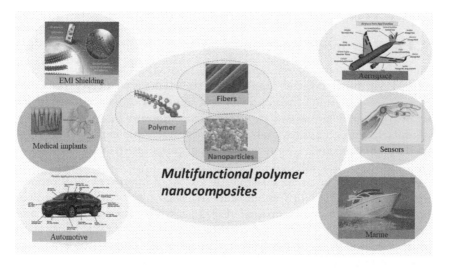

Figure 10.3 Applications of multifunctional polymer nanocomposites (MPNCs).

However, a lot of scope is available to develop MPNCs that are efficient tribo-materials, economically viable, and targeted for net-zero emissions. Besides this, it can also be inferred that although smart MPNCs have revolutionized the composite industry, green MPNCs are yet to be discovered.

REFERENCES

Aghamohammadi, H., Amousa, N., and Eslami-Farsani, R. 2021. "Recent advances in developing the MXene/polymer nanocomposites with multiple properties: A review study." *Synthetic Metals* 273: 116695. https://doi.org/10.1016/j.synthmet.2020.116695.

Almahri, S., Schneider, J., Schiffer, A., and Kumar, S. 2022. "Piezoresistive sensing performance of multifunctional MWCNT/HDPE auxetic structures enabled by additive manufacturing." *Polymer Testing* 114: 107687. https://doi.org/10.1016/j.polymertesting.2022.107687.

Al-Oqla, Faris M., Sapuan, S. M., Anwer, T., Jawaid, M., and Hoque, M. E. 2015. "Natural fiber reinforced conductive polymer composites as functional materials: A review." *Synthetic Metals* 206: 42–54. https://doi.org/10.1016/j.synthmet.2015.04.014.

Amurin, L. G., Felisberto, M. D., Ferreira, F. L. Q., Soraes, P. H. V., Oliveira, P. N., Santos, B. F., Valeriano, J. C. S., de Miranda, D. C., and Silva, G. G. 2022. "Multifunctionality in ultra high molecular weight polyethylene nanocomposites with reduced graphene oxide: Hardness, impact and tribological properties." *Polymer* 240: 124475. https://doi.org/10.1016/j.polymer.2021.124475.

Andrew, J. J., Alhashmi, H., Schiffer, A., Kumar, S., and Deshpande, V. S. 2021. "Energy absorption and self-sensing performance of 3D printed CF/PEEK cellular composites." *Materials and Design* 208: 109863. https://doi.org/10.1016/j.matdes.2021.109863.

Chan, J. X., Wong, J. F., Petrů, M., Hassan, A., Nirmal, U., Othman, N., and Ilyas, R. A. 2021. "Effect of nanofillers on tribological properties of polymer nanocomposites: A review on recent development." *Polymers* 13 (17): 1–47. https://doi.org/10.3390/polym13172867.

Delides, C. G. 2016. "Everyday life applications of polymer nanocomposites." Technological Educational Institute of Western Macedonia, no. October: 1–8.

Dong, M., Zhang, H., Tzounis, L., Santagiuliana, G., Bilotti, E., and Papageorgiou, D. G. 2021. "Multifunctional epoxy nanocomposites reinforced by two-dimensional materials: A review." *Carbon* 185: 57–81. https://doi.org/10.1016/j.carbon.2021.09.009.

Dorigato, A., and Fredi, G. 2022. "Special issue 'Investigation of polymer nanocomposites' performance'." *Molecules* 27: 1180. https://doi.org/10.3390/molecules27041180.

Gul, S., Kausar, A., Muhammad, B., and Jabeen, S. 2016. "Research progress on properties and applications of polymer/clay nanocomposite." *Polymer—Plastics Technology and Engineering* 55 (7): 684–703. https://doi.org/10.1080/03602559.2015.1098699.

Japić, D., Kulovec, S., Kalin, M., Slapnik, J., Nardin, B., and Huskić, M. 2022. "Effect of expanded graphite on mechanical and tribological properties of polyamide 6/glass fibre composites." *Advances in Polymer Technology* 2022: 9974889. https://doi.org/10.1155/2022/9974889.

Li, Y., Man, Z., Lin, X., Wei, L., Wang, H., and Lei, M. 2022. "Role of well-dispersed carbon nanotubes and limited matrix degradation on tribological properties of flame-sprayed PEEK nanocomposite coatings." *Journal of Tribology* 144 (1): 1–12. https://doi.org/10.1115/1.4050733.

Nisha, M. S., Mullai Venthan, S., Senthil Kumar, P., and Singh, D. 2022. "Tribological properties of carbon nanotube and carbon nanofiber blended polyvinylidene fluoride sheets laminated on steel substrates." *International Journal of Chemical Engineering* 2022: 3408115. https://doi.org/10.1155/2022/3408115.

Novikov, I. V., Krasnikov, D. V., Vorobei, A. M., Zuev, Y. I., Butt, H. A., Fedorov, F. S., Gusev, S. A., et al. 2022. "Multifunctional elastic nanocomposites with extremely low concentrations of single-walled carbon nanotubes." *ACS Applied Materials and Interfaces* 14 (16): 18866–18876. https://doi.org/10.1021/acsami.2c01086.

Ogbonna, V. E., Popoola, P. I., Popoola, O. M., and Adeosun, S. O. 2021. "A review on recent advances on improving polyimide matrix nanocomposites for mechanical, thermal, and tribological applications: Challenges and recommendations for future improvement." *Journal of Thermoplastic Composite Materials*. https://doi.org/10.1177/08927057211007904.

Osman, A., Elhakeem, A., Kaytbay, S., and Ahmed, A. 2022. "A comprehensive review on the thermal, electrical, and mechanical properties of graphene-based multi-functional epoxy composites." *Advanced Composites and Hybrid Materials* 5: 547–605. https://doi.org/10.1007/s42114-022-00423-4.

Padhan, M., Marathe, U., and Bijwe, J. 2020. "Surface topography modification, film transfer and wear mechanism for fibre reinforced polymer composites—An overview." *Surface Topography: Metrology and Properties* 8 (4): 043002.

Padhan, M., Marathe, U., and Bijwe, J. 2021. "Exceptional performance of bi-directionally reinforced composite of PEEK manufactured by commingling technique using poly(p-phenylene-benzobisoxazole) (PBO) fibers." *Composites Science and Technology* 218: 109125. https://doi.org/10.1016/j.compscitech.2021.109125.

Padhan, M., Marathe, U., Bijwe, J., Raja, A. K., and Trivedi, P. D. 2021. "Functionalization of spherical alumina nano-particles for enhancing the performance of PAEK-based composites." *Applied Surface Science* 562: 150107. https://doi.org/10.1016/j.apsusc.2021.150107.

Panda, J. N., Bijwe, J., and Pandey, R. K. 2019. "Optimization of graphite contents in PAEK composites for best combination of performance properties." *Composites Part B: Engineering* 174: 106951. https://doi.org/10.1016/j.compositesb.2019.106951.

Panda, J. N., Bijwe, J., and Pandey, R. K. 2020. "On the significant tribo-potential of PAEK based composites and their dry bearings." *Tribology International* 142: 105994. https://doi.org/10.1016/j.triboint.2019.105994.

Panda, J. N., Bijwe, J., and Pandey, R. K. 2021. "Particulate PTFE as a super-efficient secondary solid lubricant in PAEK composites for exceptional performance in adhesive wear mode." *Composites Part C: Open Access* 4: 100110. https://doi.org/10.1016/j.jcomc.2021.100110.

Qi, Z., Liu, H., He, T., Wang, J., and Yan, F. 2022. "Friction and transfer behavior of alumina nanoparticle-reinforced carbon fiber/polyoxymethylene composites in the deep sea environment." *Tribology International* 170: 107516. https://doi.org/10.1016/j.triboint.2022.107516.

Rasul, M. G., Kiziltas, A., Hoque, M. S. B., Banik, A., Hopkins, P. E., Tan, K. T., Arfaei, B., and Shahbazian-Yassar, R. 2022. "Improvement of the thermal conductivity and tribological properties of polyethylene by incorporating functionalized boron nitride nanosheets." *Tribology International* 165: 107277. https://doi.org/10.1016/j.triboint.2021.107277.

Saba, N., Jawaid, M., Sultan, M. T. H., and Alothman, O. 2017. "Hybrid multifunctional composites—Recent applications." *Hybrid Polymer Composite Materials: Applications* 2017: 151–167. https://doi.org/10.1016/B978-0-08-100785-3.00005-X.

Sengwa, R. J., and Dhatarwal, P. 2022. "Toward multifunctionality of PEO/PMMA/MMT hybrid polymer nanocomposites: Promising morphological, nanostructural, thermal, broadband dielectric, and optical properties." *Journal of Physics and Chemistry of Solids* 166: 110708. https://doi.org/10.1016/j.jpcs.2022.110708.

Shinde, N. G., and Patel, D. M. 2022. "Polymeric nanocomposite of PEEK and nickel incorporated ZnO: Synthesis, charachterization and tribological behavior." *Materials Today: Proceedings* 56: 3727–3733. https://doi.org/10.1016/j.matpr.2021.12.474.

Teijido, R., Ruiz-Rubio, L., Echaide, A. G., Vilas-Vilela, J. L., Lanceros-Mendez, S., and Zhang, Q. 2022. "State of the art and current trends on layered inorganic-polymer nanocomposite coatings for anticorrosion and multi-functional applications." *Progress in Organic Coatings* 163: 106684. https://doi.org/10.1016/j.porgcoat.2021.106684.

Ul Ain, Q., Sehgal, R., Wani, M. F., and Singh, M. K. 2021. "An overview of polymer nanocomposites: Understanding of mechanical and tribological behavior." *IOP Conference Series: Materials Science and Engineering* 1189 (1): 012010. https://doi.org/10.1088/1757-899x/1189/1/012010.

Zhang, M., Li, Y., Su, Z., and Wei, G. 2015. "Recent advances in the synthesis and applications of graphene-polymer nanocomposites." *Polymer Chemistry* 6 (34): 6107–6124. https://doi.org/10.1039/c5py00777a.

Zhang, Y., He, X., Cao, M., Shen, X., Yang, Y., Yi, J., Jipeng G., et al. 2021. "Tribological and thermo-mechanical properties of TiO_2 nanodot-decorated Ti_3C_2/epoxy nanocomposites." *Materials* 14 (10): 2509. https://doi.org/10.3390/ma14102509.

Chapter 11

Synthesis of nanomaterial coatings for various applications

*Mohd Rafiq Parray, Himanshu Kumar,
Amritbir Singh, Bunty Tomar, and S. Shiva*
Indian Institute of Technology, Jammu

CONTENTS

11.1 Introduction	167
11.2 Nanomaterials synthesis techniques	169
11.2.1 Plasma spray	170
11.2.2 Cold spray	171
11.2.3 Sputtering	172
11.2.4 Electroplating	173
11.2.5 Chemical vapor deposition	174
11.2.6 Spin coatings	175
11.3 Area of applications	176
11.3.1 Tribology	176
11.3.2 Defense	178
11.3.3 Biomedical	180
11.3.4 Micro-electronics	180
11.3.5 Energy	182
11.3.6 Food industry	182
11.4 Conclusion	183
References	184

11.1 INTRODUCTION

Recently, nanomaterials have received a lot of interest from researchers owing to the ease with which they can be rebuilt, their unique features, and the wide range of applications they may be used in. The assimilation of nanomaterials in coatings greatly enhances the component properties, including wear resistance, resistance to chemicals and corrosive environment, abrasion resistance, and good electrical and mechanical properties [1]. Many types of nanomaterials have been developed to date and are being used in a variety of industrial sectors. The transition metals of groups III and IV of the periodic table were commonly used in the form of their oxides, carbides, and nitrides, along with thermoset resin to develop super hard coatings. These coatings' gas permeability is greatly enhanced by the incorporation of nano-sized clay particles. The thermal resistance of these coatings can be enhanced by adding carbon nanotubes (CNT) and carbon fibers to the base material [2]. The addition of nano-sized metal particles such as silver oxide helps in improving the antibacterial properties of the coatings. A particular grade of nano-filler named nano-container is used to impart the self-healing and anti-corrosive properties in the coatings during adverse times of mechanical rupture and potential of hydrogen (pH) change [3]. Nano-containers have a volume of nano-size and contain active healing agents and inhibitors, including nano-silica [4,5], clay nanotubes [6], nano-ceria [7], and zirconia nano-spheres [8]. This is a significant reason for the nanomaterials to be used in the high-end application coatings.

Based on the perspective of applications, various materials, including titanium-based, carbon-based, tin-based, polymer-based, silica-based, and so on, are employed for the deposition of nanomaterials coatings. Carbon nanomaterials are considered useful in various applications in engineering and medical fields [9–11]. Three types of carbon-based nanomaterials are being significantly tested and applied in the field of medicine, including carbon nanofibers, CNTs, and graphene. Among these, carbon nanomaterials and CNTs are the most diversified and investigated nanomaterials and are applied in various medical applications, including biosensors, drug supply systems, and scaffolds for nerve tissue regeneration [12–14]. Recently, the application of CNTs has been extended into the field of fabrication of electrodes for nerve tissue stimulation [15]. In various researches, CNTs have been proved to improve neuronal performance, boost neuron differentiation, stimulate neuronal outgrowth, and provide better neuronal functioning and signaling [16]. Nerve tissue regeneration is also another potential application for these nanomaterials. These carbon nanomaterials are coated on the nerve tissue–regenerating electrodes made of metals, such as titanium and stainless steel. Neuron stimulation and recording are ineffective with these metal electrodes because of scar development and inadequate contact with body tissues in the central nervous system. [17]. Therefore, researchers are constantly developing new techniques for modifying the surface of these metal electrodes, and CNTs are fit in this respect and are widely being used in this application.

Graphene-based nanomaterials (GBNs) have proven to be a better alternative for disinfection and preventing the virus spread as compared to conventional methods [18,19]. Recently, the use of GBNs has been influential in developing adequate personal protective equipment (PPE) kits, 3D printed masks and surgical equipment, surface protecting coatings, and efficient diagnostic devices to minimize the spread of the SARS-CoV-2 virus. Graphene-based antiviral nano-coating effectively sterilizes and minimizes viruses' survival time on various surfaces [20–22]. Graphene, along with its derivatives such as graphene oxide, shows a variety of antimicrobial properties against many viruses. Considering the given benefits of GBNs, the research on graphene nano-coatings has increased significantly in the last decade, primarily in the field of antiviral and antibacterial studies [23,24].

Nanomaterials are suitable additives in the paint coatings to add antimicrobial properties other than preventing corrosion and enhancing the appearance. The presence of nano-silica titanium dioxide in the paint enhances the adhesion and impact toughness of the coating [25]. The addition of silica nanoparticles is beneficial for increasing the strength of the coating. Titanium oxide nanoparticles have several advantages as they induce antimicrobial properties in the coating [26,27]. Research has proved that titanium oxide improves the mechanical strength [28] and corrosion resistance [29] of the coating, and also imparts the self-cleaning action to it [30]. Many other materials, including SnO_2 [31], ZnO [32], and WO_3 [33], are used as nanomaterial coating for several industrial and commercial applications.

Many techniques have been accomplished for preparing and depositing nanomaterials on any given substrate. It includes electroplating [34], sputtering [35], chemical vapor deposition [36], physical vapor deposition [37], plasma spray [34], spin coating [38], and cold spray [39]. Recently, a lot of attention has been given to the field of additive manufacturing for fabricating nanomaterials in end-use products [40], as this technology provides several benefits over all conventional manufacturing routes [41]. As each synthesis technique possesses some benefits and drawbacks, the selection of deposition and fabrication methods largely depends upon user application. For example, the cold spray technique allows the deposition of coatings at relatively lower temperatures as compared to the melting point of the nanomaterials being deposited. Thus, this deposition method can be employed for fabrication of coatings of materials prone to higher oxidation and phase transition [1].

Nanomaterial coatings not only serve the purpose of covering the substrate surface but also perform several functions such as antimicrobial and antifouling properties due to the presence of several nano-fillers inside them [26]. Nanomaterial coatings are used in a variety of disciplines, including tribology, energy, military, biomedical, and electronics, due to their unique physical, chemical, and mechanical properties [42]. This chapter presents an overview of synthesizing methods and areas of potential applications for nanomaterial coatings, considering the importance of nanomaterials in industrial and commercial applications.

11.2 NANOMATERIALS SYNTHESIS TECHNIQUES

Nanomaterial developments in recent years have captivated the scientific community's interest owing to their many functional capabilities like sensing, electrical, optical, and magnetic properties, among others. Figure 11.1 depicts the numerous techniques that may be used to

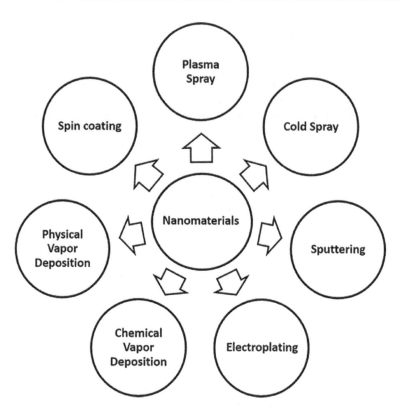

Figure 11.1 A schematic representation of the deposition method for nanomaterials.

synthesize nanomaterials. The synthesis techniques for nanomaterials are explored in detail in the following sections, including their advantages and disadvantages.

11.2.1 Plasma spray

Plasma spray is a well-established thermal spray technique widely used for thermal barrier and wear resistance applications [43,44]. The plasma spray consists of a plasma gun, and argon (Ar) and helium (He) are the primary and secondary plasma gases. The metallic powder was fed through the hopper via argon (Ar) as a carrier gas. When a molten powder particle strikes a surface at supersonic speed, it flattens and solidifies.. Figure 11.2 depicts a schematic illustration of the plasma spray. The nanomaterials are suitable candidates for the plasma spray due to their higher melting temperature (~16,000°C), which facilitates the melting of ceramic nanoparticles [45]. The nano-ceramics particles such as nano-carbides and nitrides improve the hardness and wear resistance of the substrate [46]. Plasma spray is a versatile technique to deposit the alumina, zirconia, and yttria stabilized

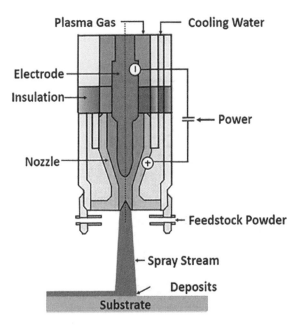

Figure 11.2 The basic diagram of plasma spray technique.

zirconia (YSZ) nano-coatings for the various applications [34]. Plasma spray process has many advantages like it is a single-step process. Nano-sized ceramic grains or particles are formed during spray processing itself. There is no need for post-processing such as heat treatment or calcination. Besides, the plasma spray has a higher capital cost, it is difficult to coat the small-sized bores, and it has higher maintenance cost.

11.2.2 Cold spray

Cold spray (CS) is a versatile thermal spray technique used to deposit the temperature-sensitive materials below the melting point. It utilizes the preheated propulsion gas to feed the nano-crystalline powder at supersonic speed through a de-Laval nozzle. A schematic representation of the CS process is shown in Figure 11.3. CS, a low-temperature technique, has shown remarkable capability for producing high-performance metal coatings containing nanostructures. The nanostructured coatings can be produced by three different steps using cold spray. In the first place, nano-crystallization may occur at the interfaces between particles and between a coating and its substrate owing to dynamic recrystallization, and this can lead to nanostructured grains. Second, nano-crystalline powders are used as the feedstock for cold spray because they preserve the nanostructure of the powders. Thirdly, the nanostructured coatings can be obtained via depositing the

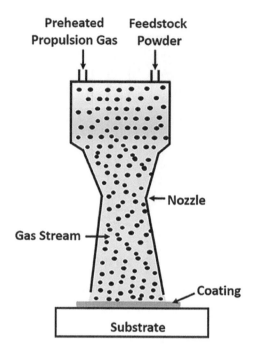

Figure 11.3 The schematic diagram of cold spray deposition.

nanomaterial-reinforced metal matrix composite coatings using CS. For thermal barrier applications, a coating of nano-crystalline NiCoCrAlTaY alloy utilizing CS has recently been developed [39]. This coating has excellent oxidation resistance.. CS technique combined with shot peening is also utilized to produce the nano-crystalline materials through the grain refining and plastic deformation mechanism [39]. Furthermore, Liu et al. produced the nano-crystalline Cu coating with improved hardness from the nano-crystalline powder [47]. Solid-state technology is a major benefit of the CS technique, which produces a wide range of distinct coating properties. The primary drawback is the coating's reduced ductility as a result of plastic deformation.

11.2.3 Sputtering

Sputtering is a method of physical vapor deposition in which the material is ejected from the target as the cathode and coated on the substrate as the anode by bombardment by gaseous ions. The expelled atoms from the target material are propelled by the plasma gas in order to deposit it on the substrate. Furthermore, a sputtering gas is used in the chamber to enhance the reactivity of the plasma present in the deposition chamber. A schematic of the sputtering process is shown in Figure 11.4. The sputtering

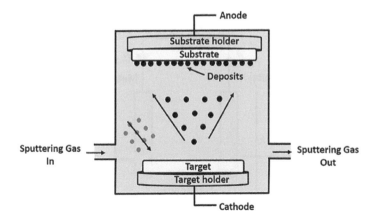

Figure 11.4 A schematic illustration of sputtering technique.

is widely used to deposit the nano-coatings on the tool, engine parts, and magnetic and anti-reflection coatings. There are mainly four types of sputtering processes in practice. DC sputtering utilizes a direct current (DC) voltage of (500–1,000) V and Ar as sputtering gas. Firstly, Ar^+ ion strike on the target material and ejected atom deposited on the cathode in a low-pressure plasma environment. Secondly, radio frequency (RF) sputtering uses a high-frequency alternating field instead of DC. The ions and electrons are alternately accelerated in both directions by the alternating current that leads to oscillating electrons in the plasma region and colliding with the Ar atoms results in higher sputtering rate. Thirdly, magnetron sputtering (MS) utilizes the combination of an electric field and magnetic field on cathode, resulting in more ionization of the target materials and higher deposition rate. Panepinto et al. developed the TiO_2-based nanostructured photo-anodes for the energy storage application using the MS technique [35]. Furthermore, nano-crystalline Si thin film was developed using sputtering for the photovoltaic application [48]. The sputtering has various advantages such as it is capable of depositing a wide range of materials such as alloys, metals, composites, and insulators. It has dense and homogeneous deposition over large substrates. Moreover, it has some drawbacks, such as the ultraviolet (UV) rays of generated plasma and ion bombardment may cause substrate damage.

11.2.4 Electroplating

Electroplating is an efficient technique for depositing the nanomaterials. It is based on the principles of electrochemical phenomena (redox reaction), namely, those involved with the reduction of electroactive material on the anode and the deposition of that material on the cathode surface. A schematic illustration of the electroplating process is shown in

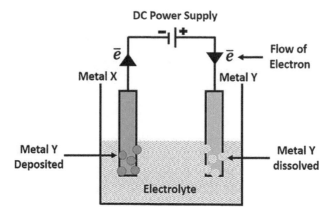

Figure 11.5 Basic representation of electroplating technique.

Figure 11.5. Electroplating is widely used in anticorrosion nano-coatings, solar photovoltaic application, thermoelectric thin film, and supercapacitors [49–52]. Electroplating has various advantages such as using electroplating to preserve specific artifacts from corrosion may be quite beneficial. It also enhances shock and heat resistance properties, making it an ideal method for protecting the metallic components. Electroplated metallic components exhibit low friction when they are in contact with each other, leading to significantly reduced wear. It is also beneficial in the electronic and solar cell applications to make the object magnetic and conductive. Moreover, it has some disadvantages such as it is only suitable for conductive materials; waste items generated during electroplating must be appropriately disposed of in order to avoid contaminating the environment.

11.2.5 Chemical vapor deposition

Chemical vapor deposition (CVD) is a process that utilizes a chemical reaction to deposit a solid thin layer on the substance from a heated substrate due to vapor solid reaction. A schematic representation of the working principle is illustrated in Figure 11.6. It is widely used in the semiconductor industry to produce the thin film. There are different types of CVD processes. It is classified based on the chemical reaction initiated. Based on the operating pressure, it can be classified as follows: low-pressure CVD, atmospheric pressure CVD, and ultra-vacuum CVD. Furthermore, based on the substrate temperature, it can be classified as hot wall CVD and cold wall CVD, and based on physical characteristics of vapor, it can be classified as direct liquid injection and aerogel-assisted CVD. It is widely used for nanofabrication of various applications, such as nano-coatings for semiconductor chemical sensor, anticorrosion applications, oxidation resistance coatings, nanocomposite membrane for water purification application, and thermal

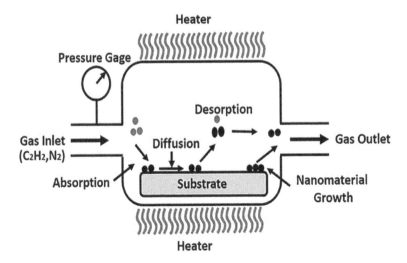

Figure 11.6 A schematic illustration of chemical vapor deposition.

barrier coatings [53–57]. The CVD process has various benefits, such as it can be applied on a broad range of materials such as metals, glass, ceramics, and metal alloys. CVD coating exhibits higher adhesion strength and is suitable for high-temperature applications. The precursor gas used can be optimized based on the desired area of application. Moreover, it has several drawbacks, such as it is not suitable for the temperature-sensitive materials and limitation to component size based on reaction chamber capacity.

11.2.6 Spin coatings

Spin coating is the versatile technique to deposit a thin polymeric film on the substrate. In this process, the semi-solid coating material is ejected through the dispenser nozzle and coated on the spinning substrate by centrifugal force. A schematic diagram of the spin-coating process is shown in Figure 11.7a. There are four steps involved in the whole process; the first step is to apply the coating material through the dispenser nozzle; secondly, spin up the substrate at a desirable higher speed; and thirdly, spin off the substrate at a constant rate and finally evaporation and solidification, as shown in Figure 11.7b. The thickness of the coating depends upon various factors, such as spinning speed, spinning time, fluid viscosity, and substrate temperature. The spin coating is widely applicable in various applications, such as nanostructured hydrophobic coating for solar cell application, electronic devices, nanodevice in solar power application, and optoelectronic device [58–61]. Spin coating is beneficial for achieving thin uniform multilayer coatings. However, it has various drawbacks, such as wastage of material during the spinning process and post-processing is required to improve the coating crystallinity and remove the air entrapment.

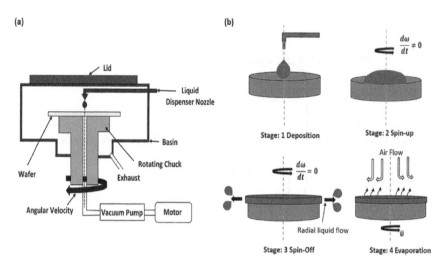

Figure 11.7 (a) Schematic diagram of spin-coating technique and (b) steps involved in spin-coating process.

11.3 AREA OF APPLICATIONS

As per the description of a nanoparticle, it has negligible dimensions, which indicates that any of its measurements should be less than 100 nanometers. One-dimensional examples include wires, rods, and nanofibers, while two-dimensional examples include films, slabs, multilayers, and network nanostructures. To clarify this further, a sphere or cluster of nano-phase materials with zero dimensions is portrayed as a point-like particle that is determined by the three dimensions of the nanomaterials [61]. Nanomaterials may be used in a wide range of applications, including but not limited to those mentioned in Figure 11.8, on account of their unique physical, chemical, and mechanical capabilities. Furthermore, these applications are individually discussed in detail in this section.

11.3.1 Tribology

Energy conservation is one of the most pressing issues in scientific advancement today, and it remains a significant challenge. For example, roughly one-third of the gasoline used by vehicles is spent on overcoming friction. Even a modest 20% decrease in friction is predicted to significantly reduce actual economic expenses in terms of energy use and environmental advantages. Tribology optimizations and improvements, such as controlling friction and wear, are increasingly recognized as key strategies for energy efficiency not only in macro-scaled moving assemblies, such as pumps, compressors, and turbines, but also in micro-scaled/nano-scaled technologies

Synthesis of nanomaterial coatings for various applications 177

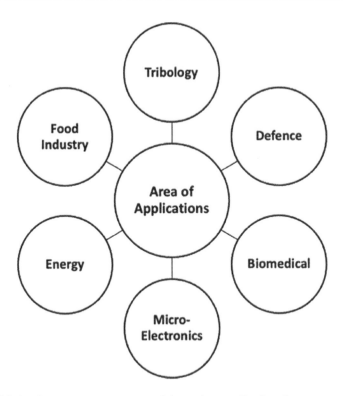

Figure 11.8 A schematic representation of the various applications for nanomaterials.

such as micro-/nano-electromechanical systems (MEMS/NEMS) devices [42,62]. A machine's frictional losses account for the majority of its energy losses. The tribological properties depend on the friction and wear that happen when two surfaces that touch each other move relative to each other. Lubricants are applied to moving (sliding) surfaces to minimize the amount of wear and the temperature generated by such surfaces. Lubricants are responsible for the creation of a thin film between the contact pairs, and this film formation is what stops the surfaces from wearing down. The coefficient of friction and the wear scar diameter (WSD) are two metrics that may be used to evaluate the performance of a lubricant [63].

There are various low-cost nano-powders produced through self-propagating high-temperature synthesis (SHS) that may be used as friction modifiers and solid lubricants for plastic or liquid lubricating materials. These nano-powders come in many different forms and can be used for a wide range of applications. They may also be used to create different compact tribo-technical building materials (ceramic and composite) using ex situ technologies, such as solid-phase powder metallurgy by sintering from powder and liquid-phase technologies with the introduction of nano-powders into matrix material melt [64].

Various studies discussed the current developments in friction modifiers for liquid lubricants; it was said that there are three different kinds of friction modifiers: organic friction modifiers, metal-organic compounds (mostly molybdenum organic compounds), and nano-powders. The use of microparticles and nanoparticles of various nature (polymers, metals, ceramics, and so on) as friction modifiers is now the subject of extensive research in tribological centers. These powder additives bring about a significant improvement in the wear resistance properties of lubricants and bring about a reduction in the friction coefficient at high temperatures and loads. Because a reduction in particle size leads to an improvement in that particle's capacity to be kept in oil without precipitation and to a reduction in wear intensity, nano-dimensional modifiers are the most effective powder modifiers. It was found that ceramic tribo-technical materials are attractive because of their light weight, high hardness and rigidity, heat resistance and corrosion resistance. They are very important in gas turbine engines for high-speed ball bearings made of silicon nitride or sialon or for tightening materials made of silicon carbide and are intended for use at high temperatures and aggressive media. In the past, these materials were made by sintering or hot pressing ceramic granules together. Reduced particle size, the use of nano-powders, and the creation of nanostructured ceramics have all been demonstrated to increase tribological characteristics significantly [64].

For the majority of technologically relevant solid–solid interfaces, contact occurs at various asperities. A sharp atomic force microscope (AFM) or friction force microscope (FFM) tip gliding over a surface mimics a single touch of this kind. However, asperities exist in a variety of forms. The AFM and FFM are used to explore a wide range of wear processes, comprising surface quality, adherence, abrasion, wear, penetration, material transfer detection, and interface lubrication [65].

11.3.2 Defense

Using nanotechnology-based materials is not a brand-new concept. Nanotechnology was being employed in a number of sectors before the name "nanotechnology" was created, including the arts and steel manufacturing. Nanotechnology-infused nanomaterials and frameworks are being developed by all of the world's major armed forces. In the present scenario, nanotechnology research focuses on enhancing healthcare offices and supplying light weight solid and multi-useable materials as armors in a network-driven warfare region [66].

Nanomaterials provide the military with an abundance of new possibilities, which is without a doubt a significant benefit. Many civilian applications for nano-advancements are being developed, and some of them may find their way into the military in the near future. The development of national capabilities in this area is moving forward at a rapid pace. Even in India, where nanotechnology is seen as fundamentally important, there has been a constant investment. The Indian legislature established the Nanoscience and Technology Initiative (NSTI) to conduct nanomaterials research in the country. Since then, India has achieved a lot of development. The Defense Research and Development Organization (DRDO) is doing a wide range of research in nanotechnology in order to increase its use in the defense sector. NBC (nuclear, organic, and chemical) attack assurance devices, stealth and disguise, sensors, and high-vitality weapons have all been major focus areas. Figure 11.9 depicts the various applications of nanomaterials in the field of defense. The following nanotechnology applications are driving research and development efforts in the military sector, particularly in the areas of nanostructures and nanocomposites [67,68].

- Protective apparel that isn't heavy but nonetheless effective.
- Fabrics that are resistant to ballistics.
- Fabrics with self-cleaning nanofibers for chemical and biological warfare defense.
- Customized suits with interchangeable textiles for better thermal regulation and interchangeable camouflage are examples of adaptive suits.
- Integrating body and brain sensing, environmental and situational awareness into a smart suit or helmet.

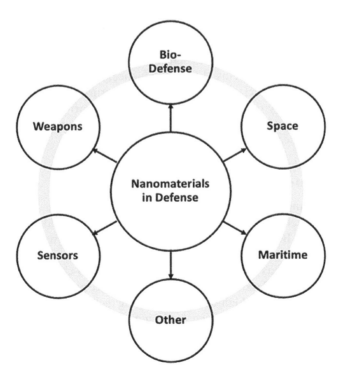

Figure 11.9 A schematic representation of the defense applications for nanomaterials.

11.3.3 Biomedical

The rising utilization of nanomaterials in biomedicine has sparked the emergence of a new hybrid science known as nano-biotechnology. In nano-biotechnology, nanomaterials have a wide range of uses, including diagnostics, medication delivery systems, prosthetics, and implants. Biological systems, which are also nanoscale, are a natural fit for nanoscale materials in biomedical equipment. Among the components often used in the production of these nanotechnology items are inorganic and metallic nanoparticles, nanotubes, lipid membranes, and conductive materials. Bio-specific molecules may be conjugated with nanoparticles by the use of chemical or physical processes that take advantage of specific biological events, such as the binding of antibodies to antigens or the binding of receptors to ligands [69–71]. It is suggested that nanostructures may be designed to better respond to certain applications by manipulating their surface qualities. These nanoparticles and nano-patterned flat surfaces are used in diagnostics, biosensors and bio-imaging devices, drug delivery systems, and bone-substituting implants, with an emphasis on inorganic (metallic and metal oxide) as well as organic (CNTs and liposomes). Nanoparticles of metal oxide have been used in the construction of a number of medical devices. The magnetic properties of iron oxide have a variety of therapeutic and diagnostic applications, including contrast agents for magnetic resonance imaging (MRI), magnetic particle imaging, and ultrasonic methods such as magneto-motive ultrasound, photoacoustic imaging, and magnetic particle hyperthermia. The nanoparticles' large surface area promotes the rapid adsorption of plasma proteins. As a consequence, the chemical constituents' topology of particle surface, as well as their mixture (wettability, surface-free energy), modify the particles' interaction with various substances and regulate their eventual application [70,72]. Broad area of nanomaterials in biomedical applications is presented in Figure 11.10.

11.3.4 Micro-electronics

Advances in nanomaterials for microelectronic and optical devices have been a growing subject in recent years, owing to their flexibility and low weight for everyday usage. In order to meet customer needs for lightweight, compact, dependable, and multifunctional electronic or communication equipment, packaging density has long been a major focus in micro-electronics and photonics technologies. Nanotechnology and nanoscience play an essential role in the development of ultra-high-density packaging technology. Throughout the last decade, researchers have been exploring and studying nanotechnology, such as CNTs, nanoparticles, molecular self-assembly, nano-impressions, and sensors. Aiming for system-level components (chip–chip or chip–package connections, thin-film conductors, dielectrics, passives, and encapsulants) leading to nanomodules is the goal of electronic packaging scientists.

Synthesis of nanomaterial coatings for various applications 181

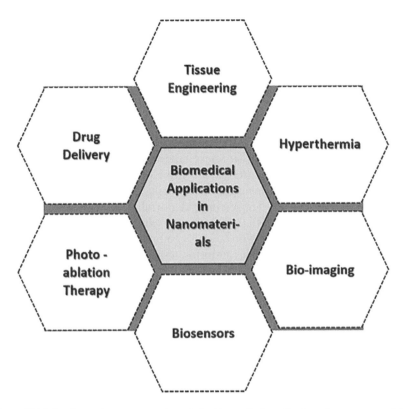

Figure 11.10 A schematic representation of the biomedical applications for nanomaterials.

As a result of this focus, researchers anticipate to see a number of commercial applications within this decade. On system-level applications, however, nanotechnology implementation calls for a solid foundation in nanoscience and engineering [73]. Nanotechnology has made significant contributions to the development of key improvements in computers and electronics, which have resulted in systems that are more efficient, smaller, and more portable, and which are able to handle and store ever-increasing volumes of information [74]. These ever-evolving applications include the following:

- Transistors, the fundamental switches that power all contemporary computers, have become even smaller by the emergence of nanotechnology.
- Computers will "boot" quickly using magnetic random access memory (MRAM). MRAM uses nanometer-scale magnetic tunnel junctions to store data fast and allow resume-play.
- Quantum dots are increasingly being used in ultra-high definition screens and TVs to provide more brilliant colors while being more energy efficient.

- Flexible semiconductor nano-membranes have been created for smartphone and e-reader displays. Other nanoparticles like graphene and cellulosic nanomaterials allow wearable and "tattoo" sensors, sewable photovoltaics, and rollable electronic paper.
- Other electronic and computer products.

11.3.5 Energy

The most pressing issue of the 21st century is energy. A reliance on non-renewable fossil fuels causes not just environmental problems but also significant long-term ramifications for the global economy and civilization. Devices capable of converting and storing energy (CSE) are urgently needed to power energy-deficit sectors like portable electronics (e.g., mobile phones, camcorders, laptops), transportation (electric cars, hybrid vehicles), and even stationary applications. Some common electrochemical energy storage systems (ECS) include batteries (especially Li-ion batteries) and supercapacitors (as well as other types of photo electrochemical water splitting cells). Nanostructured materials are critical to the functioning of these energy devices. According to some experts, ECS system advances might be made possible by new discoveries in nanomaterial chemistry [75,76]. Nanomaterials may be used in a wide variety of ways to generate, convert, store, and save energy, as demonstrated in Figure 11.11.

11.3.6 Food industry

Nanotechnology has changed the food sector by devising revolutionary delivery mechanisms for nano-formulated agrochemicals, enhancing nutritional qualities, and creating new bioactive goods. Nano-emulsions boost oral bioavailability, while liposomes carry hydrophobic compounds. In the case of packaged foods, their processing and preservation are dependent on nanomaterial devices. Foods using nanoparticles or nano-encapsulation have a longer lifespan as they are shielded from moisture, gases, and lipids. They provide more efficient ways for transporting bioactive compounds. Food safety and quality are also ensured using a variety of nanomaterials and nano-sensors. This application masks flavors and tastes. Nanotechnology delivers controlled-release, water-insoluble dietary components and supplements. Nanotechnology in food processing improves texture, encapsulates food additives, creates novel tastes and experiences, and controls scent release in nutritional supplements [77–79].

Nano-biotechnology is advancing at a rapid pace, but it is not without its obstacles. To deal with the dangers connected with the use of nanomaterials and nano-devices in the food business, environmental consequences and safety problems must be a top focus in research. Finally, sophisticated nanomaterials and nano-devices have already started a revolution in the food business, despite the fact that nanotechnology is still in its infancy.

Synthesis of nanomaterial coatings for various applications 183

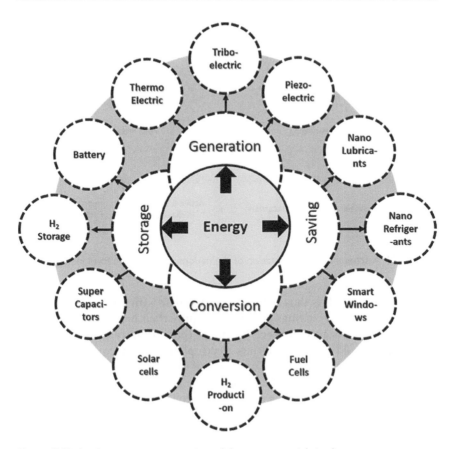

Figure 11.11 A schematic representation of the nanomaterials in the energy sector.

Nanotechnology still has a lot of promise, and new uses in the food business are being investigated. Security and safety concerns are becoming more apparent, and they will need to be properly evaluated and handled in the future [77,78,80]. As shown in Figure 11.12, nanotechnology has the potential to be used in the food chain for storing food, monitoring food quality, and packaging material.

11.4 CONCLUSION

This chapter concludes that the nanomaterials in coatings have grown rapidly in the last several years. Numerous kinds of nanoparticles were prepared and are used in a wide range of applications. Nanomaterials exhibit a wide variety of distinctive properties such as large specific surface area, magnetism effects, antimicrobial properties, high electrical and thermal

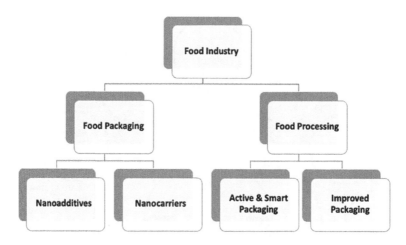

Figure 11.12 A schematic representation of the nanomaterials in the food industry.

conductivities, etc. It is thus critical to get familiar with the production process of nanomaterial coatings in order to acquire their benefits. The creation of nanomaterials is accomplished by any of these primary methods such as plasma spray, cold spray, sputtering, electroplating, chemical vapor deposition, and spin coating. The pros and cons of each synthesis method are distinctive. In addition, the notable difference in attributes associated with the various processing methods is evident. Moreover, there are several significant uses for nanomaterials in various sectors such as biomedical, defense, tribology, and energy. In each application regime mentioned previously, nanomaterials are used in a variety of ways. Hence, this chapter paves the way for the enthusiastic community of scientists to improve the properties of the existing materials by the addition of the nanomaterial coatings using different synthesis approaches.

REFERENCES

[1] P. Nguyen Tri, T. A. Nguyen, S. Rtimi, and C. M. Ouellet Plamondon, "Nanomaterials-based coatings: An introduction." In: Nguyen P.T., Rtimi S., Ouellet-Plamondon C., editors, *Nanomaterials Based Coatings*. Amsterdam, Netherlands: Elsevier, pp. 1–7, 2019.

[2] T. Nardi, S. Rtimi, C. Pulgarin, and Y. Leterrier, "Antibacterial surfaces based on functionally graded photocatalytic Fe_3O_4 @TiO_2 core-shell nanoparticle/epoxy composites," *RSC Adv.*, vol. 5, no. 127, pp. 105416–105421, 2015, doi: 10.1039/C5RA19298F.

[3] D. Pedrazzoli, A. Dorigato, and A. Pegoretti, "Monitoring the mechanical behaviour of electrically conductive polymer nanocomposites under ramp and creep conditions," *J. Nanosci. Nanotechnol.*, vol. 12, no. 5, pp. 4093–4102, 2012, doi: 10.1166/jnn.2012.6219.

[4] A. Keyvani, M. Yeganeh, and H. Rezaeyan, "Application of mesoporous silica nanocontainers as an intelligent host of molybdate corrosion inhibitor embedded in the epoxy coated steel," *Prog. Nat. Sci. Mater. Int.*, vol. 27, no. 2, pp. 261–267, 2017, doi: 10.1016/j.pnsc.2017.02.005.

[5] C. Zea, R. Barranco-García, J. Alcántara, B. Chico, M. Morcillo, and D. de la Fuente, "Hollow mesoporous silica nanoparticles loaded with phosphomolybdate as smart anticorrosive pigment," *J. Coatings Technol. Res.*, vol. 14, no. 4, pp. 869–878, 2017, doi: 10.1007/s11998-017-9924-7.

[6] S. Manasa A. Jyothirmayi, T. Siva, B. V. Sarada, M. Ramakrishna, S. Sathiyanarayanan, K. V. Gobi, and R. Subasri "Nanoclay-based self-healing, corrosion protection coatings on aluminum, A356.0 and AZ91 substrates," *J. Coatings Technol. Res.*, vol. 14, no. 5, pp. 1195–1208, 2017, doi: 10.1007/s11998-016-9912-3.

[7] I. Kartsonakis, I. Daniilidis, and G. Kordas, "Encapsulation of the corrosion inhibitor 8-hydroxyquinoline into ceria nanocontainers," *J. Sol-Gel Sci. Technol.*, vol. 48, no. 1–2, pp. 24–31, 2008, doi: 10.1007/s10971-008-1810-4.

[8] A. Chenan, S. Ramya, R. P. George, and U. Kamachi Mudali, "Hollow mesoporous zirconia nanocontainers for storing and controlled releasing of corrosion inhibitors," *Ceram. Int.*, vol. 40, no. 7, pp. 10457–10463, 2014, doi: 10.1016/j.ceramint.2014.03.016.

[9] H. He, L. A. Pham-Huy, P. Dramou, D. Xiao, P. Zuo, and C. Pham-Huy, "Carbon nanotubes: Applications in pharmacy and medicine," *Biomed Res. Int.*, vol. 2013, pp. 1–12, 2013, doi: 10.1155/2013/578290.

[10] M. Notarianni, J. Liu, K. Vernon, and N. Motta, "Synthesis and applications of carbon nanomaterials for energy generation and storage," *Beilstein J. Nanotechnol.*, vol. 7, pp. 149–196, 2016, doi: 10.3762/bjnano.7.17.

[11] F. Lu, L. Gu, M. J. Meziani, X. Wang, P. G. Luo, L. M. Veca, L. Cao, and Y.-P. Sun, "Advances in bioapplications of carbon nanotubes," *Adv. Mater.*, vol. 21, no. 2, pp. 139–152, 2009, doi: 10.1002/adma.200801491.

[12] N. Yang, X. Chen, T. Ren, P. Zhang, and D. Yang, "Carbon nanotube based biosensors," *Sensors Actuators B Chem.*, vol. 207, pp. 690–715, 2015, doi: 10.1016/j.snb.2014.10.040.

[13] J. V. Veetil and K. Ye, "Tailored carbon nanotubes for tissue engineering applications," *Biotechnol. Prog.*, vol. 25, no. 3, pp. 709–721, 2009, doi: 10.1002/btpr.165.

[14] B. Dineshkumar, K. Krishnakumar, A. R. Bhatt, D. Paul, J. Cherian, A. John, and S. Suresh "Single-walled and multi-walled carbon nanotubes based drug delivery system: Cancer therapy: A review," *Indian J. Cancer*, vol. 52, no. 3, p. 262, 2015, doi: 10.4103/0019-509X.176720.

[15] F. Vitale, S. R. Summerson, B. Aazhang, C. Kemere, and M. Pasquali, "Neural stimulation and recording with bidirectional, soft carbon nanotube fiber microelectrodes," *ACS Nano.*, vol. 9, no. 4, pp. 4465–4474, 2015, doi: 10.1021/acsnano.5b01060.

[16] A. Fabbro, M. Prato, and L. Ballerini, "Carbon nanotubes in neuroregeneration and repair," *Adv. Drug Deliv. Rev.*, vol. 65, no. 15, pp. 2034–2044, 2013, doi: 10.1016/j.addr.2013.07.002.

[17] X. Luo, C. L. Weaver, D. D. Zhou, R. Greenberg, and X. T. Cui, "Highly stable carbon nanotube doped poly(3,4-ethylenedioxythiophene) for chronic neural stimulation," *Biomaterials*, vol. 32, no. 24, pp. 5551–5557, 2011, doi: 10.1016/j.biomaterials.2011.04.051.

[18] S. Talebian, G. G. Wallace, A. Schroeder, F. Stellacci, and J. Conde, "Nanotechnology-based disinfectants and sensors for SARS-CoV-2," *Nat. Nanotechnol.*, vol. 15, no. 8, pp. 618–621, 2020, doi: 10.1038/s41565-020-0751-0.

[19] E. V. R. Campos, A. E. S. Pereira, J. L. de Oliveria, L. B. Carvalho, M. Guilger-Casagrande, R. de Lima, and L. F. Fraceto, "How can nanotechnology help to combat COVID-19? Opportunities and urgent need," *J. Nanobiotechnol.*, vol. 18, no. 1, p. 125, 2020, doi: 10.1186/s12951-020-00685-4.

[20] A. K. Srivastava, Y. Kumar, and P. K. Singh, "A rule-based monitoring system for accurate prediction of diabetes," *Int. J. E-Health Med. Commun.*, vol. 11, no. 3, pp. 32–53, 2020, doi: 10.4018/IJEHMC.2020070103.

[21] V. Palmieri and M. Papi, "Can graphene take part in the fight against COVID-19?," *Nano Today*, vol. 33, p. 100883, 2020, doi: 10.1016/j.nantod.2020.100883.

[22] S. Bhattacharjee, R. Joshi, A. A. Chughtai, and C. R. Macintyre, "Graphene modified multifunctional personal protective clothing," *Adv. Mater. Interfaces*, vol. 6, no. 21, p. 1900622, 2019, doi: 10.1002/admi.201900622.

[23] K. Krishnamoorthy, K. Jeyasubramanian, M. Premanathan, G. Subbiah, H. S. Shin, and S. J. Kim, "Graphene oxide nanopaint," *Carbon N. Y.*, vol. 72, pp. 328–337, 2014, doi: 10.1016/j.carbon.2014.02.013.

[24] R. Pemmada, X. Zhu, M. Dash, Y. Zhou, S. Ramakrishna, X. Peng, V. Thomas, S. Jain, and H. S. Nanda, "Science-based strategies of antiviral coatings with viricidal properties for the COVID-19 like pandemics," *Materials (Basel)*, vol. 13, no. 18, p. 4041, 2020, doi: 10.3390/ma13184041.

[25] J. Verma, A. S. Khanna, R. Sahney, and A. Bhattacharya, "Super protective antibacterial coating development with silica-titania nano core-shells," *Nanoscale Adv.*, vol. 2, no. 9, pp. 4093–4105, 2020, doi: 10.1039/D0NA00387E.

[26] A. Auyeung, M. A. Casillas-Santana, G. A. Martinez-Castanon, Y. N. Slavin, W. Zhao, J. Asnis, U. O. Hafeli, and H. Bach, "Effective control of molds using a combination of nanoparticles," *PLoS One*, vol. 12, no. 1, p. e0169940, 2017, doi: 10.1371/journal.pone.0169940.

[27] G. B. Hwang, A. Patir, E. Allan, S. P. Nair, and I. P. Parkin, "Superhydrophobic and white light-activated bactericidal surface through a simple coating," *ACS Appl. Mater. Interfaces*, vol. 9, no. 34, pp. 29002–29009, 2017, doi: 10.1021/acsami.7b05977.

[28] J. Verma and A. Bhattacharya, "Development of coating formulation with silica-titania core-shell nanoparticles against pathogenic fungus," *R. Soc. Open Sci.*, vol. 5, no. 8, p. 180633, 2018, doi: 10.1098/rsos.180633.

[29] R. Solano, D. Patiño-Ruiz, and A. Herrera, "Preparation of modified paints with nano-structured additives and its potential applications," *Nanomater. Nanotechnol.*, vol. 10, p. 184798042090918, 2020, doi: 10.1177/1847980420909188.

[30] S. Pal, V. Contaldi, A. Licciulli, and F. Marzo, "Self-cleaning mineral paint for application in architectural heritage," *Coatings*, vol. 6, no. 4, p. 48, 2016, doi: 10.3390/coatings6040048.

[31] H. Fu, Y. You, S. Wang, and H. Chang, "SnO2 nanomaterial coating microfiber interferometer for ammonia concentration measurement," *Opt. Fiber Technol.*, vol. 68, p. 102819, 2022, doi: 10.1016/j.yofte.2022.102819.

[32] M. Poloju, N. Jayababu, and M. V. Ramana Reddy, "Improved gas sensing performance of Al doped ZnO/CuO nanocomposite based ammonia gas sensor," *Mater. Sci. Eng. B*, vol. 227, pp. 61–67, 2018, doi: 10.1016/j.mseb.2017.10.012.

[33] S. B. Kulkarni, Y. H. Navale, S. T. Navale, F. J. Stadler, N. S. Ramgir, and V. B. Patil, "Hybrid polyaniline-WO3 flexible sensor: A room temperature competence towards NH3 gas," *Sensors Actuators B Chem.*, vol. 288, pp. 279–288, 2019, doi: 10.1016/j.snb.2019.02.094.

[34] J. Karthikeyan, C. C. Berndt, J. Tikkanen, S. Reddy, and H. Herman, "Plasma spray synthesis of nanomaterial powders and deposits," *Mater. Sci. Eng. A*, vol. 238, no. 2, pp. 275–286, 1997, doi: 10.1016/S0921-5093(96)10568-2.

[35] A. Panepinto and R. Snyders, "Recent advances in the development of nanosculpted films by magnetron sputtering for energy-related applications," *Nanomaterials*, vol. 10, no. 10, p. 2039, 2020, doi: 10.3390/nano10102039.

[36] Y. Chen and B. Wang, "Effects of deposition parameters on structures and photoluminescence of MoO3—Nanomaterials grown by CVD," *Opt. Mater. (Amst).*, vol. 92, pp. 150–155, 2019, doi: 10.1016/j.optmat.2019.04.010.

[37] A. Krishna, E. T. Gecil, L. S. Aravinda, N. S. Kumar, K. N. Reddy, N. Balashanmugam, and M. R. Sankar, "Synthesis and thermal simulations of novel encapsulated CNT multifunctional thin-film based nanomaterial of SiO_2-CNT and TiN-CNT by PVD and PECVD techniques for thermal applications," *Diam. Relat. Mater.*, vol. 109, p. 108029, 2020, doi: 10.1016/j.diamond.2020.108029.

[38] J. Wang, Y. Wei, Y. Xu, Q. Wang, H. Lu, L. Qiu, and J. Zhu, "Photoluminescence and electroluminescence properties of aligned CsPbBr3 nanowire films prepared by off-center spin-coating," *Synth. Met.*, vol. 267, p. 116481, 2020, doi: 10.1016/j.synthmet.2020.116481.

[39] D. Guo, Y. Wang, R. Fernandez, L. Zhao, and B. Jodoin, "Cold spray for production of in-situ nanocrystalline MCrAlY coatings—Part I: Process analysis and microstructure characterization," *Surf. Coatings Technol.*, vol. 409, p. 126854, 2021, doi: 10.1016/j.surfcoat.2021.126854.

[40] N. V. Challagulla, V. Rohatgi, D. Sharma, and R. Kumar, "Recent developments of nanomaterial applications in additive manufacturing: A brief review," *Curr. Opin. Chem. Eng.*, vol. 28, pp. 75–82, 2020, doi: 10.1016/j.coche.2020.03.003.

[41] B. Tomar, S. Shiva, and T. Nath, "A review on wire arc additive manufacturing: Processing parameters, defects, quality improvement and recent advances," *Mater. Today Commun.*, vol. 31, p. 103739, 2022, doi: 10.1016/j.mtcomm.2022.103739.

[42] W. Zhai, N. Srikanth, L. B. Kong, and K. Zhou, "Carbon nanomaterials in tribology," *Carbon N. Y.*, vol. 119, pp. 150–171, 2017, doi: 10.1016/j.carbon.2017.04.027.

[43] H. Kumar, C. Kumar, S. G. K. Manikandan, M. Kamaraj, and S. Shiva, "Laser re-melting of atmospheric plasma sprayed high entropy alloy," *Adv. Eng. Mater.*, 2022, pp. 105–127, 2022.

[44] H. Kumar, G. A. Bhaduri, S. G. K. Manikandan, M. Kamaraj, and S. Shiva, "Influence of annealing on microstructure and tribological properties of AlCoCrFeNiTi high entropy alloy based coating," *Met. Mater. Int.*, 2022, doi: 10.1007/s12540-022-01264-y.

[45] H. Kumar, G. A. Bhaduri, S. G. K. Manikandan, M. Kamaraj, and S. Shiva, "Microstructural characterization and tribological properties of atmospheric plasma sprayed high entropy alloy coatings," *J. Therm. Spray Technol.*, vol. 31, pp. 1956–1974, 2022, doi: 10.1007/s11666-022-01422-z.

[46] S. Siegmann, M. Leparoux, and L. Rohr, "The role of nano-particles in the field of thermal spray coating technology," *Proc. SPIE*, vol. 5824, p. 224, 2005, doi: 10.1117/12.605225.

[47] J. Liu, H. Cui, X. Zhou, X. Wu, and J. Zhang, "Nanocrystalline copper coatings produced by cold spraying," *Met. Mater. Int.*, vol. 18, no. 1, pp. 121–128, 2012, doi: 10.1007/s12540-012-0014-1.

[48] O. Shekoofa, J. Wang, D. Li, Y. Luo, C. Sun, Z. Hao, Y. Han, B. Xiong, L. Wang, and H. Li, "Nano-crystalline thin films fabricated by Si-Al co-sputtering and metal induced crystallization for photovoltaic applications," *Sol. Energy*, vol. 173, pp. 539–550, 2018, doi: 10.1016/j.solener.2018.07.077.

[49] F. Nasirpouri, K. Alipour, F. Daneshvar, and M.-R. Sanaeian, "Electrodeposition of anticorrosion nanocoatings." In Susai Rajendran, Tuan ANH Nguyen, Saeid Kakooei, Mahdi Yeganeh, Yongxin Li, editors, *Corrosion Protection at the Nanoscale*, Netherlands, Elsevier, 2020, pp. 473–497.

[50] P. K. Kannan, S. Chaudhari, S. R. Dey, and M. Ramadan, "Progress in development of czts for solar photovoltaics applications," In Abdul-Ghani Olabi, editors, *Encyclopedia of Smart Materials*, Netherlands, Elsevier, 2022, pp. 681–698.

[51] M. Hong, J. Zou, and Z.-G. Chen, "Synthesis of thermoelectric materials," In Ranjan Kumar, Ranber Singh, editors, *Woodhead Publishing Series in Electronic and Optical Materials*, New Delhi, India, Woodhead Publishing, 2021, pp. 73–103.

[52] R. Wang and J. Wu, "Structure and basic properties of ternary metal oxides and their prospects for application in supercapacitors." In Deepak P. Dubal, Pedro Gomez-Romero, editors, *Metal Oxides in Supercapacitors*, Netherlands, Elsevier, 2017, pp. 99–132.

[53] M. Mittal, S. Sardar, and A. Jana, "Nanofabrication techniques for semiconductor chemical sensors," In Chaudhery Mustansar Hussain, Suresh Kumar Kailasa, editors, *Handbook of Nanomaterials for Sensing Applications*, Netherlands, Elsevier, 2021, pp. 119–137.

[54] A. Behera, P. Mallick, and S. S. Mohapatra, "Nanocoatings for anticorrosion," In Susai Rajendran, Tuan ANH Nguyen, Saeid Kakooei, Mahdi Yeganeh, Yongxin Li, editors, *Corrosion Protection at the Nanoscale*, Netherlands, Elsevier, 2020, pp. 227–243.

[55] O. Sakai and T. Takahashi, "Improvement of oxidation resistance of Silicon Nitride sintered body by CVD Si3N4 coating," In N. Mizutani, K. Akashi, T. Kimura, S. Ohno, M. Yoshimura, T. Maruyama, et al., editors, *Advanced Materials '93*, Netherlands, Elsevier, 1994, pp. 269–271.

[56] S. L. Sonawane, P. K. Labhane, and G. H. Sonawane, "Carbon-based nanocomposite membranes for water purification," In Bharat Bhanvase, Shirish Sonawane, Vijay Pawade, Aniruddha Pandit, editors, *Handbook of Nanomaterials for Wastewater Treatment*, Netherlands, Elsevier, 2021, pp. 555–574.

[57] K. V. Madhuri, "Thermal protection coatings of metal oxide powders," In Yarub Al-Douri, editors, *Metal Oxide Powder Technologies*, Netherlands, Elsevier, 2020, pp. 209–231.

[58] A. Mishra, N. Bhatt, and A. K. Bajpai, "Nanostructured superhydrophobic coatings for solar panel applications," In Phuong Nguyen Tri, Sami Rtimi, Claudiane M. Ouellet Plamondon, editors, *Nanomaterials-Based Coatings*, Netherlands, Elsevier, 2019, pp. 397–424.

[59] P. He, J. Cao, H. Ding, X. Zhao, and Z. Li, "Electronic devices based on solution-processed two-dimensional materials," In Eui-Hyeok Yang, Dibakar Datta, Junjun Ding, Grzegorz Hader, editors, *Synthesis, Modeling, and Characterization of 2D Materials, and Their Heterostructures*, Netherlands, Elsevier, 2020, pp. 351–384.

[60] Z. Shariatinia, "Perovskite solar cells as modern nano tools and devices in solar power energy," In Sheila Devasahayam, Chaudhery Mustansar Hussain, editors, *Nano Tools and Devices for Enhanced Renewable Energy*, Netherlands, Elsevier, 2021, pp. 377–427.

[61] A. Nayfeh and N. El-Atab, "Agglomeration-based nanoparticle fabrication," In Ammar Nayfeh, Nazek El-Atab, editors, *Nanomaterials-Based Charge Trapping Memory Devices*, Netherlands, Elsevier, 2020, pp. 133–153.

[62] K. Chowdary, A. Kotia, V. Lakshmanan, A. H. Elsheikh, and M. K. A. Ali, "A review of the tribological and thermophysical mechanisms of bio-lubricants based nanomaterials in automotive applications," *J. Mol. Liq.*, vol. 339, p. 116717, 2021, doi: 10.1016/j.molliq.2021.116717.

[63] S. M. Alves, B. S. Barros, M. F. Trajano, K. S. B. Ribeiro, and E. Moura, "Tribological behavior of vegetable oil-based lubricants with nanoparticles of oxides in boundary lubrication conditions," *Tribol. Int.*, vol. 65, pp. 28–36, 2013, doi: 10.1016/j.triboint.2013.03.027.

[64] A. P. Amosov, "Nanomaterials of SHS technology for tribological applications: A review," *Russ. J. Non-Ferrous Met.*, vol. 58, no. 5, pp. 530–539, 2017, doi: 10.3103/S1067821217050029.

[65] B. Bhushan, "Nanotribology, nanomechanics and nanomaterials characterization," *Philos. Trans. R. Soc. A Math. Phys. Eng. Sci.*, vol. 366, no. 1869, pp. 1351–1381, 2008, doi: 10.1098/rsta.2007.2163.

[66] A. Singh, S. Dubey, and H. K. Dubey, "Nanotechnology : The Future Engineering," no. May, 2019.

[67] P. Sharma, N. Bhardwaj, and V. Kumar, "Defence applications of nanotechnology: Development and strategies," *Eur. J. Mol. Clin. Med.*, vol. 07, no. 07, pp. 4310–4316, 2020.

[68] A. Lele, "Role of nanotechnology in defence," *Strateg. Anal.*, vol. 33, no. 2, pp. 229–241, 2009, doi: 10.1080/09700160802518700.

[69] D. Liu, F. Yang, F. Xiong, and N. Gu, "The smart drug delivery system and its clinical potential," *Theranostics*, vol. 6, no. 9, pp. 1306–1323, 2016, doi: 10.7150/thno.14858.

[70] V. S. Saji, H. C. Choe, and K. W. K. Yeung, "Nanotechnology in biomedical applications: A review," *Int. J. Nano Biomater.*, vol. 3, no. 2, p. 119, 2010, doi: 10.1504/IJNBM.2010.037801.

[71] A. P. Ramos, M. A. E. Cruz, C. B. Tovani, and P. Ciancaglini, "Biomedical applications of nanotechnology," *Biophys. Rev.*, vol. 9, no. 2, pp. 79–89, 2017, doi: 10.1007/s12551-016-0246-2.

[72] H. Liu, J. Zhang, X. Chen, X.-S. Du, J.-L. Zhang, G. Liu, and W.-G. Zhang, "Application of iron oxide nanoparticles in glioma imaging and therapy: From bench to bedside," *Nanoscale*, vol. 8, no. 15, pp. 7808–7826, 2016, doi: 10.1039/C6NR00147E.

[73] C.-P. Wong, W. Lin, L.-B. Zhu, H.-J. Jiang, R.-W. Zhang, Y. Li, and K.-S. Moon, "Nano materials for microelectronic and photonic packaging," *Front. Optoelectron. China*, vol. 3, no. 2, pp. 139–142, 2010, doi: 10.1007/s12200-010-0009-9.

[74] National Nanotechnology Initiative, "No Title.".

[75] M.-R. Gao, Y.-F. Xu, J. Jiang, and S.-H. Yu, "Nanostructured metal chalcogenides: Synthesis, modification, and applications in energy conversion and storage devices," *Chem. Soc. Rev.*, vol. 42, no. 7, p. 2986, 2013, doi: 10.1039/c2cs35310e.

[76] Y.-J. Wang, D. P. Wilkinson, and J. Zhang, "Noncarbon support materials for polymer electrolyte membrane fuel cell electrocatalysts," *Chem. Rev.*, vol. 111, no. 12, pp. 7625–7651, 2011, doi: 10.1021/cr100060r.

[77] M. Shafiq, S. Anjum, C. Hano, I. Anjum, and B. H. Abbasi, "An overview of the applications of nanomaterials and nanodevices in the food industry," *Foods*, vol. 9, no. 2, p. 148, 2020, doi: 10.3390/foods9020148.

[78] N. Durán and P. D. Marcato, "Nanobiotechnology perspectives. Role of nanotechnology in the food industry: A review," *Int. J. Food Sci. Technol.*, vol. 48, no. 6, pp. 1127–1134, 2013, doi: 10.1111/ijfs.12027.

[79] C. Chellaram, G. Murugaboopathi, A. A. John, R. Sivakumar, S. Ganesan, S. Krithika, and G. Priya, "Significance of nanotechnology in food industry," *APCBEE Procedia*, vol. 8, pp. 109–113, 2014, doi: 10.1016/j.apcbee.2014.03.010.

[80] Y.-H. Cho and O. G. Jones, "Assembled protein nanoparticles in food or nutrition applications," *Adv. Food Nutr. Res.*, vol. 88, pp. 47–84, 2019.

Chapter 12

Corrosion mitigation using polymeric nanocomposite coatings

Avi Gupta, Jaya Verma, and Deepak Kumar
Indian Institute of Technology, Delhi

CONTENTS

12.1 Introduction: Background and concept 191
12.2 Corrosion protection mechanisms 193
    12.2.1 Anodic passivation 193
    12.2.2 Cathodic protection 194
    12.2.3 Solution (electrolytic) inhibition 194
    12.2.4 Active corrosion inhibition 194
12.3 Nanostructured polymeric coatings for corrosion mitigation 194
12.4 Nano-core-shell-doped state-of-the-art polymeric coatings for corrosion prevention 196
    12.4.1 Development of core-shell nanoparticles 196
    12.4.2 Development of nano-core-shell-incorporated polymer coatings 198
    12.4.3 Corrosion response 198
12.5 Future scope 199
References 200

12.1 INTRODUCTION: BACKGROUND AND CONCEPT

Corrosion's financial cost is a major worry for the world. The National Association of Corrosion Engineers estimates that corrosion cost the United States 276 billion dollars annually in 2002 (Koch et al. 2002). The cost of corrosion in the US shipping sector alone was estimated at 2.7 billion dollars per year in the same research. According to research by the National Academies, corrosion accounts for losses of between $2 trillion and $4 trillion each decade in the United States alone (National Research Council 2011). Thirty-plus years later, the economic expenses of an old infrastructure remain among the biggest burdens on the country (Winkleman et al. 2011). According to recent estimates, infrastructure, automobile transportation, homeland security, and utility-related corrosion losses total $200 billion in Europe and about $300 billion in the United States (Böhm 2014). Although it is challenging, if not impossible, to estimate the indirect costs associated with decreased productivity brought on by disruptions, holdups, failures, and a decline in readiness (the nation's capability to reply to situations regimentally or otherwise), some accounts place them at levels that are at least equal to the direct costs and possibly more, demonstrating the seriousness and extent of this problem (National Research Council 2011; Winkleman et al. 2011). With the ubiquity of diverse steels as essential structural components, the redox interaction between oxygen (O) and iron (Fe) in the presence of water has substantial relevance. In the presence of water, the redox reaction between Fe and O (Dennis et al. 2015) has significant relevance given the prevalence of various steels as key structural components.

$$Fe\ (s) \rightarrow Fe^{2+}(aq) + 2e^- \tag{9.1}$$

$$O_2\ (g) + 2H_2O\ (l) + 4e^- \rightarrow 4OH^- \tag{9.2}$$

Overall reaction:

$$2Fe^{2+}(aq) + 4OH^- \rightarrow 2Fe\ (OH)_2 \tag{9.3}$$

$$4Fe\ (OH)_2\ (s) + O_2\ (g) \rightarrow 2Fe_2O_3 \times H_2O\ (s) + 2H_2O\ (l) \tag{9.4}$$

The electroneutrality between the cathodic and anodic corrosion cells on the metal surface is maintained by ion transport in an electrolytic medium, which explains why metals deteriorate more quickly in conditions with acid rain, downhole hydraulic fracturing, or exposure to seawater. In such severe conditions, more care must be taken to protect structural metals, because structural metal alloys often have uneven surface microstructures and exposed crystallographic facets, making them even more prone to corrosion. Due to a significant exothermic enthalpic contribution, corrosion reactions are depicted in equations (9.3) and (9.4) from a thermodynamic perspective; as a result, efforts to inhibit corrosion concentrate mainly on obstructing charge and mass transport channels to obstruct the half-reactions of corrosion shown in equations (9.1) and (9.2).

The term hybrid nanocomposite–based coatings, in which "hard" nanoparticles are dispersed inside "soft matter" type matrix, where both of the constituents contribute to the global functioning, is the main theme of this chapter. With a focus on the interfaces of the harder and softer components and the "interphase" area formed inside the polymeric matrix, these composite materials provide significant freedom of designing and are studied on their own right as material systems displaying unique features. Since nanoparticles have a high surface-to-volume ratio; this suggests that more atoms are available for interactions with the polymeric matrix. In a crux, this means that the characteristics of the host polymeric matrix may be extremely substantially changed even for a very tiny volume of additional nanoparticulate fillers. Unique synergies may be created by the change of the polymeric matrix, which is not possible with big, microscopic fillers.

12.2 CORROSION PROTECTION MECHANISMS

One of the best methods for preventing corrosion is by using protective coatings. Nanocomposite coatings' simple implementation of the modular design idea makes it especially suitable with the inclusion of specialized features as needed for a given application. Instead of striving to be exhaustive in our treatment of coating principles, we have tried to present illuminating examples and capture a discipline that is experiencing a vigorous rejuvenation in light of new prospects. The following methods, in order, are used through coatings to prevent metal from corroding: anodic passivation, cathodic protection, solution (electrolytic) inhibition, and active corrosion inhibition (Figure 12.1; Figueira et al. 2015; Hughes et al. 2010; Presuel-Moreno et al. 2008; Zheludkevich et al. 2005).

12.2.1 Anodic passivation

The anodic passivation method forms a passivating layer on top of the metal surface, inhibits redox reactions by creating an ion-impervious or ion-selective barrier, which stops corrosion. This approach, which is extensively employed, involves fusing a passivation layer with a natural oxide layer to create a bipolar precipitate. Alternately, during the initial anodization or conversion coating stage, additional layers of porous and denser oxides are also deposited nearby, substantially limiting ion movement.

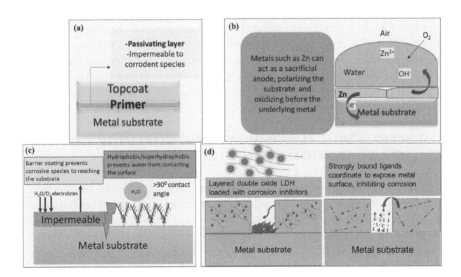

Figure 12.1 Different corrosion inhibition techniques, including anodic passivation (a), cathodic protection (b), barrier protection (c), and active corrosion inhibition (d). (Based on Dennis et al. 2015.)

12.2.2 Cathodic protection

Cathodic protection is attained by coating the substrate with a further electropositive metal that polarizes it and acts as the sacrificial anode. Zinc, magnesium, and aluminium alloys have conventionally been the basis for metallurgical coatings that provide cathodic protection of base metals like steels and Al alloys because their electrochemical potentials allow for preferential oxidation of the cladding over the substrate. In order to sufficiently polarize the substrate to counteract pitting corrosion while circumventing "overprotection," which might have additional negative consequences like hydrogen embrittlement or alkaline attack, the redox couple between the sacrificial anode and the substrate must be carefully coupled.

12.2.3 Solution (electrolytic) inhibition

Electrolytic inhibition slows down the process of corrosion by obstructing the transit of ions between the cathode and anode by using diffusion barriers and a low-ionic-conductivity matrix. By producing compact, ion-impermeable barriers, these coatings aim to either fundamentally reduce the amount of solution/electrolyte that may reach the surface of metal or augment the tortuousness of the electrolyte transport channels.

12.2.4 Active corrosion inhibition

Active corrosion inhibition addresses the inevitability of coating failure by mixing components that are selectively released on breakdown of coating so that they can rebuild a protective wall at the metal surface. For the purposes of this discussion, active corrosion inhibition is distinguished from the more general perception of "self-healing," which incorporates, for instance, the incorporation of particles of monomers/catalysts within containers that leads to reconstitution of polymeric organic coatings even without direct corrosion inhibition.

12.3 NANOSTRUCTURED POLYMERIC COATINGS FOR CORROSION MITIGATION

Recent developments in coating design have made an effort to combine different types of functioning for inhibiting corrosion in a synergetic manner inside a solo coating scheme in order to extend the lifespan of the structural components and to provide a shield against stimulating corrosive environments (Kendig and Buchheit 2003; Koch et al. 2002; Presuel-Moreno et al. 2008). The complex topography of materials, which includes both hard (nanoparticles/fillers) and soft matter (polymer phase) and hybrid structures, has been explored in response to the wide range of corrosion protection demands in an effort to broaden the functionality options. Numerous of these coating formulations were motivated by previously unheard-of possibilities made possible by advancements in experimental tools, high-throughput material assessment techniques based on informatics, and techniques for the fabrication of nanomaterials.

The qualities of each individual component are synergistically combined in multi-component hybrid nanocomposites to perform equivalent or greater than the summation of their parts (Figure 12.2). Other kinds of inclusions dispersed in the continuous phase include species of molecules, salts, pore-forming hybrid inorganic–organic particles, assembled polyelectrolytes, and capsules made of polymer. However, the majority of these materials are made of inorganic "hard" particles that are dispersed throughout a polymeric matrix. The inorganic component frequently aids in boosting mechanical strength, impact resistance, and electroactivity, while the organic component provides flexibility, compliance with the substrate, and permits suitability for topcoats (Metroke et al. 2002; Zheludkevich et al. 2005).

Figure 12.3 shows types of hybrid composites with variable extent of inclusion size, dispersion as well as the effects on the development of the "interphase" area. It is conceivable to include several plasticity and toughening processes that operate at various spatial and temporal scales to increase both toughness and strength by alteration of this interphase area, choice of the nano-filler, and the polymer matrix. Furthermore, these improvements in mechanical qualities are crucial for the lifespan and toughness of coatings intended to limit corrosion (Ritchie 2011).

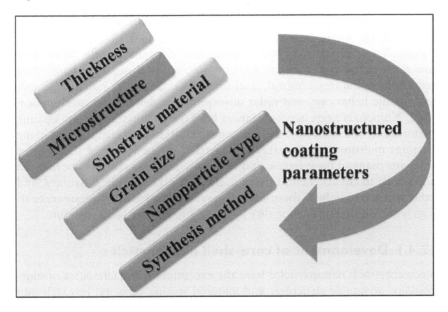

Figure 12.2 Several factors affecting the performance of nanostructured coatings. (Based on Farooq et al. 2022.)

12.4 NANO-CORE-SHELL-DOPED STATE-OF-THE-ART POLYMERIC COATINGS FOR CORROSION PREVENTION

Multifunctional coatings adapt well to the outside environment and can provide on-demand protection for the coated item. The triggering conditions including moisture, temperature, pH, active ions, and mechanical damage are just a few of the triggers that cause them to react. These reactions prevent chemical and physical damages to the material by releasing corrosion-inhibitor chemicals that have been stored or implanted on demand. The concept of multifunctional surfaces is a subject of considerable academic and commercial interest because of the wide areas of application that can be affected by their creation, including self-cleaning, anti-icing, liquid separation, corrosion resistance, drag reduction, oil-repellent, antibacterial, adhesion architecture coatings, antifouling paints for the marine industry, etc. (Junyan et al. 2015; Kota et al. 2014). Coatings have become lighter, cheaper, more durable, and more adaptable because of advancements in nanotechnology and manufacturing techniques. The design, preparation, and characterization of organic coatings have advanced significantly recently, making them more intelligent and adaptable. The functional and multifunctional coatings may fulfil the most demanding demands of users, particularly in high-tech sectors, by developing new qualities utilizing appropriate fillers and matrix as well as acceptable surface treatment procedures.

Organic coatings are seldom investigated, despite the fact that they are the most practical and affordable way to prevent corrosion on equipment and buildings, sometimes even adding an aesthetic look (Bagale et al. 2020; Nguyen-Tri et al. 2018). These coatings are designed to offer innovative functions such as antibacterial, self-cleaning, self-healing, super-hard, solar reflecting, pollutants removal, anti-ultraviolet (anti-UV) radiation, super hydrophilic behaviour, and radar absorption for desired end-user applications. Numerous aggressive substances have the potential to disrupt coating performance (Buskens et al. 2016; Deng et al. 2011; Yang et al. 2019). Giving coatings multifunctional character is a relatively new way of reducing or resolving many of these issues, and as a consequence, coatings now respond adequately to environmental changes, mechanical or chemical damage, and other threats. In light of these advantages, it is possible to create smart coatings with additional functionality by using a nano-core-shell structure.

12.4.1 Development of core-shell nanoparticles

Since core-shell nanoparticles have the exceptional characteristics of morphology, adaptable structure, and material storing capacity, research into core-shell microparticle/nanoparticle applications could ultimately result in significant breakthroughs. Core-shell particle synthesis can be divided into two steps: first, the core is formed, followed by formation of shell or

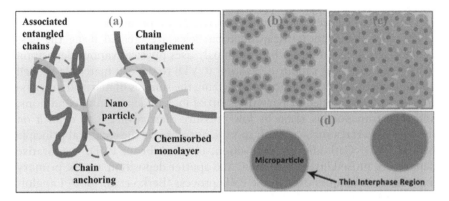

Figure 12.3 Schematic representation of the interphase area. (a) The area of the interphase where a nanoparticle makes contact with the chains of a polymer matrix shows the various possible interactions. (b) A single-walled unfunctionalized carbon nanotube/polymer composite's interphase region, showing sections with and without nanotubes. (c) Hybrid single-walled carbon nanotubes that have been functionalized, revealing the bigger interphase region that has fundamentally changed the composite. (d) The extremely narrow interphase zone, which leaves most of the host matrix unaltered, is shown in a conventional composite system based on the inclusion of microparticles. (Based on Dennis et al. 2015.)

deposition of shell on top of the already built core. The numerous synthesis approaches can be divided into four groups based on distinct principles: solid-phase reactions, liquid-phase reactions, gas-phase reactions, and mechanical mixing (Chen et al. 2020).

In the solid-phase reaction process, the shell precursors and the core materials are effectively mixed, followed by calcination according to a formulation to produce ultrafine-coated particles. Solid-phase reaction techniques are frequently used to synthesize materials that require phase shift during the synthesis process in order to acquire particular features via calcination at high temperatures, which cannot be done using wet chemistry techniques.

The phrase "liquid-phase reaction" describes chemical processes that deposit surface modifiers or coatings on preformed particles while occurring in a moist environment. In the lab and in business, the liquid-phase reaction method is employed for creating core-shell nanoparticles/microparticles since it requires less sophisticated equipment and operates at lower reaction temperatures. Some of the widely employed liquid-phase reaction procedures are the sol-gel method, hydrolysis method, electrochemical method, hydrothermal method, and emulsion polymerization (Marini et al. 2008; Yujun et al. 2006). The core components and shell precursors were dissolved in solutions for the sol-gel process, and subsequent hydrolysis reactions of the precursors led to the production of active monomers. Subsequently, the active monomers polymerize to form sols, the surface of the core materials is next covered with gels with particular spatial arrangements. After drying and heating, the necessary core-shell particles are produced.

The core-shell nanoparticles/microparticles are created using the gas-phase reaction technique, which primarily uses physical vapour deposition (PVD) and chemical vapour deposition (CVD) to deposit materials of shell on the target particles' surfaces. Vaporizing the sources of material into gaseous molecules, atoms, or partially ionized ions under vacuum conditions, and then transporting them by low-pressure plasma or gas to deposit on the core material's surfaces with specific purposes, is the technical principle behind PVD. Cathodic arc deposition, electron beam PVD, evaporative deposition, pulsed laser deposition, and sputter deposition are the primary deposition techniques used in the PVD process (Berkó et al. 1999; Langlois et al. 2015; Óvári et al. 2010).

According to the mechanical mixing technique, convection and diffusion may equally adsorb two or more particles on the surfaces of the modified nanoparticles with high adsorption or adhesion qualities at a particular temperature. After that, the particles are tightly mixed to cover the target particles' surfaces with the influence of external forces (gravity, mechanical, and aerodynamic force), completing the surface modification or recombination (He et al. 2015; Heidarpour et al. 2009).

12.4.2 Development of nano-core-shell-incorporated polymer coatings

The developed core-shell nanoparticles are then dispersed in a selective polymer media (matrix) and applied onto the surface to be protected. There are several coatings that are resistant to corrosion, including urethanes, latexes, epoxies, and silicone alkyd. One such effective process to develop SiO_2-ZnO core-shell nanoparticles-doped polyurethanes was reported and found effective in mitigation of corrosion of the steel substrate (Verma et al. 2022). The schematic in Figure 12.4 elaborates the steps that were involved in the fabrication of these coatings.

12.4.3 Corrosion response

High surface energies of the functionalized or non-functionalized core-shell nanoparticles have been held responsible for the strong adhesion to the polymer matrix. This yields highly densified coating structures, and as a consequence, this characteristic minimizes corrosion reactions by narrowing the channels through which the corrosive electrolyte can travel while passing through the coating system. When testing SiO_2-ZnO core-shell nanoparticles incorporating polyurethane coatings (PU/CS (Peptz)) on steel substrates, Verma et al. (2022) noted a similar corrosion prevention mechanism. The specimen Peptz was the most effective against corrosion of all those studied. It exhibited the lowest corrosion current (3.36×10^{-8} A)

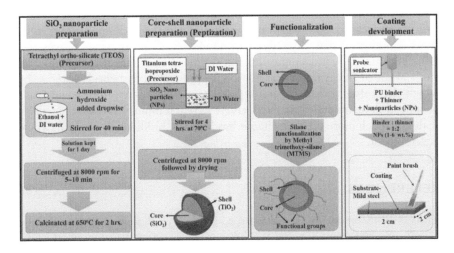

Figure 12.4 The development of nano-core-shell-technology-assisted polyurethane coating. (Based on Verma et al. 2022.)

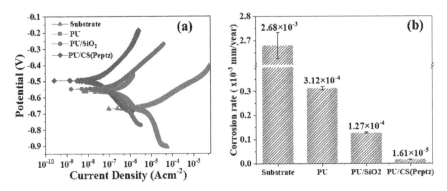

Figure 12.5 (a) Taffel plots and (b) variation of corrosion rate for substrate material and the coatings. (Based on Verma et al. 2022.)

and the highest corrosion potential (0.49 V) (Figure 12.5a). It also showed the lowest rate of corrosion (1.65×10^{-5} mm/year; Figure 12.5b). The reason why PU/CS (Peptz) exhibits much higher corrosion performance than other coatings is due to TiO$_2$'s maximum degree of dispersion in PU matrix leading to highly densified coating structure to reduce electrolytic permeability.

12.5 FUTURE SCOPE

Although the development of corrosion-resistant coatings with nanostructured inclusions has increased recently, we are still just at the beginning of the process in certain ways. Improved multiscale modelling, testing and formulation with higher throughput, *in situ* studies of processes of corrosion, integrating computational materials science and engineering, and swift adoption of new discoveries in nanoscience and polymer processing are all required to explore the vast multidimensional parameter space. The application of integrated computational approaches to hybrid systems, in which electrolyte transport, crack propagation, plasticity, and mass transport need to be modelled over extremely varied temporal and spatial length scales, is much more difficult than it has been for predictively enabling alloy development.

REFERENCES

Bagale, U. B., Sonawane, S. H., Bhanvase, B., Hakke, V. S., Kakunuri, M., Manickam, S., and Sonawane, S. S. 2020. "Multifunctional coatings based on smart nanocontainers." In Abdel Salam Hamdy Makhlouf, Nedal Yusuf Abu-Thabit, editors, *Advances in Smart Coatings and Thin Films for Future Industrial and Biomedical Engineering Applications*, Elsevier, Amsterdam, 135–162.

Berkó, A., Klivényi, G., and Solymosi, F. 1999. "Fabrication of Ir/TiO$_2$ (110) planar catalysts with tailored particle size and distribution." *Journal of Catalysis* 182(2): 511–514.

Böhm, S. 2014. "Graphene against corrosion." *Nature Nanotechnology* 9(10): 741–742.

Buskens, P., Burghoorn, M., Mourad, M. C. D., and Vroon, Z. 2016. "Antireflective coatings for glass and transparent polymers." *Langmuir* 32(27): 6781–6793.

Chen, H., Zhang, L., Li, M., and Xie, G. 2020. "Synthesis of core-shell micro/nanoparticles and their tribological application: A review." *Materials* 13(20): 4590.

Deng, X., Mammen, L., Zhao, Y., Lellig, P., Mullen, K., Li, C., Butt, H.-J., and Vollmer, D. 2011. "Transparent, thermally stable and mechanically robust superhydrophobic surfaces made from porous silica capsules." *Advanced Materials* 23(26): 2962–2965.

Dennis, R. V., Patil, V., Andrews, J. L., Aldinger, J. P., Yadav, G. D., and Banerjee, S. 2015. "Hybrid nanostructured coatings for corrosion protection of base metals: A sustainability perspective." *Materials Research Express* 2(3): 32001.

Farooq, S. A., Raina, A., Mohan, S., Singh, R. A., Jayalakshmi, S., and Haq, M. I. U. 2022. "Nanostructured coatings: Review on processing techniques, corrosion behaviour and tribological performance." *Nanomaterials* 12(8): 1323.

Figueira, R. B., Silva, C. J. R., and Pereira, E. V. 2015. "Organic-inorganic hybrid sol-gel coatings for metal corrosion protection: A review of recent progress." *Journal of Coatings Technology and Research* 12(1): 1–35.

He, W., Lv, Y., Zhao, Y., Xu, C., Jin, Z., Qin, C., and Yin, L. 2015. "Core-shell structured gel-nanocarriers for sustained drug release and enhanced antitumor effect." *International Journal of Pharmaceutics* 484(1–2): 163–171.

Heidarpour, A., Karimzadeh, F., and Enayati, M. H. 2009. "In situ synthesis mechanism of Al$_2$O$_3$-Mo nanocomposite by ball milling process." *Journal of Alloys and Compounds* 477(1–2): 692–695.

Hughes, A. E., Cole, I. S., Muster, T. H., and Varley, R. J. 2010. "Designing green, self-healing coatings for metal protection." *NPG Asia Materials* 2(4): 143–151.

Junyan, L., Li, W., Jingxian, B., and Ling, H. 2015. "Durable superhydrophobic/highly oleophobic coatings from multi-dome SiO_2 nanoparticles and fluoroacrylate block copolymers on flat substrates." *Journal of Materials Chemistry A* 3(40): 20134–20144.

Kendig, M. W., and Buchheit, R. G. 2003. "Corrosion inhibition of aluminum and aluminum alloys by soluble chromates, chromate coatings, and chromate-free coatings." *Corrosion* 59(5): 379–400.

Koch, G. H., Brongers, M. P. H., Thompson, N. G., Virmani, Y. P., and Payer, J. 2002. *Corrosion Cost and Preventive Strategies in the United States*. Washington, DC: Federal Highway Administration.

Kota, A. K., Kwon, G., and Tuteja, A. 2014. "The design and applications of superomniphobic surfaces." *NPG Asia Materials* 6(7): e109.

Langlois, C., Benzo, P., Arenal, R., Benoit, M., Nicolai, J., Combe, N., Ponchet, A., and Casanove, M. J. 2015. "Fully crystalline faceted Fe-Au core-shell nanoparticles." *Nano Letters* 15(8): 5075–5080.

Marini, M., Toselli, M., Borsacchi, S., Mollica, G., Geppi, M., and Pilati, F. 2008. "Facile synthesis of core-shell organic-inorganic hybrid nanoparticles with amphiphilic polymer shell by one-step sol-gel reactions." *Journal of Polymer Science Part A: Polymer Chemistry* 46(5): 1699–1709.

Metroke, T. L., Kachurina, O., and Knobbe, E. T. 2002. "Spectroscopic and corrosion resistance characterization of amine and super acid-cured hybrid organic-inorganic thin films on 2024-T3 aluminum alloy." *Progress in Organic Coatings* 44(3): 185–199.

National Research Council. 2011. "Research Opportunities in Corrosion Science and Engineering."

Nguyen-Tri, P., Nguyen, T. A., Carriere, P., and Xuan, C. N. 2018. "Nanocomposite coatings: Preparation, characterization, properties, and applications." *International Journal of Corrosion* 2018: 4749501.

Óvári, L., Berko, A., Balazs, N., Majzik, Z., and Kiss, J. 2010. "Formation of Rh–Au core–shell nanoparticles on TiO2 (110) surface studied by STM and LEIS." *Langmuir* 26(3): 2167–2175.

Presuel-Moreno, F., Jakab, M. A., Tailleart, N., Goldman, M., and Scully, J. R. 2008. "Corrosion-resistant metallic coatings." *Materials Today* 11(10): 14–23.

Ritchie, R. O. 2011. "The conflicts between strength and toughness." *Nature Materials* 10(11): 817–822.

Verma, J., Gupta, A., and Kumar, D. 2022. "Steel protection by SiO_2/TiO_2 core-shell based hybrid nanocoating." *Progress in Organic Coatings* 163: 106661.

Winkleman, A., Svedberg, E. B., Schafrik, R. E., and Duquette, D. J. 2011. "Preventing corrosion from wearing our future away." *Advanced Materials & Processes* 169(3): 26–32.

Yang, F., Li, C., Xu, W., and Cai, Z. 2019. "Multifunctional antifogging coatings based on ZrO2 and SiO2 nanoparticles by spray-spin-blow layer-by-layer assembly." *Journal of Materials Research* 34(22): 3827–3836.

Yujun, W., Ying, X., Fei, C., and Guangsheng, L. 2006. "Preparation of core-shell glass bead/polysulfone microspheres with two-step sol-gel process." *Journal of Applied Polymer Science* 99(6): 3365–3369.

Zheludkevich, M. L., Miranda Salvado, I., and Ferreira, M. G. S. 2005. "Sol-gel coatings for corrosion protection of metals." *Journal of Materials Chemistry* 15(48): 5099–5111.

Chapter 13

Sustainable nanomaterial coatings for anticorrosion
A review

Sangeeta Das and Preetam Bezbarua
Girijananda Chowdhury Institute of Management and Technology

Shubhajit Das
National Institute of Technology, Arunachal Pradesh

CONTENTS

13.1 Introduction	203
13.2 Methods to prepare anti-corrosive coatings	205
13.2.1 Chemical vapour deposition (CVD)	206
13.2.2 Physical vapour deposition (PVD)	207
13.2.3 Spray coating	207
13.2.4 Sol-gel	207
13.2.5 Electrodeposition	208
13.2.6 Laser cladding	208
13.3 Types of anti-corrosive nanocoatings	209
13.3.1 Metallic nanocoatings	209
13.3.2 Ceramic nanocoatings	210
13.3.3 Polymeric nanocoatings	210
13.3.4 Multifunctional anti-corrosive nanocoatings	211
13.4 Conclusion	211
References	212

13.1 INTRODUCTION

In the ongoing and evolving industries, the metals along with their alloys have been the mainstay. They are widely used in construction, electronics, water treatment industries, etc. However, metals under harsh climatic and environmental conditions are seen to be thermodynamically unstable and tend to undergo corrosion. It has been found that the economic losses faced by the industries account to billions of dollars per year due to the corrosive nature of metals. Corrosion inhibitors are used to prevent this phenomenon because they soak up on metal surfaces and can react with corrosion products such as ions of metals to generate a layer that functions as a barrier to reduce corrosion. These inhibitors are easy to use, economically affordable and efficient. However, they come with certain limitations such as they can hydrolyse and decompose during the process (Ahmed, 2020). Hence, there is a need for optimizing the physical and chemical properties of these coatings to achieve better environment-friendly properties for different applications. Nanocoatings based on nanotechnology are used to overcome the disadvantages of traditional coatings that provide better properties for different special applications, as shown in Figure 13.1.

Fine particle size and high surface area along with different nanoscale properties are the desirable properties of nanocoatings. The fine-sized particles fill and block the spaces on the metal surfaces preventing the diffusion of corrosive materials. Good adhesion properties of nanocoatings are the result of high surface area of nanoparticles that improves the lifespan of the coatings. The nanocoatings are more resistant to corrosion and wear due to improved mechanical, chemical, and electronic properties of nanomaterials. In a recent study, researchers have developed various high-performance anti-corrosive coatings by incorporating uniformly distributed nano-fillers within the matrix medium of the coating materials (Kar, 2019). There are mainly three different types of anti-corrosive nanocoatings, viz., metallic nanocoatings, ceramic nanocoatings, and nanocomposite coatings. Titanium oxide, tantalum oxide, alumina, zirconia, graphene-based nanocoatings, etc. come under ceramic nanocoatings. Metallic or polymeric matrices, including different nanomaterials, constitute nanocomposite coatings. Furthermore, a special type of nano-filler known as nanocontainer is added to coating formulation for anticorrosion and self-healing properties. Such coatings are a combination of passive matrix material and active nanocontainers packed

Figure 13.1 Applications of nanomaterial-based coatings (Tri et al., 2019).

with corrosion inhibitors or other active agents. The nanocontainer shells should have controlled permeability, which is triggered by specific conditions, and should remain stable during coating application and treatment (Shchukina et al., 2019). A few applications of anti-corrosive nanocoatings include electronic devices, eye glasses, windows, air filters, various industrial machineries, automotive parts, bone fixation materials and implants, protection of wood materials, etc. This chapter gives an insight on the various types of anti-corrosive coatings along with different fabrication techniques.

13.2 METHODS TO PREPARE ANTI-CORROSIVE COATINGS

Corrosion can be prevented using different techniques considering process cost and performance. The different methods of corrosion prevention include (Abdeen et al., 2018):

i. *Selection of material*: Either the material is relatively non-reactive in the galvanic series or can form a protective oxide layer in a particular environment.
ii. *Adjustment of environmental conditions*: This includes addition of inhibitors, adjustment of surrounding pH and temperature, reduction in the amount of sulphur, oxygen and chloride, reducing flow velocity and cleaning of contaminants.
iii. *Modification of surface*: Surface modification can be done by employing physical barriers such as films and coatings to lessen fractures and cracks.
iv. *Cathodic protection*: In this method, a power source is used or a more active anodic material is attached to the structure to be protected. The corrosion current is suppressed and is forced to flow to the metal to protect it.

The most widely used preventive measure of corrosion is the application of coating due to the availability of a wide variety of coating materials and techniques to deposit coating for specific applications. In the present days, nanocoatings that have been formed by incorporating nanomaterials in coatings are ultrafine microstructures that have one component in nanoscale. The nanomaterials comprise nanoparticles (zero-dimensional); nanotubes, nanowires and nanorods (one-dimensional); and nanoplatelet, nanosheets and nanofilms (two-dimensional). The nanoparticles in nanocoatings fill the voids due to their fine sizes and prevent any corrosive elements from dispersing into the surface of the substrate. Also, the grain boundaries of nanocoatings possess high density due to which it adheres better to the substrate and increases the longevity of the coating (Abdeen et al., 2018).

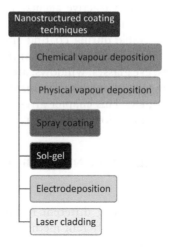

Figure 13.2 Techniques to produce nanostructured coatings.

The two broad nanocoating-deposition techniques are physical vapour deposition (PVD) and chemical vapour deposition (CVD). Apart from these two traditional processes, there exist other methods like spray coating, electrodeposition, laser cladding, sol-gel method, etc. The nanostructured coating methods are shown in Figure 13.2. Substrate preparation is an important step in coating techniques that includes cleaning and chemical modification of the substrate. The cleanliness of the substrate surface (free from contamination, oxides/scales, etc.), surface defects (pores, scratches), roughness, etc. decide the quality of coating on a substrate (Farooq et al., 2022).

13.2.1 Chemical vapour deposition (CVD)

It is the most widely used coating technique that coats a wide variety of metallic and ceramic compounds, like elements, metals and their alloys and intermetallic compounds. The deposition through CVD process takes place due to a chemical reaction between volatile precursors and the surface of the substrate. As a result of this chemical reaction, a solid phase is formed and gets deposited on to the substrate. There is a possibility of occurrence of different reactions on the substrate that are critically influenced by the substrate temperature. The classification of CVD process includes atmospheric pressure CVD, metal-organic CVD, low-pressure CVD, laser CVD, photochemical vapour deposition, chemical beam epitaxy, plasma-assisted CVD and plasma enhanced CVD (Makhlouf, 2011). The low-pressure CVD and metal-organic CVD are the two most widely used methods for nanocoating fabrication (Sishkovsky & Lebedev, 2011).

13.2.2 Physical vapour deposition (PVD)

This coating technique involves the transfer of coating material on an atomic level in vacuum. The solid or liquid material converts into vapour phase, which is then condensed to form a solid and dense film. Due to the formation of thin coating by PVD, there is a necessity for multilayer coatings. The PVD coating can easily get removed from the components that are subjected to elevated rate of wear causing surface abrasion. The corrosion resistance qualities of the components are weakened as a result of this abrasion, making them more sensitive to corrosive media. For the evaporation to take place in PVD, the thermal energy necessary for it may be supplied by electron beam, heating wire, laser beam, molecular beam, etc. The atoms of the source material get heated to its evaporation point using this thermal energy. The atoms that have been vaporized pass through the vacuum and accumulate on the substrate (Fotovvati et al., 2019).

13.2.3 Spray coating

This coating technique consists of a heat source and a wire or powder coating material. The heat source melts the coating material that is sprayed with a high velocity on to the substrate. The different types of spray coating processes are HVOF (high-velocity oxy-fuel) spraying, plasma spraying, combustion flame spraying, two-wire electric arc spraying, cold spray, etc. The preparation of the substrate is an essential step for perfect bonding of the coating materials on to the substrate. The substrate must be properly cleaned, roughened, masked and preheated before the application of spray coating. Using spray coating, a wide variety of coating materials, viz. metals, metal alloys, ceramics, plastics, cermets, etc. can be deposited on the substrate. The processing cost of spray coating is comparatively less due to rapid spray rate and high deposition efficiency. A wide variety of coating thicknesses is possible using this technique. In spite of all the advantages of spray coating, it contains pores that allow the movement of gases and/or liquids through the pores to the coating-substrate interface. Post-processing techniques like fusion, sintering and hot isostatic pressing can reduce the pores. For coating complex geometries, the spray gun and the substrate need sophisticated modifications (Tucker, 2013).

13.2.4 Sol-gel

This technique can produce coatings with improved properties to produce functional systems by taking advantage of size-adjustable properties of nano-sized components. The sol-gel coatings with corrosion-resistant properties can be used for a wide range of applications from decorative items to high-performance engineering applications. These coatings can be fabricated and applied at low temperatures and are also environment friendly. The sol-gel technique involves an oxide network formation through condensation reaction of molecular precursors in a liquid medium. The precursors for sol-gel reactions are mainly metal alkoxides or a combination of metal alkoxides and polymers that can participate in sol-gel reactions. Initially, polymerization takes place through hydrolysis at the metal–alkoxy linkage yielding alcohol and new reactants, hydroxylated metal centres, M-OH, where M refers to a network-forming element such as Si, Ti, Zr, etc. Thereafter, condensation or three-dimensional propagation occurs when hydroxylated species condense to form oxypolymers (M-O-M). The solid particles dispersed in any liquid medium, known as sols, join together to form a network of particles and get converted into gel. The viscosity of the gel is infinite and finally becomes static and this process is known as gelation. When the sol of coating precursors gets converted into gel, a coating film on the substrate is formed. For better adhesion of coating to the substrate, the sol must be applied on the substrate before the starting of gelation process (Pathak & Khanna, 2012).

13.2.5 Electrodeposition

This technique is based on the principle of electrolysis process that is used to deposit solids or condensed materials in different forms like films, coatings, etc. Electrodeposition can be used to deposit a wide range of materials, from metals, alloys to semiconductors and polymers in different sizes and dimensions. A simple electrode position cell consists of anode and cathode submerged in an electrolyte containing ions of the metal or solid to be deposited that is connected through a power supply. For a reaction to take place, an external driving force in the form of current and voltage is applied. A redox process, which occurs in a controlled setting and generates a rugged substance on the cathode, is known as electrodeposit. The factors that influence the electrode position of nanocoatings include two current modes direct or pulse, nanoparticle size and distribution, morphology, electrolyte parameters such as ion source, particle load, surfactants, pH, agitation and temperature (Nasirpouri et al., 2020).

13.2.6 Laser cladding

This technique is used to achieve good metallurgical bonding and less pores in comparison to other coating techniques. In laser cladding, the laser beam is emitted from nozzles of different shapes that come in contact with various powder mixtures and melts to generate a powder coating deposit on the substrate. Carbon steel, stainless steel, and non-ferrous alloy metals may be used as substrate materials, while metal components (Chrome, Nickel, Cobalt and Fe base) and earthenware powders can be used as clad materials (WC, SiC, TiC, Al_2O_3, ZrO_2). Laser energy, laser beam spot diameter, laser scanning velocity, work-piece motion velocity, pre-powder layer thickness, powder feed rate, and nozzle angle are the major characteristics of the laser cladding process (Singh et al., 2020).

13.3 TYPES OF ANTI-CORROSIVE NANOCOATINGS

The metals and alloys subjected to harsh environmental conditions can be well-protected using anti-corrosive nanocoatings. Nanocomposite coatings generally consist of two different phases, viz., one amorphous phase and one nanocrystalline phase or may be two different nanocrystalline phases. In recent years, nanocomposite coatings make use of nanoscience and nanotechnologies to design improved surface protective coatings possessing superior durability and performance at significantly lower cost. The substrate surface is basically altered using nanocomposite coatings to produce a new material having either improved or novel properties compared to the original substrate (Yasin et al., 2020). The properties and the lifespan of the substrates on which the coatings are applied are significantly improved. Based on the required engineering applications, the anti-corrosive nanocomposite coatings are categorized as metallic nanocoatings, ceramic nanocoatings and polymeric nanocoatings. In addition to this, a few anticorrosive nanocoatings also possess multifunctional and smart properties like self-healing, self-cleaning, stimuli responsiveness and corrosion sensing (Ulaeto et al., 2020). Nanocomposite coatings consist of matrix as the main material and filler as the secondary material. Nanocomposite coatings improved the mechanical properties greatly along with increased corrosion resistance. Nanocomposite coatings act as an efficient barrier by forming a more effective passivation layer (Emmanuel et al., 2021).

13.3.1 Metallic nanocoatings

Metal-based matrixes with a secondary reinforcing phase of metallic, ceramic, or polymeric material are utilized in metallic matrix composite coatings. Metallic nanocomposite coatings are made with both a metallic matrix and a strengthening phase, or at least one of them, and have a distinct nanometre (1–100 nm) length scale. Depending on the application, the reinforcing elements are injected into the metallic matrix as powders, nanowires, nanoparticles, nanotubes, or nanosheets. Al_2O_3, SiO_2, TiO_2, SiC, Cr_2O_3, WC, carbon nanotubes (CNTs), graphene sheets, and other materials with diameters or length scales ranging from micrometres (m) to nanometres (n) are examples of reinforcing materials (nm; Yasin et al., 2020).

The metallic nanocoatings are basically used for barrier effect and cathodic protection. Preferably, the metallic nanocoatings possess long life and highly protective efficiency due to its self-passivation nature and have lower potential than substrates. The techniques used to develop metallic nanocoatings are diffusion-bonded cladding, electro deposition, vapour deposition, flame spraying, etc. The metal surface should be free from porosity and various defects to avoid localized effect on the metal surface due to galvanic effects (Behera et al., 2020). Wadullah et al. (2022) studied the characteristics, and the structure of Nb_2O_5 nanocoating thin film for biological applications was examined. They found that the Nb_2O_5 coating is free of impurities, defects or micro-cracks and has the morphology of spherical nanoparticles.

13.3.2 Ceramic nanocoatings

Ceramic matrix composite (CMC) is developed by dispersing fibres into a ceramic matrix. With the advances in nanotechnology, the traditional reinforced fibres (in micro level) are replaced by a fibre (in nanometric level) that leads to the development of ceramic matrix nanocomposites (CMNC). Reinforced nanomaterials like graphene and CNT improves the ceramic structure at nano level to make them suitable for specific use in the field of automobile, industrial and aerospace engineering (Sharma et al., 2022). The function of barrier protection of coatings can be enhanced by using ceramic-based coatings as they are chemically non-reactive with the environment and alloy substrates. Due to the advantageous properties of ceramic coatings such as phase stability at high temperature, high melting point, resistance to oxidation and corrosion, etc. numerous researches have been carried out in the field of ceramic coatings as diffusion barrier layer (Kirubaharan & Kuppusami, 2020).

Ceramic nanocoatings can be used to coat the most widely used alloy in any industrial applications, i.e., steel, in order to prevent it from corrosion and wear. The ceramic nanocoatings can be developed using plasma electric oxidation (PEO) method. The PEO coatings on steels increase the surface hardness, and the coating structure is compact with good coating thickness, which enhances the load-bearing capacity and wear resistance (Attarzadeh et al., 2021).

13.3.3 Polymeric nanocoatings

Polymeric nanocomposites are created primarily by spreading nanoscale filler into a polymeric matrix. Alternating polymeric and inorganic layers result in a well-ordered layered architecture. Normally, just a few molecular layers of polymer may be intercalated in these materials. Gas-phase methods are the most often used techniques for roll-to-roll deposition of thin layers. These techniques enable the coating of flexible substrates with nanoscale inorganic or polymeric layers ranging in thickness from nanometres to hundreds of nanometres. Based on the deposition processes, gas-phase methods are classified as PVD or CVD. PVD procedures are performed in high vacuum and include transferring the solid coating material into the gas or vapour phase, followed by condensation on top of the substrate (Müller et al., 2017).

Nanocoatings have been shown to be quite useful in surface functionalization to give certain characteristics such as antibacterial, self-healing, gas barrier, etc. It has also shown wide application in fields like biomedicine, solar panels, packaging, etc. Polymer nanocomposites have the potential to increase the performance of existing automotive technologies such as engines and powertrains, exhaust systems and catalytic converters, paints and coatings, and tyres. Intelligent packaging, surface biocides, active packaging, silver nanoparticles as antimicrobial agents, nutrition and nutraceuticals, nanosensors and analysis for the detection of food relevant analytes (gases, tiny organic molecules and food-borne pathogens) and bioplastics are all examples of nanocoatings with exceptional properties (Vasile, 2018).

13.3.4 Multifunctional anti-corrosive nanocoatings

There are numerous available anti-corrosive nanocoatings, such as graphene oxide nanopaints, superhydrophobic polymer nanocoatings, etc. Graphene oxide (GO) nanopaint is created by integrating GO sheets in alkyd resin milled with a ball mill. The paint contains GO as a pigment, alkyd resin as a binder, thickening agent, hydrating agent, inner coat drier, outer coat drier, thinner, and stabilizer, and other additives (Krishnamoorthy et al., 2014). In general, a paint formulation consists of two main components: (i) pigmentation and (ii) binder. Pigment is a component that may provide the colour of the paint, whereas binder is a material that acts as the film-forming agent in the paint (Wijewardane & Goswami, 2012). These nanocoatings have shown effective inhibition of corrosion and bacterial growth. Furthermore, they have nominal environmental footprint (Krishnamoorthy et al., 2012).

Superhydrophobic polymer nanocoatings were prepared by Gu et al. (2014), by binding 1H, 1H, 2H, 2H perfluorodecyltriethoxysilane (PFDTS) onto OH-functionalized CNTs, which created superhydrophobic hybrid membranes with fire retardant capabilities. Under extreme conditions, the PFDTS/CNT membranes were discovered to be superhydrophobic with high oil/water separation efficiency. De et al. (2015) used a CVD technique at a low temperature (1,000°C) with no external catalysts to generate superhydrophobic multiwalled carbon nanotube (MWCNT) coatings on stainless steel. There are other superhydrophobic polymer nanocoatings as well as silica-based superhydrophobic nanocoatings, ZxO-based superhydrophobic nanocoatings, etc. (Das et al., 2018). These nanocoatings are mainly employed as anti-corrosive film on corrosive surfaces. Along with that, it has also shown properties like anti-icing, anti-fogging, self-cleaning, etc. These further increase the value of the nanocoatings.

13.4 CONCLUSION

Corrosion has been a major hurdle in the metal industry. Economic losses incurred by the industries account to billions of dollars due to the corrosive nature of the metals. In order to bring down this occurrence, the use of corrosion prevention methods is employed. Inhibitors have been widely used as a corrosion prevention measure. These inhibitors can bind to the metal surface and react with corrosion products like metal ions to generate a protective layer. To optimize the physical and chemical properties of the inhibitors, nanocoatings based on nanotechnology are used. Nanotechnology enhances the mechanical, chemical and electronic properties of the inhibitors. There are various methods to prepare anti-corrosive coatings such as CVD, PVD, spray coating, sol-gel, etc. There are different types of nanocoatings as well, namely, metallic nanocoatings, ceramic nanocoatings, polymeric nanocoatings, etc. The different methods of preparation are employed according to the requirement and the condition of the surroundings. This helps keep the maintenance cost to a minimum. The different forms of nanocoatings have their own forms of application in various fields such as automobile, food packaging, etc. It is seen that nanocoatings have proven to be a powerful preventive measure against corrosion and their research and development are being carried out to further fabricate various high-performance anti-corrosive coatings.

REFERENCES

Abdeen, D. H., Hachach, M. E., Koc, M., & Atieh, M. A. (2018). A review on the corrosion behaviour of nanocoatings on metallic substrates. *Materials* 12, 210. https://doi.org/10.3390/ma12020210

Ahmed, A. F. (2020). Applications of nanomaterials in corrosion protection coatings and inhibitors. *Corrosion Reviews* 38(1), 67–86. https://doi.org/10.1515/corrrev-2019-0011

Attarzadeh, N., Molaei, M., Babaei, K., & Fattah-Alhosseini, A. (2021). New promising ceramic coatings for corrosion and wear protection of steels: A review. *Surfaces and Interfaces* 23, 100997. https://doi.org/10.1016/j.surfin.2021.100997

Behera, A., Mallick, P., & Mohapatra, S. S. (2020). Nanocoatings for anticorrosion. *Corrosion Protection at the Nanoscale* 2020, 227–243. https://doi.org/10.1016/B978-0-12-819359-4.00013-1

Das, S., Kumar, S., Samal, S. K., Mohanty, S., & Naak, S. K. (2018). A review on superhydrophobic polymer nanocoatings: Recent development and applications. *Industrial & Engineering Chemistry Research* 57, 2727–2745. https://doi.org/10.1021/acs.iecr.7b04887

De Nicola, F., Castrucci, P., Scarselli, M., Nanni, F., Cacciotti, I., & De Crescenzi, M. (2015). Super-hydrophobic multi-walled carbon nanotube coatings for stainless steel. *Nanotechnology* 26(14), 145701–145707. https://www.researchgate.net/publication/273639485_Super-Hydrophobic_Multi-Walled_Carbon_Nanotube_Coatings_for_Stainless_Steel

Emmanuel, O. A., Fayomi, O. S. I., Agboola, O., Ayoola, A. A., Oloke, O. C., & Amusan, L. M. (2021). Short review on nanocomposite coating advances in the industry. *IOP Conference Series: Materials Science and Engineering* 1107, 012069. https://doi.org/10.1088/1757-899X/1107/1/012069

Farooq, S. A., Raina, A., Mohan, S., Arvind Singh, R., Jayalakshmi, S., & Irfan UlHaq, M. (2022). Nanostructured coatings: Review on processing techniques corrosion behaviour and tribological performance. *Nanomaterials* 12, 1323. https://doi.org/10.3390/nano12081323

Fotovvati, B., Namdari, N., & Dehghanghadikolaei, A. (2019). On coating techniques for surface protection: A review. *Journal of Manufacturing and Materials Processing* 3, 28. https://doi.org/10.3390/jmmp3010028

Gu, J., Xiao, P., Chen, J., Liu, F., Huang, Y., Li, G., Chen, T., & Zhang, J. (2014). Robust preparation of superhydrophobic polymer/carbon nanotube hybrid membranes for highly effective removal of oils and separation of water-in-oil emulsions. *Journal of Material Chemsitry A* 2, 15268–15272. https://pubs.rsc.org/en/content/articlelanding/2014/ta/c4ta01603c/unauth

Kar, P. (2019). Chapter 8—Anticorrosion and antiwear. *Nanomaterials-Based Coatings* 2019, 195–236. https://doi.org/10.1016/B978-0-12-815884-5.00008-9

Kirubaharan, A. M. K., & Kuppusami, P. (2020). Chapter 16—Corrosion behavior of ceramic nanocomposite coatings at nanoscale. *Corrosion Protection at the Nanoscale Micro and Nano Technologies* 2020, 295–314. https://doi.org/10.1016/B978-0-12-819359-4.00016-7

Krishnamoorthy, K., Jeyasubramanian, K., Premanathan, M., Subbiah, G., Shin, H. S., & Kim, S. J. (2014). Graphene oxide nanopaint. *Carbon* 72, 328–337. https://doi.org/10.1016/j.carbon.2014.02.013

Krishnamoorthy, K., Veerapandian, M., Zhang, L. H., Yun, K., & Kim, S. A. K. (2012). Antibacterial efficiency of graphene nanosheets against pathogenic bacteria via lipid peroxidation. *The Journal of Physical Chemistry* 116(32), 17280–17287. https://doi.org/10.1021/jp3047054

Makhlouf, A. S. H. (2011). Chapter 1—Current and advanced coating technologies for industrial applications. *Nanocoatings and Ultra-Thin Films* 2011, 3–23. https://doi.org/10.1533/9780857094902.1.3

Müller, K., Bugnicourt, E., Latore, M., Jorda, M., Sanz, Y. E., Lagaron, J. M., Miesbauer, O., Bianchin, A., Hankin, S., Bölz, U., Pérez, G., Jesdinszki, M., Lindner, M., Scheuerer, Z., Castelló, S., & Schmid, M. (2017). Review on the processing and properties of polymer nanocomposites and nanocoatings and their applications in the packaging automotive and solar energy fields. *Nanomaterials* 7(4), 74. https://doi.org/10.3390/nano7040074

Nasirpouri, F., Alipour, K., Daneshvar, F., & Sanaeian, M. (2020). Chapter 24—Electrodeposition of anticorrosion nanocoatings. *Corrosion Protection at the Nanoscale* 2020, 473–497. https://doi.org/10.1016/B978-0-12-819359-4.00024-6

Pathak, S. S., & Khanna, A. S. (2012). Chapter 12—Sol-gel nanocoatings for corrosion protection. *Corrosion Protection and Control Using Nanomaterials* 2012, 304–329. https://doi.org/10.1533/9780857095800.2.304

Sharma, N., Saxena, T., Alam, S. N., Ray, B. C., Biswas, K., & Jha, S. K. (2022). Ceramic-based nanocomposites: A perspective from carbonaceous nanofillers. *Materials Today Communications* 31, 103764. https://doi.org/10.1016/j.mtcomm.2022.103764

Shchukina, E., Wang, H., & Shchukin, D. G. (2019). Nanocontainer-based self-healing coatings: Current progress and future perspectives. *Chemical Communications* 55, 3859–3867. https://doi.org/10.1039/C8CC09982K

Singh, S., Goyal, D., Kumar, P., & Bansal, A. (2020). Laser cladding technique for erosive wear applications: A review. *Materials Research Express* 7, 012007. https://doi.org/10.1088/2053-1591/ab6894

Sishkovsky, I. V., & Lebedev, P. N. (2011). Chapter 3—Chemical and physical vapor depositionmethods for nanocoatings. *Nanocoatings and Ultra-Thin Films* 2011, 57–77. http://doi.org/10.1533/9780857094902.1.57

Tri, P. N., Nguyen, T. A., Ritmi, S., & Plamondon, C. M. O. (2019). Chapter 1—*Nanomaterials-based coatings*: An introduction. Nanomaterials-Based Coatings 2019, 1–7. https://doi.org/10.1016/B978-0-12-815884-5.00001-6

Tucker, R. C. (2013). *Introduction to Coating Design and Processing. ASM Handbook Volume 5A Thermal Spray Technology*. ASM International, Materials Park, OH. https://www.asminternational.org/documents/10192/1849770/05348G_Sample.pdf, accessed on 15.06.22.

Ulaeto, S. B., Pancrecious, J. K., Ajekwene, K. K., Mathew, G. M., & Rajan, T. P. D. (2020). Chapter 25—Advanced nanocoatings for anticorrosion. In: *Corrosion Protection at the Nanoscale Advanced Nanomaterials Series: Micro & Nano Technology Books*, edited by Susai Rajendran, Tuan Anh Nguyen, Saeid Kakooei, Mahdi Yeganeh, & Yongxin Li. 1st ed., Amsterdam: Elsevier, 499–509. http://doi.org/10.1016/B978-0-12-819359-4.00025-8

Vasile, C. (2018). Polymeric nanocomposites and nanocoatings for food packaging: A review. *Materials* 11(10), 1834. https://doi.org/10.3390/ma11101834

Wadullah, H. M., Mohammad, M. T., & Abdulrazzaq, T. K. (2020). Structure and characteristics of Nb_2O_5 nanocoating thin film for biomedical applications. *Materials Today Proceedings* 62(6), 3076–3080. https://doi.org/10.1016/j.matpr.2022.03.229

Wijewardane, S., & Goswami, D. (2012). A review on surface control of thermal radiation by paints and coatings for new energy applications. *Renewable and Sustainable Energy Reviews* 16, 1863–1873. https://doi.org/10.1016/j.rser.2012.01.046

Yasin, G., Arif, M., Mehtab, T., Shakeel, M., Khan, M. A., & Khan, W. Q. (2020). Metallic nanocomposite coatings. *Corrosion Protection at the Nanoscale* 2020, 245–274. http://doi.org/10.1016/b978-0-12-819359-4.00014-3

Chapter 14

Effect of nano-additives lubricant on the dynamic performance of textured journal bearing

Deepak Byotra, Sanjay Sharma, and Arun Bangotra
Shri Mata Vaishno Devi University

Rajeev Kumar Awasthi
Sardar Beant Singh State University

CONTENTS

14.1 Introduction	216
14.2 Lubrication governing equations	219
14.2.1 Non-dimensional form of Reynold's equation	220
14.2.2 Calculation of lubricant film thickness	220
14.2.3 Finite element method	220
14.2.4 Boundary conditions	222
14.3 Mathematical viscosity model	223
14.4 Solution procedure	224
14.5 Steady-state performance parameters	224
14.5.1 Load-carrying capacity	224
14.5.2 Coefficient of friction	225
14.6 Dynamic performance parameters	225
14.6.1 Stiffness coefficients of fluid film	225
14.6.2 Fluid film damping coefficients	225
14.6.3 Stability parameters	226
14.6.3.1 Threshold speed	227
14.6.3.2 Critical mass	228
14.6.3.3 Whirl frequency ratio	228
14.7 Conclusion	228
Nomenclature	229
Greek letter	230
Non-dimensional parameters	230
Vectors and matrices	231
References	231

DOI: 10.1201/9781003306276-14

14.1 INTRODUCTION

In the modern days, the necessity of having high-speed and high-load carrying conditions encourages researchers to devise a mechanism that can overcome the shortcomings of bearings operating under such conditions and provide an improved solution. Hydrodynamic journal bearings are important components of machines operating at high speed like compressors, turbines, etc. and find their applications in various industries due to low friction, high Load carrying capacity (LCC), low noise, and better dynamic behavior. The bearing operates at high speed, and heat developed in the lubricant causes high shear rates in the lubrication, thereby increasing the temperature and reducing the viscosity, which ultimately affect the bearing performance. The advantages of hydrodynamic bearings like long life, minimal maintenance, and damping can reduce vibration significantly. Various methods of lubricating the bearings are found in diverse applications. In order to overcome these challenges, many studies have been done on bearings with surface texturing to improve the LCC. The extensive use of texturing on the bearing is a productive method to improve its static and dynamic performance characteristics. It also leads to an increase in both static and dynamic losses due to friction. To keep the losses at a minimum, the location of the texture, number of textures, and texture depth are important parameters and need to be optimal. Different textures such as cylindrical, square, elliptical, rectangular, chevron, etc. are being used by various researchers to improve the bearings' performance. These surface textures with varying geometries and profiles like micro-dimples are formed by various techniques on the bearings surface. The repeated use of any mechanical components/parts results in the generation of high heat and material erosion, which not only has negative impacts on the environment but also disturbs the ecological balance, resulting in the failure of the concept of green tribology. The present study aims to find stability and dynamic performance parameters to optimize the design of texture.

Tala et al. [1] found the effect of texture location on bearing performance and observed that the important characteristics can be enhanced by proper orientation of texture on contact surface. Hamdavi et al. [2] investigated the effect of bearing partially textured on the LCC, etc. to find that partial texturing has a good effect on the performance of bearings. Kango et al. [3] studied negative texturing at various bearing locations' surface and found that negative texture improves the performance of bearing in comparison to full textured bearing surface. Matele and Pandey [4] studied the bearing partially textured with geometry and location, and found that stability margin for circular texture is maximum and for square texture, it is found to be the minimum. Kango et al. [5] studied the effect of spherical texturing and observed the reduced friction and temperature. Manser et al. [6] compared the texture profiles and geometry of hydrodynamic bearings and found enhancement in the LCC with the benchmark data. Singh and Awasthi [7] investigated the influence of various shapes and observed that spherically textured bearing exhibits the best performance compared to other textured surfaces. Shinde et al. [8] investigated conical bearing with partial texturing and found that bearing with partial texture results in improvement in fluid film pressure (FFP), whereas enhancement in LCC in comparison with conical shape hydrodynamic journal bearing (CSHJB) has maximum decrease in coefficient of friction (COF). Cupillard et al. [9] investigated the influence of surface texturing on friction and LCC using computational fluid dynamics (CFT) and observed that the COF can be reduced by using a suitable geometry of texture. Sharma et al. [10] studied chevron-shaped texture for investigating the performance improvement of bearing and found that texturing improves the performance of bearing when the texture is formed in the pressure-increasing region. Kango and Sharma [11] studied the combined effect in texture using positive roughness and found that roughness increases the load capacity. Shinde et al. [12] studied the bearing performance without considering the cavitation effect and observed the development of maximum pressure and observed the power losses due to friction in comparison to other micro-textures. Rasep et al. [13] studied the textured surface performance by taking mineral and vegetable oil and found improvement in the journal bearing performance. Tala et al. [14] studied the hydrodynamic effects of texture shapes in the bearing and observed that cubic textures enhance the bearing performance in comparison to the other geometries of the bearing. Shinde and Pawar [15] studied the performance of smooth bearing surface and observed that the film pressure, LCC, etc. improved due to micro-textures. Singh and Awasthi [16] studied the effect of cylindrical texturing on the characteristics of bearing and observed that effect of textures is better if the textures are formed in the upperward zone. Tauviqirrahman et al. [17] studied the combined effects of boundary slip as well as texture in bearings and observed that the high eccentricity ratios, texturing, and slip reduce the bearing performance. Kulkarni et al. [18] investigated the micro-textures on the bearing surface by applying chemical machining process and observed that micro-texturing on bearings be manufactured using this type of machining. Guha [19] investigated the roughness in the dynamic bearing characteristics and observed that bearings' stability degrade with increasing roughness parameters. Parkins [20] studied non-linear characteristics of the eight coefficients indicate flexibility in bearing using realistic criteria and observed the measured non-linearity coefficient is significant at eccentricity ratios higher than 0.78. Feng et al. [21] found the effect of water-lubricated bearing by applying misaligned thermo-hydrodynamic model and found that turbulence increases the LCC while the thermo-hydrodynamic (THD) effect decreases. Rho et al. [22] investigated the synchronous control of bearings and observed the imbalance in the output of the system, which is lowered by synchronous control in bearings. Leung et al. [23] observed the effect of a spherical bearing and observed that geometry and nanoparticles in lubricants improved bearing performance. Sheeja and Prabhu [24] investigated the plain bearing and found good impact of non-Newtonian effect, thermo-hydrodynamic effect for friction, and negligible eccentricity ratio and lubricant flow rate. Crosby and Chetti [25] investigated the influence of lubricated bearings and observed that the influence of couple stress found high bearing performance. Xu et al. [26] studied the LCC and damping of journal bearing and found that the thermal effects in oils have a better effect in misaligned bearings. Sun et al. [27] investigated the cavitation and, journal whirl with 3D computational fluid dynamics and by using mesh deformation technique and observed that the whirl frequency increases and the stability of bearing decreases and this method can be used to evaluate the static and dynamic characteristics of journal bearing. Kim et al. [28] studied the global analysis method and found that the stiffness coefficient and damping coefficients display better agreement with the existing literature. Arif et al. [29] investigated the influence of location of slip zones on the dynamic stability and found that in lubricants rheology, the stability is improved by increasing the eccentricity ratio. Kostrzewsky et al. [30] investigated eccentricity and dynamic coefficients against sommerfeld number of a three-lobe journal bearing and found good agreement for Sommerfeld numbers and increasing disagreement at higher Sommerfeld numbers. Nair et al. [31] studied characteristics of journal bearing with nanoparticle lubricants and found an increase in LCC. Sadiq et al. [32] studied the influence of bio-oil and mineral-based oil and found that pressure for base oils is observed to be more than those of soybean oil, etc. Jang and Yoon [33] investigated characteristics of herringbone grooved journal bearing with plain sleeve and plain journal bearing with herringbone grooved sleeve under dynamic load and observed that a herringbone grooved journal bearing with plain sleeve generates less torque than that of plain journal bearing with herringbone grooved sleeve. Flack et al. [34] investigated the stiffness and damping coefficient using harmonic excitations and reported that dynamic stiffness and damping values for different data are within 13% at worst and good within 3%. Wei et al. [35] studied an oil film force model and found that the tolerances in bearing diameter and width and tolerance form a negative effect. Mongkolwongrojn and Arunmetta [36] investigated the characteristics of bearings and found that soybean oil strongly influences the characteristics of bearings. Mandal et al. [37] studied the nanofluid-lubricated bearings and the study showed that the LCC is increased with nanoparticles compared to other conventional engine oils. Rho and Kim [38] investigated the stability of bearings and found that the control of bearings can be used for improvement in stability of whirl amplitude. Chetti [39] investigated the effect of four-lobe bearing and found that couple stresses improve the characteristics of four-lobe bearing compared to those lubricated with newtonian fluids. Bouzidane and

Thomas [40] investigated the characteristics of the bearing and found that the bearing has better characteristics and stability due to its high stiffness and damping. Wang et al. [41] studied the texturing effect on the stability of bearings and found that for partial texture along the bearing circumferential direction, the dimple depth influences the stability of the bearing. Awasthi et al. [42] found that bearing stability can be improved by providing different shaped texturing on the surface of bearings. Kalakada et al. [43] developed a viscosity model using regression analysis and by adding the nano-additives in the base oil and found an enhanced bearing performance. Also, Khatri and Sharma [44], Sharma and Kushare [45], and Chandrawat and Sinhasan [46] found that the stability of bearing can be enhanced with surface texturing of various shapes.

After studying the literature survey, it is observed that the textured surface can considerably increase the bearing performance and also nanoparticles' addition in base lubricants further improved the performance of the journal bearing. As a result of which, combined effect of textures on the bearing surface and nanoparticles' addition in the lubricants can be further realized to enhance the performance of journal bearing.

14.2 LUBRICATION GOVERNING EQUATIONS

Figure 14.1 represents various shapes of the textures formed in the journal bearing. Considering non-dimensional governing Reynolds equation for laminar flow for iso-viscous Newtonian lubricant, given by Sharma et al. [10] as under:

$$\frac{\partial}{\partial x}\left(F_2 \frac{\partial p_l}{\partial x}\right) + \frac{\partial}{\partial y}\left(F_2 \frac{\partial p_l}{\partial y}\right) = U\left[\frac{\partial}{\partial x}\left\{h - \frac{F_1}{F_0}\right\}\right] + \frac{\partial h}{\partial t} \quad (14.1)$$

where F_0, F_1, and F_2 are viscosity functions which are defined by:

$$F_0 = \int_0^h \frac{dz}{\mu} \quad (14.2)$$

$$F_1 = \int_0^h z \frac{dz}{\mu} \quad (14.3)$$

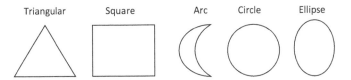

Figure 14.1 Various shapes of texture.

$$F_2 = \left(\int_0^h \left(\frac{z^2}{\mu} dz - \frac{F_1}{F_0} \frac{z\,dz}{\mu} \right) \right) \tag{14.4}$$

14.2.1 Non-dimensional form of Reynold's equation

Non-dimensional parameters are substituted in the Reynolds equation to convert it into the non-dimensional form by considering incompressible flow, laminar and iso-viscous Newtonian lubricant, as given by Manser et al. [6] as:

$$\frac{\partial}{\partial \alpha}\left(\bar{h}^3 \bar{F}_2 \frac{\partial \bar{p}_l}{\partial \alpha} \right) + \frac{\partial}{\partial \beta}\left(\bar{h}^3 \bar{F}_2 \frac{\partial \bar{p}_l}{\partial \beta} \right) = \bar{\Omega}\left[\frac{\partial}{\partial \alpha}\left\{ \left(1 - \frac{\bar{F}_1}{\bar{F}_0}\right) \bar{h} \right\} \right] + \frac{\partial \bar{h}}{\partial \bar{t}} \tag{14.5}$$

Here $\bar{F}_0$, $\bar{F}_1$ and $\bar{F}_2$ are defined as:

$$\bar{F}_0 = \int_0^1 \frac{1}{\bar{u}} d\bar{z} \tag{14.6}$$

$$\bar{F}_1 = \int_0^1 \frac{\bar{z}}{\bar{u}} d\bar{z} \tag{14.7}$$

$$\bar{F}_2 = \int_0^1 \frac{\bar{z}}{\bar{u}} \left(\bar{z} - \frac{\bar{F}_1}{\bar{F}_0} \right) d\bar{z} \tag{14.8}$$

14.2.2 Calculation of lubricant film thickness

The variation in fluid film thickness in a bearing having a texture in the inner surface given by Sharma et al. [10]:

$$\bar{h} = (1 - \bar{X}_j \cos\alpha - \bar{Z}_j \sin\alpha) + \bar{h}_t \tag{14.9}$$

where $\bar{X}_j$ and $\bar{Z}_j$ are journal coordinates representing the center position.

$\bar{h}_t$ is the variation in the thickness in the fluid film having textures, as reported by Manser et al. [47] and Qiu et al. [48].

14.2.3 Finite element method

The Reynolds equation is solved by using finite element method (FEM). The numerical-based research studies are performed by MATLAB for finding different parameters in the bearing. For discretization of the lubricant regime in two-dimensional flow, four-noded elements are employed and the pressure in the element is expressed by Awasthi et al. [42] as:

$$\bar{p} = \sum_{j=1}^{4} N_j \bar{P}_j \tag{14.10}$$

$$\alpha = \sum_{j=1}^{n} N_j \alpha_j \quad \text{and} \quad \beta = \sum_{j=1}^{n} N_j \beta_j$$

$$\frac{\partial \bar{p}}{\partial \alpha} = \sum_{j=1}^{n} \frac{\partial N_j}{\partial \alpha} \bar{p}_j \quad \text{and} \quad \frac{\partial \bar{p}}{\partial \beta} = \sum_{j=1}^{n} \frac{\partial N_j}{\partial \beta} \bar{p}_j \tag{14.11}$$

where N_j = the shape function of the element, while n = number of nodes in each element.

Substituting pressure from Equation (14.11), the Equation (14.5) given by Awasthi et al. [42] is stated as:

$$\frac{\partial}{\partial \alpha}\left[\bar{h}^3 \bar{F}_2 \frac{\partial}{\partial \alpha}\left(\sum_{j=1}^{4} N_j \bar{P}_j\right)\right] + \frac{\partial}{\partial \beta}\left[\bar{h}^3 \bar{F}_2 \frac{\partial}{\partial \beta}\left(\sum_{j=1}^{4} N_j \bar{P}_j\right)\right]$$

$$-\bar{\Omega}\left[\frac{\partial}{\partial \alpha}\left\{\left(1-\frac{\bar{F}_1}{\bar{F}_0}\right)\bar{h}\right\}\right] - \frac{\partial \bar{h}}{\partial \bar{t}} = R^e \tag{14.12}$$

where R^e is known as the residue, which is minimized by using Galerkin's technique given by Awasthi et al. [42]:

$$\iint_{\Omega^e} N_i R^e d\alpha\, d\beta = 0 \tag{14.13}$$

By integrating the second-order terms of Equation (14.14) part-wise, the global system of equation in matrix form is given by Awasthi et al. [42] as:

$$[\bar{F}_M]^e \{\bar{p}_V\}^e = \{\bar{Q}_V\}^e + \Omega\{\bar{R}_H\}^e + \dot{\bar{x}}_j\{\bar{R}_{Xj}\}^e + \dot{\bar{z}}_j\{\bar{R}_{Zj}\}^e \tag{14.14}$$

In Equation (14.16), the various coefficients of matrices for the eth element can be defined as:

$$\bar{F}_{Mij}^e = \int\int_{A\ e} \left\{ \frac{\bar{h}^3}{12\bar{\mu}} \left(\frac{\partial N_{Vi}}{\partial \alpha} \frac{\partial N_{Vj}}{\partial \alpha} + \frac{\partial N_{Vi}}{\partial \beta} \frac{\partial N_{Vj}}{\partial \beta} \right) \right\} d\alpha\, d\beta$$

$$\bar{Q}_{Vi}^e = \int_{\Gamma^e} \left\{ \left(\frac{\bar{h}^3}{12\bar{\mu}} \frac{\partial \bar{p}_V}{\partial \alpha} - \frac{1}{2}\bar{\Omega}\bar{h} \right) n_1 + \left(\frac{\bar{h}^3}{12\bar{\mu}} \frac{\partial \bar{p}_V}{\partial \beta} \right) n_2 \right\} N_{Vi}\, d\bar{\Gamma}^e$$

$$\bar{R}_{Hi}^e = \int\int_{A\ e} \frac{1}{2}\bar{h}\, \frac{\partial N_{Vi}}{\partial \alpha}\, d\alpha\, d\beta \qquad (14.15)$$

$$\bar{R}_{Xji}^e = \int\int_{A\ e} \cos\alpha N_{Vi}\, d\alpha\, d\beta$$

$$\bar{R}_{Xzi}^e = \int\int_{A\ e} \sin\alpha N_{Vi}\, d\alpha\, d\beta$$

The directional cosines in each direction are represented by n_1 and n_2.

14.2.4 Boundary conditions

The boundary conditions in the domain of lubricant flow are used to solve the Reynolds equation by Sharma et al. [10]:

- The pressure is assumed as zero at external edges of all nodes of bearing boundary.

$$\bar{p}_l \big|_{\beta = \pm 1} = 0.0$$

- Assuming pressure gradient at the trailing edges as 0.0 by considering the effect of cavitation in boundary conditions.

$$\bar{p}_l = \frac{\partial \bar{p}_t}{\partial \alpha} = 0.0$$

After calculation of the vector value $\{\bar{P}_0\}$ and $\{\bar{Q}\}$, the following conditions are checked at every node:

- $\{\bar{Q}\} < 0$, nodes in cavitation zone;
- $\{\bar{P}_0\} > 0$, nodes in flow zone of the lubricant.

If these conditions are satisfied, a solution will exist and if the nodes do not meet these conditions and move from one region to the other, the computation process continues until a solution is obtained. The calculation is repeated again until the pressure difference at nodes of subsequent calculations fall below 0.1 TLM. The calculation process is stopped if the conditions of convergence are satisfied.

$$\bar{Z}_0 = \left| \frac{\left[p_t^{-m'}\right] - \left[p_t^{-(m'-1)}\right]}{\left[p_t^{-(m'-1)}\right]} \right| \times 100 \leq \text{TLM} \qquad (14.16)$$

14.3 MATHEMATICAL VISCOSITY MODEL

The mathematical viscosity model's regression model is the relation between the relative viscosity $\bar{\mu}$, temperature, and percentage of nanoparticle additives (ϕ) by weight in the base lubricant by using the experimental data. The values of relative viscosity with nanoparticle additives CuO, CeO_2, and Al_2O_3 in the base lubricant are computed for temperatures range between 30°C and 90°C and wt. concentration of nanoparticle additives 0.0%–0.5% as reported by Kalakada et al. [43] as under:

$$\bar{\mu} = \frac{\mu}{\mu_0} = e^{(K_1 - K_2 \bar{T})} \qquad (14.17)$$

where $\bar{T} = \dfrac{T}{T_{atm}}$

For CuO; $K_1 = 1.142 - 1.126\ \phi + 51.120\ \phi^2 - 6.354\ \phi^3$
           $K_2 = 1.163 - 1.041\ \phi + 4.218\ \phi^2 - 5.080\ \phi^3$
For CeO_2; $K_1 = 1.142 - 0.311\ \phi + 1.377\ \phi^2 - 2.113\ \phi^3$
           $K_2 = 1.163 - 0.431\ \phi + 1.521\ \phi^2 - 2.060\ \phi^3$
For Al_2O_3; $K_1 = 1.142 - 0.311\ \phi + 1.377\ \phi^2 - 2.113\ \phi^3$
           $K_2 = 1.163 - 0.431\ \phi + 1.521\ \phi^2 - 2.060\ \phi^3$

It has been observed that the value of $\bar{\mu}$ nanolubricants enhances when the weight percentage of nanoparticles, ϕ at any temperature enhances. The dynamic performance parameters are calculated in substitutes the relative viscosity $\bar{\mu}$ as derived from model of viscosity as shown in Equation (14.5).

The experimental data related to the viscosity of the nanoparticles-lubricant in the temperature range of 30°C–90°C is obtained by Kalakada et al. [43]. The experimental outcomes showed the relative viscosity and $\bar{\mu}$ nanoadditives-lubricant, enhances with the enhancement of weight percentage of nanoparticles at any value of temperature.

14.4 SOLUTION PROCEDURE

Dynamic performance parameters and stability of bearing having textured surfaces and operating with the addition of nanolubricants are calculated using the Reynolds equation and the boundary conditions with the help of finite element analysis and variable viscosity. The viscosity field changes with the temperature and is calculated at any temperature and %wt. fraction of nanoparticles with 0.1, 0.25, and 0.5 in lubricant by using the model of viscosity. The calculated viscosity from the model is put in Reynolds equation to determine the pressure distribution. The process continues till the criteria as mentioned in Equation (14.16) are satisfied. The various dynamic parameters like stiffness, damping coefficients, frequency ratio, and the threshold speed can be calculated at various eccentricity ratios. The dynamic parameter of bearing with texture and nanoparticles' lubricant is computed by using the procedure shown in Figure 14.2.

The bearing geometrical parameters and operating parameters like aspect ratio ($\lambda=1$) having length (50 mm) and radius (25 mm), clearance ratio ($c_r=1$), lubricant viscosity at different eccentricity ratios and depth of textures are also considered for textured bearing in circumferential and axial directions. The influence of variation in parameters and the effect of %wt. fraction of additives at various temperature ranges on dynamic performance parameters such as direct stiffness, damping, and threshold speed are computed for the stability of bearing.

14.5 STEADY-STATE PERFORMANCE PARAMETERS

14.5.1 Load-carrying capacity

The components of LCC are calculated by Khatri and Sharma [44] as:

$$\bar{F}_x = -\int_{-\lambda}^{\lambda}\int_0^{2\pi} \bar{p}_t \cos\alpha\, d\alpha\, d\beta$$

$$\bar{F}_z = -\int_{-\lambda}^{\lambda}\int_0^{2\pi} \bar{p}_t \sin\alpha\, d\alpha\, d\beta \tag{14.18}$$

Resultant film reaction and angle of attitude are calculated as:

$$\bar{W}_b = \left[\left(\bar{F}_x\right)^2 + \left(\bar{F}_z\right)^2\right]^{1/2}$$

$$\phi = \tan^{-1}\left(\frac{\bar{F}_z}{\bar{F}_x}\right) \tag{14.19}$$

14.5.2 Coefficient of friction

It is computed by using an equation given by Khatri and Sharma [44] as:

$$\bar{P}_L = \sum_{e=1}^{n_e} \int_{A^e} \left(\Omega \frac{\bar{\tau}_c}{\bar{h}} + \frac{\bar{h}}{2} \frac{\partial \bar{p}_l}{\partial \alpha} \right) dA \qquad (14.20)$$

where $\bar{\tau}_c$: Normalized shear stress couette represented by $\bar{\tau}_c = 1 + (R_e)^{0.855} R_e$ represented as $R_e = \frac{\rho U h}{\mu}$ and $\bar{\tau}_c = 1.0$, if the flow is laminar.

14.6 DYNAMIC PERFORMANCE PARAMETERS

In hydrodynamic journal bearing, a thin film of lubricant supports the radial load and offers dampening due to the effect of lubricant film squeeze. During oscillation, the journal center position is defined by two degrees of freedom in X and Z directions and we can determine four fluid film stiffness and damping coefficients, which predict the stability of bearing. With the help of these coefficients, we can calculate the stability characteristics, i.e., threshold speed, whirl frequency ratio, and critical mass.

14.6.1 Stiffness coefficients of fluid film

These are represented by Sharma et al. [10] and Khatri and Sharma [44]:

$$\bar{S}_{ij} = -\frac{\partial \bar{F}_i}{\partial \bar{q}_j}, \ (i = x, z) \qquad (14.21)$$

where i represents the force direction, while $\bar{q}_j$ represents the center position of the journal, i.e., $\bar{X}_j, \bar{Z}_j$

The Fluid Film Stiffness coefficient in matrix form can be represented as:

$$\begin{bmatrix} \bar{S}_{xx} & \bar{S}_{xz} \\ \bar{S}_{zx} & \bar{S}_{zz} \end{bmatrix} = -\begin{bmatrix} \dfrac{\partial \bar{F}_x}{\partial \bar{x}} & \dfrac{\partial \bar{F}_x}{\partial \bar{z}} \\ \dfrac{\partial \bar{F}_z}{\partial \bar{x}} & \dfrac{\partial \bar{F}_z}{\partial \bar{z}} \end{bmatrix} \qquad (14.22)$$

14.6.2 Fluid film damping coefficients

These are presented by Khatri and Sharma [44] as:

$$\bar{C}_{ij} = -\frac{\partial \bar{F}_i}{\partial \dot{\bar{q}}_j}, (i = x, z) \qquad (14.23)$$

where i represents the force direction, while $\dot{\bar{q}}_j$ defines the velocity component of the journal center.

The damping coefficient in matrix is represented as:

$$\begin{bmatrix} \bar{C}_{xx} & \bar{C}_{xz} \\ \bar{C}_{zx} & \bar{C}_{zz} \end{bmatrix} = -\begin{bmatrix} \dfrac{\partial \bar{F}_x}{\partial \bar{\dot{x}}} & \dfrac{\partial \bar{F}_x}{\partial \bar{\dot{z}}} \\ \dfrac{\partial \bar{F}_z}{\partial \bar{\dot{x}}} & \dfrac{\partial \bar{F}_z}{\partial \bar{\dot{z}}} \end{bmatrix} \quad (14.24)$$

14.6.3 Stability parameters

The linear equation for motion when the journal is disturbed is defined by Sharma and Kushare [45] as:

$$[\bar{M}_j]\{\bar{\ddot{X}}_j\} + [\bar{C}]\{\bar{\dot{X}}_j\} + [\bar{S}]\{\bar{X}_j\} = 0 \quad (14.25)$$

Equation (14.25) as stated above for journal center motion trajectories in the matrix form is written by numerically integrating the linear motion equations:

$$\begin{bmatrix} \bar{M}_j & 0 \\ 0 & \bar{M}_j \end{bmatrix}\begin{Bmatrix} \bar{\ddot{X}}_j \\ \bar{\ddot{Z}}_j \end{Bmatrix} + \begin{bmatrix} \bar{C}_{xx} & \bar{C}_{xz} \\ \bar{C}_{zx} & \bar{C}_{zz} \end{bmatrix}\begin{Bmatrix} \bar{\dot{X}}_j \\ \bar{\dot{Z}}_j \end{Bmatrix} + \begin{bmatrix} \bar{S}_{xx} & \bar{S}_{xz} \\ \bar{S}_{zx} & \bar{S}_{zz} \end{bmatrix}\begin{Bmatrix} \bar{X}_j \\ \bar{Z}_j \end{Bmatrix} = \begin{Bmatrix} 0 \\ 0 \end{Bmatrix}$$

$$(14.26)$$

The polynomial equation in a quadratic form is stated by Sharma and Kushare [30] as:

$$A_0 s^4 + A_1 s^3 + A_2 s^2 + A_3 s + A_4 = 0 \quad (14.27)$$

Here, s = complex variable

$A_0 = 1 > 0$

$A_1 = \dfrac{1}{\bar{M}_j}\left[\bar{C}_{xx} + \bar{C}_{zz}\right] > 0$

$A_2 = \dfrac{1}{\bar{M}_j^2}\left[\bar{C}_{xx}\bar{C}_{zz} + \bar{M}_j\left(\bar{S}_{xx} + \bar{S}_{zz}\right) - \bar{C}_{xz}\bar{C}_{zx}\right] > 0 \quad (14.28)$

$A_3 = \dfrac{1}{\bar{M}_j^2}\left[\bar{S}_{xx}\bar{C}_{zz} + \bar{S}_{zz}\bar{C}_{xx} - \bar{S}_{xz}\bar{C}_{zx} - \bar{S}_{zx}\bar{C}_{xz}\right] > 0$

$A_4 = \dfrac{1}{\bar{M}_j^2}\left[\bar{S}_{xx}\bar{S}_{zz} - \bar{S}_{xz}\bar{S}_{zx}\right] > 0$

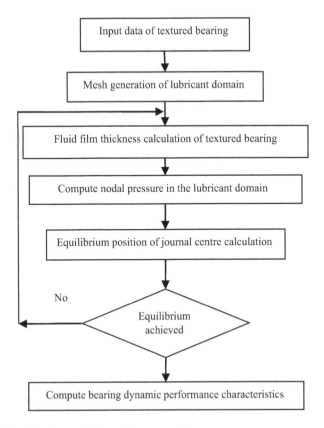

Figure 14.2 Flowchart of the solution procedure.

By using Equation (14.27) and Routh's criteria, bearing stability parameters are represented by the terms, i.e., threshold speed ($\bar{\omega}_{th}$) and critical mass ($\bar{M}_c$).

14.6.3.1 Threshold speed

Threshold speed ($\bar{\omega}_{th}$) is presented by Sharma et al. [10] as:

$$\bar{\omega}_{th} = \left[\frac{\bar{M}_c}{\bar{F}_o}\right]^{1/2} \tag{14.29}$$

where $\bar{F}_o$ = Resultant fluid film force, $\dfrac{\partial \bar{h}}{\partial \bar{t}} = 0$.

The bearing is said to be operating in the stable region when the condition, i.e., $(\bar{\omega}_j) < (\bar{\omega}_{th})$ is satisfied.

14.6.3.2 Critical mass

With the help of dynamic coefficients and Routh's Hurwitz criteria, critical mass can be defined by Sharma and Kushare [45] as:

$$\bar{M}_c = \left[\frac{\bar{G}_1}{\bar{G}_2 - \bar{G}_3} \right] \tag{14.30}$$

where

$$\bar{G}_1 = \left[\bar{C}_{xx}\bar{C}_{zz} - \bar{C}_{zx}\bar{C}_{xz} \right]$$

$$\bar{G}_2 = \frac{\left[\bar{S}_{xx}\bar{S}_{zz} - \bar{S}_{zx}\bar{S}_{xz} \right]\left[\bar{C}_{xx} + \bar{C}_{zz} \right]}{\left[\bar{S}_{xx}\bar{C}_{zz} + \bar{S}_{zz}\bar{C}_{xx} - \bar{S}_{xz}\bar{C}_{zx} + \bar{S}_{zz}\bar{C}_{xz} \right]} \tag{14.31}$$

$$\bar{G}_2 = \frac{\left[\bar{S}_{xx}\bar{C}_{xx} + \bar{S}_{xz}\bar{C}_{xz} - \bar{S}_{zx}\bar{C}_{zx} + \bar{S}_{zz}\bar{C}_{zz} \right]}{\left[\bar{C}_{xx} + \bar{C}_{zz} \right]}$$

14.6.3.3 Whirl frequency ratio

This is another important parameter to define stability. It is obtained by applying Routh's criteria and can be expressed by Chandrawat and Sinhasan [46] as:

$$\bar{\omega}^2_{whirl} = \frac{k_{eq}}{\bar{M}_c} \tag{14.32}$$

where

$$k_{eq} = \frac{\bar{S}_{xx}\bar{C}_{zz} + \bar{S}_{zz}\bar{C}_{xx} - \bar{S}_{xz}\bar{C}_{zx} + \bar{S}_{zz}\bar{C}_{xz}}{\bar{C}_{xx} + \bar{C}_{zz}}$$

Therefore,

$$\bar{\omega}^2_{whirl} = \frac{\left(\bar{S}_{xx} - k_{eq} \right)\left(\bar{S}_{zz} - k_{eq} \right) - \bar{S}_{xz}\bar{C}_{zx}}{\bar{C}_{xx}\bar{C}_{zz} - \bar{C}_{xz}\bar{C}_{zx}}$$

14.7 CONCLUSION

After review of literature, it has been observed that texturing improves the dynamic characteristics of journal bearing surfaces and thereby enhances their performance. However, the better results in terms of LCC (COF), stiffness coefficients, and dynamic coefficients, etc. are found in the partial texturing region with low to medium eccentricity ration and texture depth.

However, any further increase in the number of textures in pressure increasing and eccentricity ratio beyond the optimal values deteriorates the performance of journal bearing. The bearing performance also depends upon the textures' location and its geometry as it is observed that proper location of texture and geometry improved its performances further. The bottom profile plays an important role in increasing the bearing performance as the studies show that flat bottom surfaces significantly have an edge over the other bottom profiles. The addition of nano-additives in the lubricant was observed to be more effective in order to enhance the bearings' performance. It is also observed that the addition of TiO_2 nanoparticles even at the lower concentration improves the dynamic coefficient of bearing, including the couple stress parameters. The studies also reveal that addition of CuO nanoparticles in the standard oils shows good results by reduction in friction and anti-wear properties. The addition of nanoparticles additives like CuO, CeO_2 and Al_2O_3 in the base lubricant have shown considerable increase in LCC of journal bearing. The friction forces increase with any increase in weight fraction of nanoparticle additives in thermo-viscous and non-thermo-viscous cases. It is concluded that designers should concentrate on considering optimum texture depth, location of textures' geometry, number of textures, bottom profiles, and selection of nanoparticle additives with proper weight fraction for enhanced bearing performance, which results in improved machine efficiency and reduction in energy consumption.

NOMENCLATURE

c_r = Radial clearance in mm;

D_j = Journal diameter in mm;

e_j = Eccentricity in mm;

W_b = LCC, N;

F_X, F_z = LCC in X; Z direction, N;

F_b = Frictional Force, N;

h = Film thickness of fluid in mm;

L_b = Bearing Length, mm;

p_l = Pressure in N/mm^2;

p_{st} = Supply pressure of lubricant, N/mm^2 $(\mu_r \omega_j R_j^2 / c_r^2)$;

Q_t = Flow of lubricant, mm$^3 \cdot$ sec^{-1}

R_j, R_b = Journal radius & bearing radius, mm;

t = Time, sec;

T = Temp, °C;

W = Applied load, N;

x, y, z = Film thickness in axial and cicumfrential direction, mm;

$X_j Z_j$ = Journal coordinate;

GREEK LETTER

μ = Viscosity of lubricant with nanoparticles, Ns/m^2;

μ_0 = Viscosity of base lubricant, Ns/m^2;

ω_j = Journal rotational speed, rad/sec;

ω_{th} = Journal threshold speed, rad/sec;

λ = Aspect ratio (L_b/D_b);

ϕ_1 = Angle of attitude, rad;

ϕ = Weight concentration percentage of nano particles to base lubricant

NON-DIMENSIONAL PARAMETERS

$\bar{c}$ = c_r/R_j;

$\bar{C}_{ij}$ = Damping coefficient, $C_{ij}(c_r^3/\mu R_j^4)$;

$\bar{F}_0, \bar{F}_1, \bar{F}_2$ = Viscosity functions;

$\bar{h}$ = h/c_r;

$\bar{h}_{min}$ = h_{min}/c_r;

$\bar{p}_t, \bar{p}_c$ = $p_t/p_{st}, p_c/p_{st}$;

$\bar{p}_l, \bar{p}_{max}$ = $p_l/p_{st}, p_{max}/p_{st}$;

$\bar{t}$ = $tc_r^2 p_{st}/\mu_r R_j^2$;

$\bar{T}$ = T/T_{atm};

$\bar{F}, \bar{F}_0$ = $F/p_{st}R_j^2, F_0/p_{st}R_j^2$;

$\bar{F}_x, \bar{F}_z$ = $F_x/p_{st}R_j^2, F_z/p_{st}R_j^2$;

$\bar{S}_{ij}$ = Stiffness coefficient, $S_{ij}(c_r^3/p_sR_j^4)$;

$\bar{X}_j, \bar{Z}_j$ = $X_j/c_r, Z_j/c_r$;

$\bar{x}, \bar{z}$ = $x/c_r, z/c_r$;

α, β = $x/R_j, z/R_j$;

ε = Eccentricity ratio (e_j/c_r);

$\bar{\mu}$ = Relative viscosity, μ/μ_0

VECTORS AND MATRICES

$[\bar{F}_M]$ = Matrix of liquidity;

$[\bar{N}_M]$ = Shape function;Matrix ,

$\{\bar{p}_V\}$ = Node pressure;Vector

$\{\bar{Q}_V\}$ = Vector , Node flow;

$\{\bar{R}_X, \bar{R}_Z\}$ = Velocity of journal vectors;

$\{\bar{R}_H\}$ = Column vector hydrodynamic

REFERENCES

1. Tala-Ighil, N., Fillon, M., and Maspeyrot, P. 2011. Effect of textured area on the performances of a hydrodynamic journal bearing. *Tribology International*, 44(3), pp. 211–219.
2. Hamdavi, S., Ya, H. H., and Rao, N. 2016. Effect of surface texturing on hydrodynamic performance of journal bearing. *ARPN Journal of Engineering and Applied Sciences*, 11(1), pp. 172–176.
3. Kango, S., Singh, D., and Sharma, R. K. 2012. Numerical investigation on the influence of surface texture on the performance of hydrodynamic journal bearing. *Meccanica*, 47(2), pp. 469–482.
4. Matele, S., and Pandey, K. N. 2018. Effect of surface texturing on the dynamic characteristics of hydrodynamic journal bearing comprising concepts of green tribology. *Proceedings of the Institution of Mechanical Engineers, Part J: Journal of Engineering Tribology*, 232(11), pp. 1365–1376.

5. Kango, S., Sharma, R. K., and Pandey, R. K. 2014. Comparative analysis of textured and grooved hydrodynamic journal bearing. *Proceedings of the Institution of Mechanical Engineers, Part J: Journal of Engineering Tribology*, 228(1), pp. 82–95.
6. Manser, B., Belaidi, I., Hamrani, A., Khelladi, S., and Bakir, F. 2019. Performance of hydrodynamic journal bearing under the combined influence of textured surface and journal misalignment: A numerical survey. *Comptes Rendus Mécanique*, 347(2), pp. 141–165.
7. Singh, N., and Awasthi, R. K. 2021. Influence of texture geometries on the performance parameters of hydrodynamic journal bearing. *Proceedings of the Institution of Mechanical Engineers, Part J: Journal of Engineering Tribology*, 235(10), pp. 2056–2072.
8. Shinde, A., Pawar, P., Shaikh, P., Wangikar, S., Salunkhe, S., and Dhamgaye, V. 2018. Experimental and numerical analysis of conical shape hydrodynamic journal bearing with partial texturing. *Procedia Manufacturing*, 20, pp. 300–310.
9. Cupillard, S., Glavatskih, S., and Cervantes, M. J. 2008. Computational fluid dynamics analysis of a journal bearing with surface texturing. *Proceedings of the Institution of Mechanical Engineers, Part J: Journal of Engineering Tribology*, 222(2), pp. 97–107.
10. Sharma, S., Jamwal, G., and Awasthi, R. K. 2019. Enhancement of steady state performance of hydrodynamic journal bearing using chevron-shaped surface texture. *Proceedings of the Institution of Mechanical Engineers, Part J: Journal of Engineering Tribology*, 233(12), pp. 1833–1843.
11. Kango, S., and Sharma, R. K. 2010. Studies on the influence of surface texture on the performance of hydrodynamic journal bearing using power law model. *International Journal of Surface Science and Engineering*, 4(4–6), pp. 505–524.
12. Shinde, A., Pawar, P., Gaikwad, S., Kapurkar, R., and Parkhe, A. 2018. Numerical analysis of deterministic micro-textures on the performance of hydrodynamic journal bearing. *Materials Today: Proceedings*, 5(2), pp. 5999–6008.
13. Rasep, Z., Yazid, M. M., and Samion, S. 2021. Lubrication of textured journal bearing by using vegetable oil: A review of approaches, challenges, and opportunities. *Renewable and Sustainable Energy Reviews*, 146, p. 111191.
14. Tala-Ighil, N., Maspeyrot, P., Fillon, M., and Bounif, A. 2007. Hydrodynamic effects of texture geometries on journal bearing surfaces. In 10th International Conference on Tribology ROTRIB'07, Bucharest, Romania from 8 to 10 November 2007.
15. Shinde, A. B., and Pawar, P. M. 2017. Effect of partial grooving on the performance of hydrodynamic journal bearing. *Industrial Lubrication and Tribology*, 69(4), pp. 574–584.
16. Singh, N., and Awasthi, R. K. 2020. Influence of dimple location and depth on the performance characteristics of the hydrodynamic journal bearing system. *Proceedings of the Institution of Mechanical Engineers, Part J: Journal of Engineering Tribology*, 234(9), pp. 1500–1513.
17. Tauviqirrahman, M., Jamari, J., Wibowo, B. S., Fauzan, H. M., and Muchammad, M. 2019. Multiphase computational fluid dynamics analysis of hydrodynamic journal bearing under the combined influence of texture and slip. *Lubricants*, 7(11), p. 97.

18. Kulkarni, H. D., Rasal, A. B., Bidkar, O. H., Mali, V. H., Atkale, S. A., Wangikar, S. S., and Shinde, A. B. 2019. Fabrication of micro-textures on conical shape hydrodynamic journal bearing. *International Journal for Trends in Engineering and Technology*, 36(1), pp. 37–41.
19. Guha, S. K. 1993. Analysis of dynamic characteristics of hydrodynamic journal bearings with isotropic roughness effects. *Wear*, 167(2), pp. 173–179.
20. Parkins, D. W. 1979. Theoretical and experimental determination of the dynamic characteristics of a hydrodynamic journal bearing. *Journal of Lubrication Technology*, 101(2), pp. 129–137.
21. Feng, H., Jiang, S., and Ji, A. 2019. Investigations of the static and dynamic characteristics of water-lubricated hydrodynamic journal bearing considering turbulent, thermohydrodynamic and misaligned effects. *Tribology International*, 130, pp. 245–260.
22. Rho, B. H., and Kim, K. W. 2002. A study of the dynamic characteristics of synchronously controlled hydrodynamic journal bearings. *Tribology International*, 35(5), pp. 339–345.
23. Leung, P. S., Craighead, I. A., and Wilkinson, T. S. 1989. An analysis of the steady state and dynamic characteristics of a spherical hydrodynamic journal bearing. *Journal of Tribology*, 111(3), pp. 459–467.
24. Sheeja, D., and Prabhu, B. S. 1992. Thermal and non-Newtonian effects on the steady state and dynamic characteristics of hydrodynamic journal bearings—Theory and experiments. *Tribology Transactions*, 35(3), pp. 441–446.
25. Crosby, W. A., and Chetti, B. 2009. The static and dynamic characteristics of a two-lobe journal bearing lubricated with couple-stress fluid. *Tribology Transactions*, 52(2), pp. 262–268.
26. Xu, G., Zhou, J., Geng, H., Lu, M., Yang, L., and Yu, L. 2015. Research on the static and dynamic characteristics of misaligned journal bearing considering the turbulent and thermohydrodynamic effects. *Journal of Tribology*, 137(2), p. 024504.
27. Sun, D., Li, S., Fei, C., Ai, Y., and Liem, R. 2019. Investigation of the effect of cavitation and journal whirl on static and dynamic characteristics of journal bearing. *Journal of Mechanical Science and Technology*, 33(1), pp. 77–86.
28. Kim, J., Palazzolo, A., and Gadangi, R. 1995. Dynamic characteristics of TEHD tilt pad journal bearing simulation including multiple mode pad flexibility model. *Journal of Vibration and Acoustics*, 117(1), pp. 123–135.
29. Arif, M., Kango, S., and Shukla, D. K. 2022. Effect of slip boundary condition and non-newtonian rheology of lubricants on the dynamic characteristics of finite hydrodynamic journal bearing. *Surface Topography: Metrology and Properties*, 10(1), p. 015002.
30. Kostrzewsky, G. J., Taylor, D. V., Flack, R. D., and Barrett, L. E. 1998. Theoretical and experimental dynamic characteristics of a highly preloaded three-lobe journal bearing. *Tribology Transactions*, 41(3), pp. 392–398.
31. Prabhakaran Nair, K., Ahmed, M. S., and Al-Qahtani, S. T. 2009. Static and dynamic analysis of hydrodynamic journal bearing operating under nano lubricants. *International Journal of Nanoparticles*, 2(1–6), pp. 251–262.
32. Sadiq, M. I., Ghopa, W. A. W., Nuawi, M. Z., Rasani, M. R., and Mohd Sabri, M. A. 2022. Experimental and numerical investigation of static and dynamic characteristics of bio-oils and SAE40 in fluid film journal bearing. *Materials*, 15(10), p. 3595.

33. Jang, G. H., and Yoon, J. W. 2002. Nonlinear dynamic analysis of a hydrodynamic journal bearing considering the effect of a rotating or stationary herringbone groove. *Journal of Tribology*, 124(2), pp. 297–304.
34. Flack, R. D., Kostrzewsky, G. J., and Taylor, D. V. 1993. A hydrodynamic journal bearing test rig with dynamic measurement capabilities. *Tribology Transactions*, 36(4), pp. 497–512.
35. Wei, Y., Chen, Z., Xu, W., and Jiao, Y. 2013. Effect analysis of dimensional tolerances on the dynamic characteristics of hydrodynamic journal bearing system. *ASME International Mechanical Engineering Congress and Exposition*, 56253.
36. Mongkolwongrojn, M., and Arunmetta, P. 2002. Theoretical characteristics of hydrodynamic journal bearings lubricated with soybean-based oil. *Journal of Synthetic Lubrication*, 19(3), pp. 213–228.
37. Mandal, S. K., Bhattacharjee, B., Biswas, N., Choudhuri, K., and Chakraborti, P. 2021. Application of nanofluids on various performance characteristics of hydrodynamic journal bearing—A review. *Proceedings of the Institution of Mechanical Engineers, Part E: Journal of Process Mechanical Engineering*, 236(3), pp. 1229–1238.
38. Rho, B. H., and Kim, K. W. 2002. The effect of active control on stability characteristics of hydrodynamic journal bearings with an axial groove. *Proceedings of the Institution of Mechanical Engineers, Part C: Journal of Mechanical Engineering Science*, 216(9), pp. 939–946.
39. Chetti, B. 2010. Static and dynamic characteristics of hydrodynamic four-lobe journal bearing with couple stress lubricants. In The 7th Jordanian International Mechanical Engineering Conference (JIMEC'7), Amman-Jordan. https://www.researchgate.net/publication/267218638_Static_and_Dynamic_Analysis_of_Hydrodynamic_Four-lobe_Journal_Bearing_with_Couple_Stress_Lubricants.
40. Bouzidane, A., and Thomas, M. 2007. Equivalent stiffness and damping investigation of a hydrostatic journal bearing. *Tribology Transactions*, 50(2), pp. 257–267.
41. Wang, S., Wan, Q., and Xiong, Z. 2021. Stability analysis of hydrodynamic journal bearing with surface texture. In *2021 International Conference on Machine Learning and Intelligent Systems Engineering (MLISE)* (pp. 499–502). IEEE.
42. Awasthi, R. K., Sharma, S. C., and Jain, S. C. 2007. Performance of worn non-recessed hole-entry hybrid journal bearings. *Tribology International*, 40(5), pp. 717–734.
43. Kalakada, S. B., Kumarapillai, P. N. N., & Rajendra Kumar, R. K. 2015. Static characteristics of thermohydrodynamic journal bearing operating under lubricants containing nanoparticles. *Industrial Lubrication and Tribology*, 67(1), pp. 38–46.
44. Khatri, C. B., & Sharma, S. C. 2017. Influence of couple stress lubricant on the performance of textured two-lobe slot-entry hybrid journal bearing system. *Proceedings of the Institution of Mechanical Engineers, Part J: Journal of Engineering Tribology*, 231(3), 366–384.
45. Sharma, S. C., and Kushare, P. B. 2017. Nonlinear transient response of rough symmetric two lobe hole entry hybrid journal bearing system. *Journal of Vibration and Control*, 23(2), pp. 190–219.

46. Chandrawat, H. M., and Sinhasan, R. 1988. A study of steady state and transient performance characteristics of a flexible shell journal bearing. *Tribology International*, 21(3), pp. 137–148.
47. Manser, B., Belaidi, I., Hamrani, A., Khelladi, S., and Bakir, F. 2020. Texture shape effects on hydrodynamic journal bearing performances using mass-conserving numerical approach. *Tribology-Materials, Surfaces & Interfaces*, 14(1), pp. 33–50.
48. Qiu, M., Delic, A., and Raeymaekers, B. 2012. The effect of texture shape on the load-carrying capacity of gas-lubricated parallel slider bearings. *Tribology Letters*, 48(3), pp. 315–327.

Chapter 15

Impact of nano-lubricants on the dynamic performance of journal bearings with surface waviness

Arun Bangotra
Government Polytechnic College
Shri Mata Vaishno Devi University

Sanjay Sharma and Deepak Byotra
Shri Mata Vaishno Devi University

Rajeev Kumar Awasthi
Sardar Beant Singh State University

CONTENTS

15.1 Introduction: Background and literature	238
15.2 Lubrication governing equations	240
15.2.1 Reynolds equation in its non-dimensional form	241
15.2.2 Surface waviness's impact on lubricant film thickness	242
15.2.3 Finite element analysis	242
15.2.4 Boundary conditions	244
15.3 Mathematical viscosity model	245
15.4 Solution methodology	245
15.5 Dynamic performance parameters	247
15.5.1 Stiffness coefficients	247
15.5.2 Damping coefficients	247
15.5.3 Stability parameters	248
15.5.4 Threshold speed	249
15.5.5 Critical mass	249
15.5.6 Whirl frequency ratio	249
15.6 Conclusion	250
Nomenclature	251
Non-dimensional parameters	251
Vectors and matrices	252
References	252

DOI: 10.1201/9781003306276-15

15.1 INTRODUCTION: BACKGROUND AND LITERATURE

Hydrodynamic journal bearings are the most important components of machines that operate at very high speeds, such as motors, compressors, pumps, and turbines, because of their low friction, self-acting, high load-carrying capacity, low noise, low vibration, and superior dynamic behavior. Because of their inadequate hydrodynamic lift capacity, bearing dynamic performances and stability have been determined to be generally low. The bearing dynamic properties must be thoroughly examined for efficient hydrodynamic bearings' design with good hydrodynamic lift capacity. The prime motive for the successful operation of rotating machines is the journal's capacity to inhibit the whirl for improved stability. Furthermore, when the bearing operates at a very fast speed, heat is created in the lubricant owing to high shear rates in the lubrication layer, causing a rise in temperature and reduction in viscosity, which affect the bearing performance. Therefore, the viscosity of the lubricant can be enhanced by adding nanoparticles, which can improve the journal bearing's performance. Thus, thermohydrodynamic analysis may be used to determine how well a bearing performs while using nanoparticles as lubricants.

In addition, despite the high-precision machines that are now accessible, manufacturing flaws are unavoidable, leading to bearing surface defects, which can significantly affect the bearing's performance. Numerous studies on the impact of such flaws on the static, dynamic, and anti-whirl properties of bearings have been done since it was discovered that journal and bearing surface flaws influence bearing's performance.

In addition, adding waviness to the bearing surface is rather simple and increases the bearing's performance and stability when compared to other bearing shapes like non-circular geometries, multi-lobes, grooves, steps, etc. On the inner surface of the bearing, several wave profiles, like sinusoidal waves, can be provided. Wave bearings provide higher pressure while providing more stability as compared to plain bearings functioning under the same conditions. For evaluation of the dynamic characteristics of bearings, the linear perturbation method and non-linear method are used. The linear method calculates the fluid film forces with the help of dynamic coefficients. In the non-linear method, the Reynolds equation, which is time-dependent, is solved at every location of the journal center. The forces are determined from the pressure distribution at each instant of time for the location of the journal. Many papers are published in the literature to examine the journal bearing's stability by using both the methods.

Parkins [1] estimated dynamic coefficients by differentiation of Reynolds equation concerning minor disturbances of position and velocities of the journal center. This process takes a long time, and the outcomes are highly dependent on the perturbation amplitudes. Lund [2,3] employed an analytical differentiation of the same equation before doing a numerical computation to calculate the dynamic coefficients of bearings by using linear mathematical models. This method is very effective, and the results are extremely reliant on the amplitudes of the perturbations. Majumdar and Brewe [4] investigated the non-linear transient behavior of oil journal bearings supporting the rotor under constant and varying loads. The journal center trajectories estimated using non-linear and linear theories were compared by Tieu and Qiu [5]. The critical speed was the same for both approaches. The trajectories generated with the two approaches, however, are markedly different under large dynamic excitation.

In his analysis of three-wave bearings operating in isothermal conditions with compressible lubricants, Dimofte [6,7] found that wave journal bearings offer superior stability and, because of increased pressure generation, can support more load than plain journal bearings. Rasheed [8] evaluated the performance of the wave journal bearing with waviness in the circumferential direction and axial direction. He found that as wave amplitude increases, the load-carrying capacity increases at any waviness number in any direction. According to Wang et al. [9,10], surface waviness influences the bearing's dynamic properties, and dynamic stiffness and damping rise with wave amplitude. Yang et al. [11] observed the dynamic behavior of the bearing with waviness using the finite difference method and revealed that waviness influences the bearing stability. Gautam et al. [12] proved that when compared to circular bearings, wave bearings have better load-carrying capability and provide improved stability at high speeds. Li et al. [13] reported that the waviness effect in bearings can significantly increase the system's stability in circumferential direction, but the waviness in axial direction harms the performance. A thrust bearing with waviness was investigated by Zhao et al. [14,15] and reported the dynamic characteristics and stability. Zhuang et al. [16] demonstrated experimentally and numerically that aerostatic thrust bearing performance is influenced by the presence of concave and convex waviness, with the effect becoming more apparent as the wave amplitude increases. Yang et al. [17–19] numerically proved that bearings with surface waviness have better stiffness and stability in comparison with the rigid circular bearing at the same working parameters. In his analysis of the effects of three-dimensionally formed irregularities on bearings, Lin [20,21] noted that bearing roughness affects the pressure distribution, as it alters the wedge action in fluid flow.

It is clear from the literature review that surface waviness has a significant impact on a bearing's performance. Several other studies looked at how journal bearing performance was affected by different surface morphologies of the bearing and journal. Sharma and Kushare [22]; Jamwal et al. [23]; Sharma et al. [24]; and Khatri and Sharma [25] discovered that bearing stability can be enhanced by providing circular shaped surface texturing. In addition, there were [26–34] several geometrical flaws, including non-circularity of the bearings, different journal shapes, lobe bearings with wear, and misalignment affecting the performance of a bearing.

Lubricant is inserted between the surfaces of bearings and journals for smooth relative motion between them and reduces wear and friction. The oil film's rigidity and damping qualities have a big impact on a rotor's critical speeds and stability. The addition of the nanoparticles can increase the lubricant's viscosity. Lee et al. [35] reported a reduction in wear and friction due to the use of nano-lubricant. Furthermore, literature [36–44] established that nanoparticles like nanodiamonds, CuO, and TiO$_2$ to engine oil improve load capacity and stability parameters of the system over the base oil. According to Bangotra and Sharma [45], the action of nano-lubricants and surface waviness improved the performance of journal bearings. Over the years, several tribologists have created different mathematical models that use experimental data to compute the viscosity of lubricants including nanoparticles. Based on the experimental findings, Kalakada et al. [46,47] developed a mathematical model that will be utilized to determine the viscosity of the nano-lubricant. According to studies, adding nanoparticles can make the lubricant more viscous, which enhances the bearing's dynamic performance.

15.2 LUBRICATION GOVERNING EQUATIONS

Figure 15.1 depicts the configuration of the bearing with circumferential and axial waviness along with the various co-ordinates.

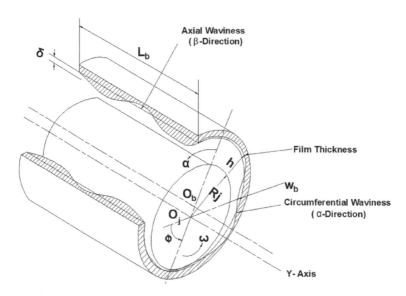

Figure 15.1 Diagram of a wave journal bearing.

Considering the Newtonian lubricant and the temperature regime in lubricants as isothermal, the two-dimensional Reynolds equation for laminar, incompressible flow, and iso-viscous Newtonian lubricant is given by Awasti et al. [26] and Tala-Ighil and Fillon [31] as follows:

$$\frac{\partial}{\partial x}\left(F_2 \frac{\partial p_l}{\partial x}\right) + \frac{\partial}{\partial y}\left(F_2 \frac{\partial p_l}{\partial y}\right) = U\left[\frac{\partial}{\partial x}\left\{h - \frac{F_1}{F_0}\right\}\right] + \frac{\partial h}{\partial t} \qquad (15.1)$$

Here, F_0, F_1 and F_2 are functions related to viscosity and are defined as:

$$F_0 = \int_0^h \frac{dz}{\mu} \qquad F_1 = \int_0^h z\frac{dz}{\mu}$$

$$F_2 = \left(\int_0^h \left(\frac{z^2}{\mu}dz - \frac{F_1}{F_0}\frac{z\,dz}{\mu}\right)\right)$$

15.2.1 Reynolds equation in its non-dimensional form

By substituting the various non-dimensional parameters in the Reynolds equation, the desired equation as given by Awasthi et al. [26] is:

$$\frac{\partial}{\partial \alpha}\left(\bar{h}^3 \bar{F}_2 \frac{\partial \bar{p}_l}{\partial \alpha}\right) + \frac{\partial}{\partial \beta}\left(\bar{h}^3 \bar{F}_2 \frac{\partial \bar{p}_l}{\partial \beta}\right) = \bar{\Omega}\left[\frac{\partial}{\partial \alpha}\left\{\left(1 - \frac{\bar{F}_1}{\bar{F}_0}\right)\bar{h}\right\}\right] + \frac{\partial \bar{h}}{\partial \bar{t}} \qquad (15.2)$$

Here $\bar{F}_0$, $\bar{F}_1$ and $\bar{F}_2$ are defined as:

$$\bar{F}_0 = \int_0^1 \frac{1}{\bar{\mu}}d\bar{z}$$

$$\bar{F}_1 = \int_0^1 \frac{\bar{z}}{\bar{\mu}}d\bar{z}$$

$$\bar{F}_2 = \int_0^1 \frac{\bar{z}}{\bar{\mu}}\left(\bar{z} - \frac{\bar{F}_1}{\bar{F}_0}\right)d\bar{z}$$

15.2.2 Surface waviness's impact on lubricant film thickness

Rasheed [8] gave the following mathematical expression for film thickness due to the influence of surface waviness, as illustrated in Figure 15.1:

$$\overline{h} = h_o + \Delta \overline{h} + \Delta \overline{h}_w \qquad (15.3)$$

$$\overline{h} = (1 - \overline{X}_j \cos\alpha - \overline{Z}_j \sin\alpha) + \Delta \overline{h} + \Delta \overline{h}_w \qquad (15.4)$$

where $\overline{X}_j$ and $\overline{Z}_j$ represent the co-ordinates of journal center under equilibrium conditions and $\Delta \overline{h}$ depicts the fluctuation in the film thickness because of dynamic effect by Sharma [27] as:

$$\Delta \overline{h} = -\overline{x} \cos\alpha - \overline{z} \sin\alpha \qquad (15.5)$$

where $\overline{x}$ and $\overline{z}$ are the journal center movements from its steady-state equilibrium condition and $\Delta \overline{h}_w$ is the change in the fluid film thickness due to waviness and is determined by the following equation with sinusoidal wave by Rasheed [8] as:

$$\Delta \overline{h}_w = \frac{\delta}{2}\left[i_1 \cos(n\alpha) + j_1 \cos(2\pi m \beta)\right] \qquad (15.6)$$

here n and m describe the number of waves in circumferential and axial directions, while i_1 and j_1 depict the direction of wave and δ is the wave amplitude. For waviness in circumferential direction, $i_1=2, j_1=0$, for axial waviness, $i_1=0, j_1=2$ and $i_1=1, j_1=1$ for waviness in both α and β directions.

15.2.3 Finite element analysis

By discretizing the lubricant flow region and combining the contributions of each element over the flow region, finite element analysis is used to find the mathematical solution to the Reynolds equation. Numerical-based research investigations are carried out to determine different hydrodynamic bearing parameters, and the strong MATLAB tool is employed for this.

Figure 15.2 shows a four-noded element used to discretize the lubricant regime in two-dimensional flow, the pressure in the element is expressed by Awasthi et al. [26] as:

$$\overline{p}_l = \sum_{j=1}^{4} N_j \overline{P}_j \qquad (15.7)$$

$$\alpha = \sum_{j=1}^{n} N_j \alpha_j \quad \text{and} \quad \beta = \sum_{j=1}^{n} N_j \beta_j$$

Nano-lubricant impact on journal bearings 243

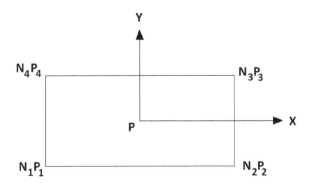

Figure 15.2 Pressure distribution in a four-noded element.

$$\frac{\partial \bar{p}_l}{\partial \alpha} = \sum_{j=1}^{n} \frac{\partial N_j}{\partial \alpha} \bar{p}_j \quad \text{and} \quad \frac{\partial \bar{p}_l}{\partial \beta} = \sum_{j=1}^{n} \frac{\partial N_j}{\partial \beta} \bar{p}_j \quad (15.8)$$

where N_j=the shape function of the element, while n=number of nodes in each element.

Substituting pressure from Equation 15.8, Equation 15.2 is written as [26]:

$$\frac{\partial}{\partial \alpha}\left[\bar{h}^3 \bar{F}_2 \frac{\partial}{\partial \alpha}\left(\sum_{j=1}^{4} N_j \bar{P}_j\right)\right] + \frac{\partial}{\partial \beta}\left[\bar{h}^3 \bar{F}_2 \frac{\partial}{\partial \beta}\left(\sum_{j=1}^{4} N_j \bar{P}_j\right)\right]$$
$$- \bar{\Omega}\left[\frac{\partial}{\partial \alpha}\left\{\left(1-\frac{\bar{F}_1}{\bar{F}_0}\right)\bar{h}\right\}\right] - \frac{\partial \bar{h}}{\partial \bar{t}} = R^e \quad (15.9)$$

where R^e is known as residue, which is minimized by using Galerkin's technique [26]:

$$\iint_{\Omega^e} N_i R^e d\alpha\, d\beta = 0 \quad (15.10)$$

By integrating the second-order terms of Equation 15.9 part-wise, the global system of equation in matrix form by Uddin and Liu [32] is:

$$\left[\bar{F}_M\right]^e \{\bar{p}_V\}^e = \{\bar{Q}_V\}^e + \Omega\{\bar{R}_H\}^e + \dot{\bar{x}}_j \{\bar{R}_{Xj}\}^e + \dot{\bar{z}}_j \{\bar{R}_{Zj}\}^e \quad (15.11)$$

In Equation 15.11, the various coefficients of matrices for the *e*th element can be defined as:

$$\bar{F}_{Mij}^e = \iint_{A\ e} \left\{ \frac{\bar{h}^3}{12\bar{\mu}} \left(\frac{\partial N_{Vi}}{\partial \alpha} \frac{\partial N_{Vj}}{\partial \alpha} + \frac{\partial N_{Vi}}{\partial \beta} \frac{\partial N_{Vj}}{\partial \beta} \right) \right\} d\alpha\, d\beta$$

$$\bar{Q}_{Vi}^e = \int_{\Gamma^e} \left\{ \left(\frac{\bar{h}^3}{12\bar{\mu}} \frac{\partial \bar{p}_V}{\partial \alpha} - \frac{1}{2} \bar{\Omega}\bar{h} \right) n_1 + \left(\frac{\bar{h}^3}{12\bar{\mu}} \frac{\partial \bar{p}_V}{\partial \beta} \right) n_2 \right\} N_{Vi}\, d\bar{\Gamma}^e$$

$$\bar{R}_{Hi}^e = \iint_{A\ e} \frac{1}{2} \bar{h}\, \frac{\partial N_{Vi}}{\partial \alpha}\, d\alpha d\beta$$

$$\bar{R}_{Xji}^e = \iint_{A\ e} \cos\alpha\, N_{Vi}\, d\alpha d\beta$$

$$\bar{R}_{Xzi}^e = \iint_{A\ e} \sin\alpha\, N_{Vi}\, d\alpha d\beta$$

n_1 and n_2 stand for the directional cosines in each direction.

15.2.4 Boundary conditions

To solve Equation 15.2, various conditions used by Khatri and Sharma [25] are:

- Pressure is considered to be zero at all nodes along the exterior margins of the bearing boundary: $\bar{p}_l \big|_{\beta=\pm 1} = 0.0$
- By taking the pressure gradient as zero at the trailing ends of the positive regions, the boundary condition considers the impact of cavitation: $\bar{p}_l = \frac{\partial \bar{p}_t}{\partial \alpha} = 0.0$

At each node, the following criteria are verified after obtaining $\{\bar{P}_0\}$ and $\{\bar{Q}\}$ using 15.11:

- If the nodes are in the cavitation zone, $\{\bar{Q}\} < 0$; if they are in the lubricating flow zone, the value is $\{\bar{P}_0\} > 0$.

The aforementioned conditions must be satisfied for solutions to exist; otherwise, nodes that do not meet them will be shifted to new regions, and computations will continue until a solution is found. Until the pressure difference at the node of the subsequent computation drops below the tolerance level of 0.1%, the calculation is repeated. The calculation procedure is terminated when the following convergence condition is met.

$$\bar{Z}_0 = \left| \frac{\left[p_t^{-m'} \right] - \left[p_t^{-(m'-1)} \right]}{\left[p_t^{-(m'-1)} \right]} \right| \times 100 \leq \text{Tolerance Limit} \qquad (15.12)$$

15.3 MATHEMATICAL VISCOSITY MODEL

By applying the experimental data, the non-dimensional regression viscosity model developed by Kalakada et al. [46,47] is:

$$\bar{\mu} = \frac{\mu}{\mu_0} = e^{(K_1 - K_2 \bar{T})} \qquad (15.13)$$

where $\bar{T} = \dfrac{T}{T_{atm}}$.

Here, $\bar{\mu}$ represents the relative viscosity, and $\bar{T}$ is the temperature of the nanoparticles. The base lubricant is taken as Society of Automotive Engineers (SAE) 15W40 engine oil with nanoparticles CuO, CeO_2, and Al_2O_3 at percentage weight fraction (ϕ) of 0.0%–0.5% at a temperature range of 30°C–90°C.

For CuO; $\quad K_1 = 1.142 - 1.126\phi + 5.120\ \phi^2 - 6.354\ \phi^3 \qquad (15.14)$

$K_2 = 1.163 - 1.041\ \phi + 4.218\ \phi^2 - 5.080\ \phi^3$

For CeO_2; $\quad K_1 = 1.142 - 0.311\phi + 1.377\ \phi^2 - 2.113\ \phi^3 \qquad (15.15)$

$K_2 = 1.163 - 0.431\ \phi + 1.521\ \phi^2 - 2.060\ \phi^3$

For Al_2O_3; $\quad K_1 = 1.142 - 0.311\phi + 1.377\ \phi^2 - 2.113\ \phi^3 \qquad (15.16)$

$K_2 = 1.163 - 0.431\ \phi + 1.521\ \phi^2 - 2.060\ \phi^3$

The relative viscosity, $\bar{\mu}$, of nano-lubricants calculated from the above equations enhances when the weight percentage of nanoparticles, ϕ, at any temperature increases. The relative viscosity calculated from the viscosity model is substituted in Equation (15.2), to calculate the pressure and the various dynamic performance parameters of the bearing.

15.4 SOLUTION METHODOLOGY

The dynamic performance parameters can be estimated by using the finite element analysis, boundary conditions, and a viscosity model with changing viscosity, as described in Section 15.3. The dynamic performance parameters and stability of bearings operating with nano-lubricants are calculated using the Reynolds equation and the boundary conditions with the help of finite element analysis and variable viscosity from the model, as mentioned in Section 15.3. The viscosity field changes with the temperature (thermo-viscous) and is calculated at any temperature and percentage concentration of CuO, CeO$_2$, and Al$_2$O$_3$ using the viscosity model. The calculated viscosity from the model is put in the Reynolds equation to determine the pressure distribution. The process is continued till the criteria as mentioned in Equation (15.12) are satisfied. The various dynamic performance parameters, such as stiffness and damping coefficients, whirl frequency ratio, and the threshold speed, can be calculated at various eccentricity ratios. The complete solution methodology and procedure is shown in Figure 15.3.

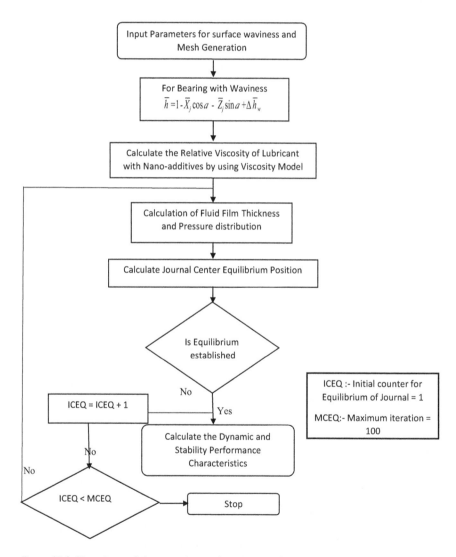

Figure 15.3 Flowchart of the complete solution procedure.

Before moving on, the mesh density must be fixed to get correct results. Any dynamic parameter says direct fluid film stiffness coefficient can be calculated using various combinations of mesh densities. The direct fluid film stiffness coefficient rises when the mesh density is increased, but the calculation time also goes up. So, the optimum value of mesh density shall be selected before proceeding to calculate the various dynamic performance parameters as discussed below.

15.5 DYNAMIC PERFORMANCE PARAMETERS

In hydrodynamic journal bearing, a thin film of lubricant supports the radial load and offers dampening because of lubricant film squeeze. During oscillation, the journal center position is defined by two degrees of freedom and we can determine four fluid film stiffness and damping coefficients, which predict the stability of the bearing. These coefficients are used to calculate the stability characteristics, i.e., threshold speed, whirl frequency ratio, and critical mass.

15.5.1 Stiffness coefficients

These are defined by Khatri and Sharma [25] as:

$$\bar{S}_{ij} = -\frac{\partial \bar{F}_i}{\partial \bar{q}_j}, (i=x,z) \tag{15.17}$$

where i represents the direction of the force, while $\bar{q}_j$ represents the position of journal center, i.e. $\bar{X}_j, \bar{Z}_j$.

The coefficients in matrix form are represented as:

$$\begin{bmatrix} \bar{S}_{xx} & \bar{S}_{xz} \\ \bar{S}_{zx} & \bar{S}_{zz} \end{bmatrix} = - \begin{bmatrix} \frac{\partial \bar{F}_x}{\partial \bar{x}} & \frac{\partial \bar{F}_x}{\partial \bar{z}} \\ \frac{\partial \bar{F}_z}{\partial \bar{x}} & \frac{\partial \bar{F}_z}{\partial \bar{z}} \end{bmatrix} \tag{15.18}$$

15.5.2 Damping coefficients

These are defined by Khatri and Sharma [25] as:

$$\bar{C}_{ij} = -\frac{\partial \bar{F}_i}{\partial \dot{\bar{q}}_j}, (i=x,z) \tag{15.19}$$

where $\dot{\bar{q}}_j$ defines the velocity component of the journal centre, i.e., $\dot{\bar{X}}_j, \dot{\bar{Z}}_j$.

The damping coefficient in matrix is represented as under:

$$\begin{bmatrix} \bar{C}_{xx} & \bar{C}_{xz} \\ \bar{C}_{zx} & \bar{C}_{zz} \end{bmatrix} = -\begin{bmatrix} \dfrac{\partial \bar{F}_x}{\partial \dot{\bar{x}}} & \dfrac{\partial \bar{F}_x}{\partial \dot{\bar{z}}} \\ \dfrac{\partial \bar{F}_z}{\partial \dot{\bar{x}}} & \dfrac{\partial \bar{F}_z}{\partial \dot{\bar{z}}} \end{bmatrix} \quad (15.20)$$

15.5.3 Stability parameters

The linear equation of motion when the journal is disturbed is defined by Sharma and Kushare [22] and Sinhasan and Goyal [30]:

$$[\bar{M}_j]\{\ddot{\bar{X}}_j\} + [\bar{C}]\{\dot{\bar{X}}_j\} + [\bar{S}]\{\bar{X}_j\} = 0 \quad (15.21)$$

Equation 15.21 as stated above for journal centre motion trajectories in matrix form is written by numerically integrating the linear motion equations:

$$\begin{bmatrix} \bar{M}_j & 0 \\ 0 & \bar{M}_j \end{bmatrix}\begin{Bmatrix} \ddot{\bar{X}}_j \\ \ddot{\bar{Z}}_j \end{Bmatrix} + \begin{bmatrix} \bar{C}_{xx} & \bar{C}_{xz} \\ \bar{C}_{zx} & \bar{C}_{zz} \end{bmatrix}\begin{Bmatrix} \dot{\bar{X}}_j \\ \dot{\bar{Z}}_j \end{Bmatrix} + \begin{bmatrix} \bar{S}_{xx} & \bar{S}_{xz} \\ \bar{S}_{zx} & \bar{S}_{zz} \end{bmatrix}\begin{Bmatrix} \bar{X}_j \\ \bar{Z}_j \end{Bmatrix} = \begin{Bmatrix} 0 \\ 0 \end{Bmatrix}$$

$$(15.22)$$

The polynomial equation in a quadratic form is stated by Sharma and Kushare [22] as:

$$A_0 s^4 + A_1 s^3 + A_2 s^2 + A_3 s + A_4 = 0 \quad (15.23)$$

Here, s = complex variable

$A_0 = 1 > 0$

$A_1 = \dfrac{1}{\bar{M}_j}\left[\bar{C}_{xx} + \bar{C}_{zz}\right] > 0$

$A_2 = \dfrac{1}{\bar{M}_j^2}\left[\bar{C}_{xx}\bar{C}_{zz} + \bar{M}_j\left(\bar{S}_{xx} + \bar{S}_{zz}\right) - \bar{C}_{xz}\bar{C}_{zx}\right] > 0$

$A_3 = \dfrac{1}{\bar{M}_j^2}\left[\bar{S}_{xx}\bar{C}_{zz} + \bar{S}_{zz}\bar{C}_{xx} - \bar{S}_{xz}\bar{C}_{zx} - \bar{S}_{zx}\bar{C}_{xz}\right] > 0$

$A_4 = \dfrac{1}{\bar{M}_j^2}\left[\bar{S}_{xx}\bar{S}_{zz} - \bar{S}_{xz}\bar{S}_{zx}\right] > 0$

By using Equation 15.23 and Routh's criteria, the stability parameters of the bearing are represented by the terms threshold speed ($\bar{\omega}_{th}$) and critical mass ($\bar{M}_c$).

15.5.4 Threshold speed

Threshold speed ($\bar{\omega}_{th}$) is defined by the following expression [22, 24]:

$$\bar{\omega}_{th} = \left[\frac{\bar{M}_c}{\bar{F}_o}\right]^{1/2} \tag{15.24}$$

where $\bar{F}_o$ = Resultant fluid film force, $\dfrac{\partial \bar{h}}{\partial \bar{t}} = 0$.

The bearing is said to be operating in the stable region when the condition, i.e., $(\bar{\omega}_j) < (\bar{\omega}_{th})$ is satisfied.

15.5.5 Critical mass

With the help of dynamic coefficients and Routh's Hurwitz criteria, critical mass can be defined by Sharma and Kushare [22] as:

$$\bar{M}_c = \left[\frac{\bar{G}_1}{\bar{G}_2 - \bar{G}_3}\right] \tag{15.25}$$

where

$$\bar{G}_1 = \left[\bar{C}_{xx}\bar{C}_{zz} - \bar{C}_{zx}\bar{C}_{xz}\right]$$

$$\bar{G}_2 = \frac{\left[\bar{S}_{xx}\bar{S}_{zz} - \bar{S}_{zx}\bar{S}_{xz}\right]\left[\bar{C}_{xx} + \bar{C}_{zz}\right]}{\left[\bar{S}_{xx}\bar{C}_{zz} + \bar{S}_{zz}\bar{C}_{xx} - \bar{S}_{xz}\bar{C}_{zx} + \bar{S}_{zz}\bar{C}_{xz}\right]}$$

$$\bar{G}_2 = \frac{\left[\bar{S}_{xx}\bar{C}_{xx} + \bar{S}_{xz}\bar{C}_{xz} - \bar{S}_{zx}\bar{C}_{zx} + \bar{S}_{zz}\bar{C}_{zz}\right]}{\left[\bar{C}_{xx} + \bar{C}_{zz}\right]}$$

15.5.6 Whirl frequency ratio

This is another important parameter to define stability. It is obtained by applying Routh's criteria and can be expressed [28] as:

$$\bar{\omega}^2_{whirl} = \frac{k_{eq}}{\bar{M}_c} \tag{15.26}$$

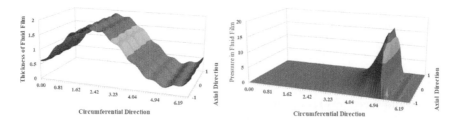

Figure 15.4 Fluid film thickness profile and pressure profile for bearings with combined surface waviness in both the directions with n=9, m=2 with 0.5% concentration of CuO at 90°C in lubricant at an eccentricity ratio of 0.8.

where

$$k_{eq} = \frac{\bar{S}_{xx}\bar{C}_{zz} + \bar{S}_{zz}\bar{C}_{xx} - \bar{S}_{xz}\bar{C}_{zx} + \bar{S}_{zz}\bar{C}_{xz}}{\bar{C}_{xx} + \bar{C}_{zz}}$$

Therefore,

$$\bar{\omega}^2_{whirl} = \frac{(\bar{S}_{xx} - k_{eq})(\bar{S}_{zz} - k_{eq}) - \bar{S}_{xz}\bar{C}_{zx}}{\bar{C}_{xx}\bar{C}_{zz} - \bar{C}_{xz}\bar{C}_{zx}}$$

Figure 15.4 displays the pressure and thickness profile of fluid film for the wave bearing and running with nano-lubricants.

15.6 CONCLUSION

The viscosity of the lubricant is increased by the addition of nanoparticles like CuO, Al_2O_3, and CeO_2. CuO nanoparticles tend to have a greater viscosity than CeO_2 and Al_2O_3 when added to the lubricant at the same weight fraction and temperature. The load capacity and friction force are higher at high eccentricity ratios with a large weight fraction of nanoparticles, 0.5% in the lubricant at high temperature, than when nanoparticles are not present in the lubricant. Also, there will be a reduction in coefficient of friction and attitude angle by using the nanoparticles at a high concentration. With the inclusion of nanoparticles, the threshold speed rise at any eccentricity ratio, but the damped frequency drops. This means that when the hydrodynamic bearing is operated with nanoparticle-containing lubricants, the stability is better than when the bearing is operated without nanoparticles. Furthermore, the wave bearing performance depends upon the wave amplitude and the number of waves in any direction. The stiffness coefficients and dynamic coefficients also increase. Moreover, the wave bearing operating with nanoparticles has better stability compared to plain bearing. Moreover, in the case of unstable fluid film, the wave bearing has a tendency that the orbit of the whirl is within the clearance space. The stability of the system can also be assessed from the locus of a journal, by plotting trajectories using the equation of motion, which presents the dynamic behaviour of the bearing system in depth.

NOMENCLATURE

h	= thickness of fluid film, mm;
h_w	= variation in film thickness due to surface waviness, mm;
L_b	= Bearing Length, mm;
p_l	= fluid film Pressure, N/mm^2;
Q_t	= Flow of lubricant, mm$^3 \cdot$ sec^{-1};
t	= time, sec;
x, y, z	= Bearing Co-ordinates, mm;
$X_j\ Z_j$	= Journal center coordinates, mm;
δ	= Wave Amplitude, mm;
μ	= Viscosity of nano-lubricant, Ns/m^2;
μ_0	= Viscosity of base lubricant, Ns/m^2;
ω_j	= Journal rotational speed, rad/sec;
ω_{th}	= Journal threshold speed

NON-DIMENSIONAL PARAMETERS

$\bar{c}$	= c_r/R_j;
$\bar{C}_{ij}$	= damping coefficient, $C_{ij}\ (c_r^3/\mu R_j^4)$;
$\bar{h}$	= h/c_r;
$\bar{h}_{min}$	= h_{min}/c_r;
$\bar{p}_t, \bar{p}_c$	= p_t/p_{st}, p_c/p_{st};
$\bar{p}_l, \bar{p}_{max}$	= p_l/p_{st}, p_{max}/p_{st};
$\bar{S}_{ij}$	= stiffness coefficient, $S_{ij}\ (c_r^3/p_s R_j^4)$;
α, β	= x/R_j, z/R_j;
$\Delta \bar{h}_w$	= h_w/c_r;
ε	= Eccentricity ratio (e_j/c_r);
$\bar{\mu}$	= Relative viscosity, μ/μ_0

VECTORS AND MATRICES

$\left[\bar{F}_M\right]$ = Matrix due to fluid liquidity;

$\left[\bar{N}_M\right]$ = Shape function Matrix;

$\{\bar{p}_V\}$ = Node pressure Vector;

$\{\bar{Q}_V\}$ = Node Flow Vector;

$\{\bar{R}_X, \bar{R}_Z\}$ = Journal Velocity Vectors;

$\{\bar{R}_H\}$ = Hydrodynamic Column vector

REFERENCES

1. Parkins DW. Theoretical and experimental determination of the dynamic characteristics of a hydrodynamic journal bearing. *Journal of Lubrication Technology.* 1979;101(2):129–137.
2. Lund JW. Review of the concept of dynamic coefficients for fluid film journal bearings. *ASME Transactions Journal of Tribology.* 1987;109(1):37–41.
3. Lund JW. Linear transient response of a flexible rotor supported in gas-lubricated bearings. *ASME Journal of Lubrication Technology.* 1976;98(1):57–65.
4. Majumdar BC, Brewe DE. Stability of a rotor supported on oil film journal bearing under dynamic load, NASA TM-102309, 1987 (AVSCOM TR 87-C-26).
5. Tieu AK, Qiu ZL. Stability of finite journal bearings—From linear and nonlinear bearing forces. *STLE Tribology Transactions.* 1995;38(3):627–635.
6. Dimofte F. Wave journal bearing with compressible lubricant—Part I: The wave bearing concept and a comparison to the plain circular bearing. *Tribology Transactions.* 1995;38(1):153–160.
7. Dimofte F. Wave journal bearing with compressible lubricant—Part II: A comparison of the wave bearing with a wave-groove bearing and a lobe bearing. *Tribology Transactions.* 1995;38(2):364–372.
8. Rasheed HE. Effect of surface waviness on the hydrodynamic lubrication of a plain cylindrical sliding element bearing. *Wear.* 1998;223(1–2):1–6.
9. Wang X, Xu Q, Huang M, Zhang L, Peng Z. Effects of journal rotation and surface waviness on the dynamic performance of aerostatic journal bearings. *Tribology International.* 2017;112:1–9.
10. Wang X, Xu Q, Wang B, Zhang L, Yang H, Peng Z. Effect of surface waviness on the static performance of aerostatic journal bearings. *Tribology International.* 2016;103:394–405.
11. Yang J, Zhu R, Yue Y, Dai G, Yin X. Nonlinear analysis of herringbone gear rotor system based on the surface waviness excitation of journal bearing. *Journal of the Brazilian Society of Mechanical Sciences and Engineering.* 2022;44(2):1–8.
12. Gautam SS, Meena L, Ghosh MK. Dynamic characteristics and stability of short wave journal bearings. *Tribology Online.* 2010;5(2):92–95.
13. Li J, Yang S, Li X, Li Q. Effects of surface waviness on the nonlinear vibration of gas lubricated bearing-rotor system. *Shock and Vibration.* 2018; 2018:16.

14. Zhao H, Choy FK, Braun MJ. Dynamic characteristics and stability analysis of a wavy thrust bearing. *Tribology Transactions*. 2005;48(1):133–139.
15. Zhao H, Choy FK, Braun MJ. Transient and steady state vibration analysis of a wavy thrust bearing. *Journal of Tribology*. 2006;128(1):139–145.
16. Zhuang H, Ding J, Chen P, Chang Y, Zeng X, Yang H, Liu X, Wei W. Effect of surface waviness on the performances of an aerostatic thrust bearing with orifice-type restrictor. *International Journal of Precision Engineering and Manufacturing*. 2021;22(10):1735–1759.
17. Yang B, Geng H, Zhou J, Yu L, Qi S, Liu X. Study on load capacity and dynamic characteristics of the wave bearing at infinite compressibility number. In *Turbo Expo: Power for Land, Sea, and Air* (Vol. 45776, p. V07BT32A014). American Society of Mechanical Engineers, Proceedings of ASME Turbo Expo 2014: Turbine Technical Conference and Exposition GT2014 June 16–20, 2014, Düsseldorf.
18. Yang B, Zhang J, Feng S, Zhao J, Geng H, Zhou J, Yu L. Parameter study on dynamic characteristics of wave journal bearings. In *2018 IEEE International Conference on Mechatronics and Automation (ICMA)* (pp. 220–226). IEEE, Changchun, 2018.
19. Yang B, Feng S, Tian J, Yu L. An investigation on the stability performance of wave journal bearing rotor system with geometry parameters. In *2019 IEEE International Conference on Mechatronics and Automation (ICMA)* (pp. 1236–1241). IEEE, Tianjin, 2019.
20. Tsann-Rong L. Steady state performance of finite hydrodynamic journal bearing with three-dimensional irregularities. *Wear*. 1994;176(1):95–102.
21. Lin TR. Hydrodynamic lubrication of journal bearings including micropolar lubricants and three-dimensional irregularities. *Wear*. 1996;192(1–2):21–28.
22. Sharma SC, Kushare PB. Nonlinear transient response of rough symmetric two lobe hole entry hybrid journal bearing system. *Journal of Vibration and Control*. 2017;23(2):190–219.
23. Jamwal G, Sharma S, Awasthi RK. The dynamic performance analysis of chevron shape textured hydrodynamic bearings. *Industrial Lubrication and Tribology*. 2019;72(1):1–8.
24. Sharma S, Jamwal G, Awasthi RK. Dynamic and stability performance improvement of the hydrodynamic bearing by using triangular-shaped textures. *Proceedings of the Institution of Mechanical Engineers, Part J: Journal of Engineering Tribology*. 2020;234(9):1436–1451.
25. Khatri CB, Sharma SC. Influence of couple stress lubricant on the performance of textured two-lobe slot-entry hybrid journal bearing system. *Proceedings of the Institution of Mechanical Engineers, Part J: Journal of Engineering Tribology*. 2017;231(3):366–384.
26. Awasthi RK, Sharma SC, Jain SC. Performance of worn non-recessed hole-entry hybrid journal bearings. *Tribology International*. 2007;40(5):717–734.
27. Sharma SC. Tribology in machine components. In Pradeep L. Menezes, Michael Nosonovsky, Sudeep P. Ingole, Satish V. Kailas, Michael R. Lovell (eds.), *Tribology for Scientists and Engineers* (pp. 821–879). Springer, New York, NY, 2013.
28. Chandrawat HM, Sinhasan R. A study of steady state and transient performance characteristics of a flexible shell journal bearing. *Tribology International*. 1988;21(3):137–148.

29. Jain D, Sharma SC. Dynamic analysis of a 2-lobe geometrically imperfect journal bearing system. *Proceedings of the Institution of Mechanical Engineers, Part J: Journal of Engineering Tribology*. 2017;231(7):934–950.
30. Sinhasan R, Goyal KC. Transient response of a two-lobe journal bearing lubricated with non-Newtonian lubricant. *Tribology International*. 1995;28(4):233–239.
31. Tala-Ighil N, Fillon M. A numerical investigation of both thermal and texturing surface effects on the journal bearings static characteristics. *Tribology International*. 2015;90:228–239.
32. Uddin MS, Liu YW. Design and optimization of a new geometric texture shape for the enhancement of hydrodynamic lubrication performance of parallel slider surfaces. *Biosurface and Biotribology*. 2016;2(2):59–69.
33. Sharma SC, Phalle VM, Jain SC. Performance of a noncircular 2-lobe multirecess hydrostatic journal bearing with wear. *Industrial Lubrication and Tribology*. 2012;64(3):171–181.
34. Rajput AK, Yadav SK, Sharma SC. Effect of geometrical irregularities on the performance of a misaligned hybrid journal bearing compensated with membrane restrictor. *Tribology International*. 2017;115:619–627.
35. Lee JH, Hwang KS, Jang SP, Lee BH, Kim JH, Choi SUS, Choi CJ. Effective viscosities and thermal conductivities of aqueous nanofluids containing low volume concentrations of Al2O3 nanoparticles. *International Journal of Heat and Mass Transfer*. 2008;51:2651–2656.
36. de Carvalho MJS, Seidl PR, Belchior CRP, Sodré JR. Lubricant viscosity and viscosity improver additive effects on diesel fuel economy. *Tribology International*. 2010;43:2298–2302.
37. Prasher R, Song D, Wang J, Phelan P. Measurements of nanofluid viscosity and its implications for thermal applications. *Applied Physics Letters*. 2006;89:133108.
38. Vijayaraghavan D, Brewe DE. Effect of rate of viscosity variation on the performance of journal bearings. *Journal of Tribology*. 1998;120(1):1–7.
39. Praveen KN, Devdatta PK, Debasmita M, Debendra KD. Viscosity of copper oxide nanoparticles dispersed in ethylene glycol and water mixture. *Experimental Thermal and Fluid Science*. 2007;32:397–402.
40. Weerapun D, Somchai W. Measurement of temperature-dependent thermal conductivity and viscosity of TiO$_2$-water nanofluids. *Experimental Thermal and Fluid Science*. 2009;33:706–714.
41. Prabhakaran Nair K, Ahmed MS, Al-Qahtani ST. Static and dynamic analysis of hydrodynamic journal bearing operating under nano lubricants. *International Journal of Nanoparticles*. 2009;2(1–6):251–262.
42. Madhusree K, Dey TK. Viscosity of alumina nanoparticles dispersed in car engine coolant. *Experimental Thermal and Fluid Science*. 2010;34:677–683.
43. Kole M, Dey TK. Viscosity of alumina nanoparticles dispersed in car engine coolant. *Experimental Thermal and Fluid Science*. 2010;34:677–683.
44. Shenoy BS, Binu KG, Pai R, Rao DS, Pai RS. Effect of nanoparticles additives on the performance of an externally adjustable fluid film bearing. *Tribology International*. 2012;45(1):38–42.
45. Bangotra A, Sharma S. Impact of surface waviness on the static performance of journal bearing with CuO and CeO$_2$ nanoparticles in the lubricant. *Industrial Lubrication and Tribology*. 2022;74(7):853–867.

46. Kalakada SB, Kumarapillai PNN, Rajendra Kumar PK. Static characteristics of thermohydrodynamic journal bearing operating under lubricants containing nanoparticles. *Industrial Lubrication and Tribology*. 2015;67(1):38–46.
47. Kalakada SB, Kumarapillai PN, Perikinalil RK. Analysis of static and dynamic performance characteristics of THD journal bearing operating under lubricants containing nanoparticles. *International Journal of Precision Engineering and Manufacturing*. 2012;13(10):1869–1876.

Chapter 16

Effect of nano-hydroxyapatite and post heat treatment on biomedical implants by sol-gel and HVOF spraying

Khushneet Singh and Sanjay Mohan
Shri Mata Vaishno Devi University

Sergey Konovalov
Siberian State Industrial University

Marcel Graf
Technische Universität Chemnitz

CONTENTS

16.1 Introduction to biotribology	258
16.2 Infection in biomedical implants	260
16.2.1 Different stages of bacterial adhesion on solid surfaces	260
16.3 Biomedical implants: Material of construction and its applications	261
16.4 Nanomaterials: Importance and its application in biomedical implants	263
16.4.1 Nano-hydroxyapatite coatings and their performance in antimicrobial treatment, adhesion strength, and osteointegration	264
16.4.2 Influence of addition of TiO_2 in HA coating	265
16.5 HVOF coating of hydroxyapatite on biomedical implants	267
16.5.1 Advantage of using nano-hydroxyapatite HVOF coating	269
16.6 Sol-gel coating of hydroxyapatite on biomedical implants	269
16.6.1 Advantage of using nano-hydroxyapatite sol-gel coating	272
16.7 Effect of post heat treatment on hydroxyapatite by using various coating techniques	273
16.7.1 Advantage of post heat treatment on HA coating	275
16.8 Market evaluation and future perspectives of biomedical industry	275
16.9 Conclusion	276
References	277

DOI: 10.1201/9781003306276-16

16.1 INTRODUCTION TO BIOTRIBOLOGY

One of the most recent areas seeking the attention of researchers in the realm of tribology is biotribology. The goal of biotribology is to acquire knowledge on the friction, lubrication, adhesion, and wear of biological systems. In 1970, Dowson coined the term "biotribology". The study of the interactions between surfaces during relative motions in the human body is known as biotribology.[1] Research in biotribology focuses on (i) functioning of the biological systems function naturally, (ii) the emergence of diseases, and (iii) the optimization of surgical procedures and devices, typically from an engineering perspective.[2] It is important to address the unique nature of different biological systems to make a wholesome and a real impact to improve quality of life.[3] In recent years, the field of biotribology has grown significantly as a result of coordinated contributions from engineers, biologists, doctors, physicists, and chemists. In order to reduce the occurrence of wear, it is necessary to simultaneously enhance the design and manufacturing of the implant as well as the surgical procedures in addition to the use of highly wear-resistant materials.[4] Figure 16.1 shows different branches of biotribology.

Bone-related injuries, such as bone cancer, osteoporosis, rheumatoid arthritis, or accidents, have affected over 90% of the world's aged population, demanding biomaterial replacement procedures for the spine, hip, and knee.[5] As an essential factor of biomedical engineering, biomaterials have drawn a lot of attention due to their ability to replace a biological part or

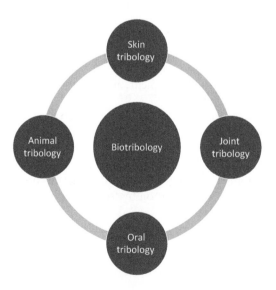

Figure 16.1 Different branches of biotribology.

function in a way that is secure, reliable, economical, and physiologically acceptable.[6] Human quality of life has significantly improved as a result of the rapid developments of biomedical engineering in modern society.[7] In addition, the implant surface won't have the necessary or suitable properties for attracting bone cells and stimulating their development. The failed implant materials will face a number of complications as a result of the revised operations. Eventually, both the expense and the length of time to recover vary considerably. Over the past several years, the worldwide medical device industries have seen a significant growth. The nanotechnology industry for implantable biomedical devices has mostly remained unexplored. Unfortunately, several biomedical implant technologies led to a wide range of new unfavourable biological reactions, such as inflammation, thrombosis, fibrosis, and a number of infections, all of which contributed to implant rejection. To meet the objectives of clinical applications, a vast variety of synthetic biomaterials, including metals, polymers, ceramics, and composites, have been developed over the past 90 years. For each application, the biomaterial has its unique set of benefits and drawbacks.[8] In other words, the characteristics of related biomaterials have an impact on their selection for a given repair or replacement situation.

According to the reports of "American Academy of Orthopaedic Surgeons," more than 120,000 total hip replacement surgeries are performed every year.[9] As a result, creating reliable, high-performing implants is extremely valuable and helpful. Biological and mechanical functions of the human body can be restored with the help of biocompatible implants, improving the quality of life. Depending on the biomedical use, the implant must sustain dynamic mechanical stresses while interacting with the nearby biological tissue for its prolonged usage. The load-bearing capacity of the implant is influenced by its bulk qualities, whereas its surface features govern the interaction with the surrounding tissue. The surface, which affects cell adhesion and behaviour influences the interaction and adsorption of different proteins. On the other hand, the body's general response to an implant is a system property that takes into account a number of variables, including surface chemistry, texture, implant mobility, biodegradation, and surgical factors. The body's limited tolerance for some dissolution products, as well as the severely corrosive environment, limits the materials that can be used in implants.

According to reports, approximately 7.7 million Americans get orthopaedic replacement surgery each year. In addition, 600,000 joint replacement operations are being conducted annually costing $3 billion.[10] However, it has been observed that most of the replacement operations were not successful because the implants might fail in some situations, even before they reach their expected average lifespan of 10–15 years. The main reasons for implant failure include stress-shielding effect, infection, tribo-corrosion, and wear debris.

16.2 INFECTION IN BIOMEDICAL IMPLANTS

Infection that develops in a time frame shorter than 3 months after surgery is referred to as postoperative infection. Delayed infection is that when an infection develops 3–24 months after surgery. Infections may also develop through haematogenous spread after 24 months of surgery; this is known as a late infection.[11] Figure 16.2 shows different categories of infection.

It has been reported that 1-month after surgery, early infections appeared to occur, and 3 weeks later, chronic infections developed into acute haematogenous infections. The type of microbe, the location of the implant, and the length of time all have an impact on the infection.[12] Complications from these infections might include persistent inflammation, implant failure, a lag in wound healing and proper tissue union, or even death.[13] Annually, 750,000 surgical site infections occur in the United States, adding 3.7 million extra hospital days and $1.6 billion in hospital expenses.[14] These infections are routinely treated with long-term antibiotic therapy, the removal of the implant, and debridement of all affected tissue.[15]

16.2.1 Different stages of bacterial adhesion on solid surfaces

Planktonic: It stands for the kind of single-celled bacteria that can be quickly found and eliminated by an individual patient's immune system or antibiotic treatment. If it hasn't been cleansed and has colonized

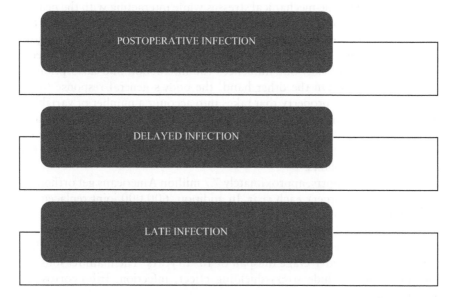

Figure 16.2 Infection in biomedical implants.

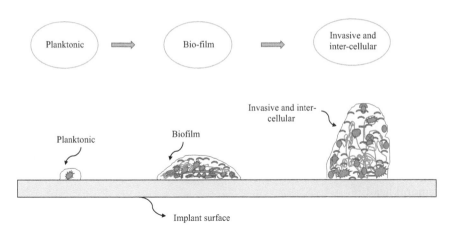

Figure 16.3 Bacterial adhesion in biomedical implant surface.

the implant or the patient's surgical site tissue, it may also serve as a source for biofilm formation.

Biofilm: When bacteria get attached to a patient's body, they create colonies that are enclosed in a structural matrix. The implanted material offers a surface for bacterial adhesion and biofilm formation.

Invasive and inter-cellular: Within the tissue of the patient, many bacteria invade, persist, and grow, as shown in Figure 16.3. To counter the effects of antibiotics and immunological reactions, the bacteria use this strategy.[16]

Bacteria control their physiological and metabolic processes to create microcolonies after first adhering successfully to the solid surfaces. Depending on the type of environment, the resulting microcolonies may include a single species or multiple species.[16,17]

According to reports, titanium oxide nanoparticles (NPs) were considered to have antibacterial properties that make them effective against bacteria, parasites, and viruses.[18] They also exhibit sporicidal activity against bacteria.[19]

16.3 BIOMEDICAL IMPLANTS: MATERIAL OF CONSTRUCTION AND ITS APPLICATIONS

Materials that have been designed and are utilized in biomedical applications include pure metals, polymers, alloys, ceramics, and composites. For orthopaedic and dental applications, biomaterials must possess a variety of properties in order to function as load-bearing implants.[20,21] The most important need for implants is that they should be biocompatible, which means they must be able to operate in the body in vivo without triggering an unfavourable reaction either locally or systemically. The body's frequently hostile environment must be tolerated by the right biomaterials, and they must also have enhanced properties like corrosion and degradation resistance, so that the implant's intended performance lifetime is not adversely impacted by the body environment. Table 16.1 shows the advantages and disadvantages of various biomaterials.

Titanium alloys have low fretting and wear resistance; thus, Co-Cr alloys are used for the modular heads because they are less prone to wear than Ti-alloys or ceramics (alumina and zirconia). The new type of modular hip prosthesis has a titanium alloy femoral stem and a Co-Cr alloy or ceramic femoral head.[22]

It's important to provide antibacterial implant coatings to avoid the colonization of the implant surface by circulating planktonic bacteria, in addition to the implant surface's capacity to support the attachment and development of protective host cells. A biofilm forms after bacterial adherence and colonization, protecting the bacteria from the host's natural defences such as leukocytes, immunoglobulins, and complements as well as against antibiotics. These elements frequently contribute to implant-related infections and failure.[16]

Table 16.1 Advantages and disadvantages of different biomaterials

S. no.	Biomaterial	Advantages	Disadvantages
1	Co-Cr and Ti- alloys	Superior Mechanical Properties	Due to the limits of bone remodelling and resorption, a high elastic modulus of Co-Cr and Ti-alloy in comparison to bone might cause stress shielding, which results in implant failure.[22] Both are bioinert and would deteriorate in the biological environment of the human body, releasing some undesirable metal ions.
2	Alumina, Calcium phosphate, Calcium silicate, Zirconia, and Titanium dioxide	High Biocompatibility and Bioactivity	Low mechanical properties, such as tensile strength and fracture toughness, are a significant design limitation.[8]
3	Polymethyl methacrylate (PMMA), Polyurethane and Polyethylene	High Resilience and Malleability	They are not suitable for load-bearing implants because of their poor mechanical qualities. The major drawbacks of PMMA and polyethylene include the extremely exothermic reaction that occurs during the polymerization process and the possible release of non-reacted monomers, which might harm the bone tissue around the PMMA implant.[8]

Furthermore, considering the advantages and disadvantages of each type of biomaterial, creating bioceramic-coated metallic implants is a workable and efficient solution. These implants are regarded as having the ability to combine the necessary biological features of bioceramics with better mechanical properties of metallic substrates. The bioceramic coating can form a barrier to prevent harmful metal ions from entering the body and can prevent corrosion of the metallic substrate. One of the simplest ways to deal with the limitations of implants is surface modification. Surface modification can be used to improve a material's bioactivity, prevent implant-related infections, and decrease or stop deterioration while maintaining the material's desirable bulk properties.

16.4 NANOMATERIALS: IMPORTANCE AND ITS APPLICATION IN BIOMEDICAL IMPLANTS

The International Organization for Standardization (ISO) has defined (i) nanomaterial (NM) as a material with any external dimension in the nanoscale or having internal structure or surface structure in the nanoscale[23] and (ii) NP as a nanoobject with all three external dimensions in the nanoscale (1–100 nm).[24] The term "nanoscale technology" describes our capacity to use our understanding of nanoscale science for the development of novel capabilities and features in goods. The use of scientific knowledge to manipulate and control matter at the nanoscale to affect properties and phenomena that depend on size and structure but are not present in individual atoms, molecules, or bulk materials is known as nanotechnology.[23] Nanotechnology is a new tech field that combines nanoscience and engineering to build useable, marketable, and economically viable technologies. The manipulation and characterization of materials on length ranging from molecule to micron are the focus of nanoscience.[25]

In recent years, scientists and engineers from all over the world have been increasingly interested in NM research. In recent years, NMs have received prime importance in a variety of implant applications, such as those that speed up the healing process and repair fractures. Due to their ease of synthesis and purification, small size, low cost, eco-friendliness, and high surface-to-volume ratio (Figure 16.4), NPs have been extensively researched as antibacterial agents and osteoblast-adhesive peptides to overcome the shortcomings of conventional coating materials, which is the most prominent factor.

In the last few decades, nanotechnology has rapidly advanced in the field of antimicrobial treatment with great effectiveness. When NPs come into contact with bacteria, a substantial quantity of bacterial cells come into contact with these NMs, which have a high surface-to-volume ratio and are highly reactive, which easily results in the destruction of cells and finally their death. NMs are sometimes known as the "wonder of modern

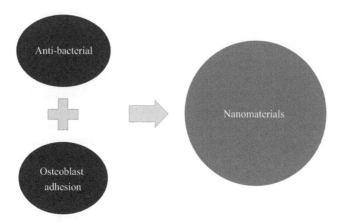

Figure 16.4 Important properties required in biomedical implants.

medicine" due to their unique characteristics and properties. After being treated with NMs, several viruses, bacteria, and fungus can be inhibited or eliminated in a short amount of time. This is due to NM's tiny size, which makes it simple for them to penetrate cell membranes and start their action within small time span[26] and this may be the reason for the rising use of NPs as an antibacterial agent. The results as reported by Goudarzi et al.[27] proved that nano-HA has greater antibacterial performance.

16.4.1 Nano-hydroxyapatite coatings and their performance in antimicrobial treatment, adhesion strength, and osteointegration

The major mineral component of human hard tissues (mostly bones and teeth) is hydroxyapatite (HA), which offers storage for the human body's calcium uptake and release.[28] HA belongs to a class of calcium phosphate–based bioceramics with a chemical formula of $Ca_{10}(PO_4)_6(OH)_2$. The word "hydroxyapatite" consists of "hydroxyl" ion and "apatite," which is the mineral name. HA possesses a hexagonal structure with the unit cell dimensions of: $a=b=0.9432$ nm and $c=0.6881$ nm.[20,29,30]

Micro- and nano-HA have been used as coatings to promote bone formation and also in drug delivery systems to distribute proteins and growth factors.[31,32] A nano-HA coating's ability to deliver and release drugs locally enables an efficient control of the infected area with less cytotoxicity.[33–35] Wang et al.'s[36] results demonstrate that the biocomposite's bending and tensile strengths improved with an increase in the content of non-dimensional HA but decreased with an increase in the content of micro-sized HA. Aksakal et al.[37] showed that coated screws, especially those with nanoscale-HA coatings, improved fixation and provided greater stability, bone ingrowth, and osteointegration than those with microscale-HA coatings. Webster et al.[38] reported that nano-HA and nanoceramics showed increased osteoblast adhesion. Ergun et al.[39] examined osteoblast adhesion on calcium phosphate NPs with different Ca/P ratios. According to their research, higher Ca/P ratios reduced the size, porosity, and average pore diameters of the calcium phosphates while increasing osteoblast adhesion. As reported by Nelson et al.,[40] the synthesis of nano-sized crystalline HA particles with controlled degrading properties can be helpful in various orthopaedic applications since they can increase osteoblast adhesion.

Human bones contain HA nanocrystals that are 5–20 nm wide and 50–60 nm long.[41,42] Since the particle size of synthetic nano-HA is close to that of natural HA, it is very suitable for regenerative medicines. HA-NMs have been the focus of continuous research and development, and have involved innovative manufacturing processes, functionalization, and characterization methods. Since HA is biologically and chemically identical to human hard tissues, it is best suited for deployment as a coating on the surface of metallic implants.[43] The HA coating not only accelerates osseointegration but also creates a strong bond between the surrounding tissues and the implant by generating strong chemical interactions.[44] Figure 16.5 shows the properties of HA. The biological attachment between the implants and the bone can be improved by HA despite its poor mechanical and tribological characteristics, which can boost the clinical success rate over the long term as compared to uncoated implants.

The most remarkable feature of HA in bone regeneration is its attractive bioactivity and its ability to have direct chemical interactions with adjoining tissues. As a result, HA-coated metallic implants could combine HA's bioactivity with the mechanical qualities of metallic substrates to create an optimum combination. For the application of HA coatings on metallic implant surfaces, a variety of procedures have been developed.

16.4.2 Influence of addition of TiO$_2$ in HA coating

Titanium dioxide (TiO$_2$), alumina (Al$_2$O$_3$), and zirconia dioxide (ZrO$_2$) are frequently used as bond coats or as composite coats in various ratios to enhance the mechanical and biological properties of HA coatings.[45,46] Due to its impact on biology and resistance to corrosion, TiO$_2$ has gained a lot of interest as a composite or bond coat material. The melting point of TiO$_2$ is 1,843°C, which is considerably less than that of Al$_2$O$_3$ and ZrO$_2$, which are above 2,000°C. A uniform HA-TiO$_2$ composite coating is more preferred due to the lower melting point of TiO$_2$. In addition, TiO$_2$ has higher thermal conductivity and diffusivity values than HA. TiO$_2$ particles hence require less thermal energy to achieve a certain temperature than HA. TiO$_2$ addition is advantageous for reducing the thermal mismatch between the HA coating and Ti substrate.[47] Liu et al.[48] found that sample coated with nano-TiO$_2$ in NiTi-alloy had better biocompatibility as compared to the uncoated one.

Figure 16.5 Properties of hydroxyapatite ($Ca_{10}(PO_4)_6(OH)_2$).

Wu et al.[49] suggested that due to its unique geometrical characteristics and excellent biocompatibility, nanostructured TiO_2 has a significant amount of potential in the biomedical field. Although the advantages of TiO_2 NPs have been well documented, more research into the underlying processes is essential for advancement. Nanostructured TiO_2 acts as a bone scaffold that increases osteoblast adhesion, proliferation, and differentiation while accelerating the creation of apatite. When compared to their conventional ones, NMs have exhibited higher levels of hardness, strength, ductility, and toughness.[50,51] These characteristics provide significant opportunities for thermal and sol-gel spraying. Nanostructured thermal spray ceramic oxide coatings have been discovered to have enhanced wear performance when compared to equivalent coatings created from ordinary ceramic oxide powder.[52–54]

Most importantly, in determining the survival of HA coating on implant surfaces, adhesion strength following in vitro or even in vivo tests is critical. The tensile adhesion and shear adhesion strength in the case of high-velocity oxy-fuel coating (HVOF) and sol-gel spraying is very high as compared to pulsed laser deposition, electrophoretic deposition, hot isostatic pressing, biomimetic coating, and sputter coating.[55] Hence, it encourages us to study both the coating techniques in order to get a better detailed review that will aid equipment manufacturers and processors in selecting materials and coatings in biomedical implants' application.

16.5 HVOF COATING OF HYDROXYAPATITE ON BIOMEDICAL IMPLANTS

Both HVOF and plasma spraying are thermal spraying coating techniques. However, in the case of plasma spraying, high plasma jet temperatures and the rapid cooling of sprayed particles[56] cause HA to degrade during flight.[57,58] Tetra-calcium phosphate (TTCP), tricalcium phosphate (TCP), amorphous calcium phosphate (ACP), and calcium oxide, which is not biocompatible, are some of the secondary phases that are formed during the degradation of HA. In body fluids, the HA is a very stable phase, while the secondary phases, in particular the ACP, have higher dissolution rates.[59] It has also been reported in the literature that the average diameter of splats in the case of plasma spray is higher in comparison to the HVOF.[60]

HVOF is being considered as an alternative to plasma spray for obtaining higher bioactive coatings. By adjusting the spraying parameter's (i.e., higher velocity and lower temperature), coating crystallinity and adhesive strength are improved. The chemical, physical, and kinematic conditions under which the particles contact the substrate during the first step of the HVOF process have a significant impact on the coating qualities.[61,62] For instance, it's ideal for the majority of the particles to deform into disk-shaped splats as they hit the substrate.[62–65] The adhesion and phase composition of the coating are determined by the melt percentage and cooling rate of the particles when they come into contact with the substrate.[66] As a result, the initial phases of deposition and the complete build-up process have an impact on the coatings' environmental stability, bioactivity, and biocompatibility.[61,67]

Munoz et al.[68] suggested that the optimized results in the HA coating obtained when the stand-off distance (SOD) is 20 cm, fuel-oxy ratio (F/O) is 0.9 and powder feed rate (PFR) is 16 g/min. Using the abovementioned set of parameters, a disk-like splat morphology was developed. However, the coating delamination is observed, when the shortest SOD (10 cm) and the highest F/O ratio (0.27) parameters are used. Table 16.2 shows the variation in the properties of several microcoatings and nanocoatings with HA as a single phase or as an reinforcement.

Table 16.2 Micro- and nano-hydroxyapatite coating by using the HVOF spraying technique

S. no.	Author	Coating material	Observations
1	Gaona et al.[69]	80 wt.% nano and TiO$_2$–20 wt.% HA; 90 wt.% nano-TiO$_2$ and 10 wt.% HA	Toughness of the coating increases with increased HA wt.% (i.e., highest in nanostructured TiO$_2$ + 20 wt% HA). However, porosity in both coatings is less than 1%. • Hardness decreases with increase in HA wt.%.
2	Henao et al.[70]	Multilayer coating 100 wt.% pure TiO$_2$ 25/75 wt.% HA/TiO$_2$ 75/25 wt.% HA/TiO$_2$ 100 wt.% HA top coat	• Microhardness of the graded coatings was 13% more than that of monolithic HA counterparts. Also, coating/substrate interfacial adhesion of graded coatings was 10% higher than monolayer HA coatings.
3	Khor et al.[61]	Hydroxyapatite (HA)	Stiffness of the coating is improved due to lower melting fraction of powder, because of higher HA content. It was observed that Young's modulus of different parts of the sprayed HA particle varies dramatically. However, the coatings' fracture toughness did not change significantly.
4	Li et al.[45]	HA + 10 vol% TiO$_2$ and HA + 20 vol% TiO$_2$	For the objective of strengthening HA coatings, it is advised that less than 20 vol.% of TiO$_2$ be added. With the addition of TiO$_2$ in HVOF HA-sprayed coatings, it strengthens the coating's shear strength, young modulus, and fracture toughness, whereas adhesion strength of HA-sprayed coatings is negatively impacted by the secondary phase of titanium.
5	Melero et al.[71]	Pure HA coating and 80–20 wt% HA-TiO$_2$	TiO$_2$ increases the coating stability by preventing HA dissolution in HVOF-sprayed coatings. The 80HA-20TiO$_2$ coating's narrow potential passive zone in polarization studies shows that the TiO$_2$ addition has a positive impact on the stability of the HA coating.
6	Lima et al.[72]	Nano-HA	High density, microstructural uniformity, and high crystallinity of HA is observed due to low flame temperature of HVOF spray coating. After 7 days of incubation in simulated body fluid, a uniform layer of apatite is formed on the HA coatings (SBF).
7	Ferrer et al.[59]	HA + TiO$_2$	HVOF is far more suitable for creating HA + TiO$_2$ coatings than atmospheric plasma spraying, and coatings were discovered to have the highest adhesion values as compared with the in vitro test.
8	Li et al.[46]	HA + TiO$_2$	The reaction product occurring around TiO$_2$ particles was found to be advantageous in improving the coating structure. In the HA/titania composite coatings, the reaction zone and phase positions were identified by TEM observation.
9	Kulpetchdara et al.[73]	Nano-HA	The research indicates that HVOF-deposited HA coatings are highly osteoconductive and may eventually replace the use of conventional HA in implant applications.

16.5.1 Advantage of using nano-hydroxyapatite HVOF coating

- Nanostructured particles in the HVOF jet have interior temperatures that are significantly lower than surface temperatures, which result in less HA degradation.
- Crystallinity of nano-HA-coating was found to be 84% and above.
- The optimized results are obtained by adjusting the oxy-fuel ratio, standoff distance and powder feed rate, i.e., only HA peaks were found, i.e., no secondary phases are present.[68]
- The nanostructured HA coating created by HVOF spraying appears to have good biocompatibility. It's been shown that HA with a nanocrystalline form has increased bioactivity.
- HVOF-sprayed nanostructure HA coating exhibits higher density and microstructural uniformity.

According to observations, the creation of various types of nanostructures as a result of the HVOF coating should significantly contribute to raising the bioactivity of the HA coating.[72] Even bioinert substances like alumina begin to demonstrate bioactivity through their interactions with osteoblasts when certain nanostructural properties, including nanoporosity, are present.[74] The HVOF-sprayed nano-HA coating demonstrated a greater binding strength as compared to plasma-spray coating.[75]

16.6 SOL-GEL COATING OF HYDROXYAPATITE ON BIOMEDICAL IMPLANTS

The sol-gel method of depositing HA is continually being researched as a potential method to overcome the limitations of implants. The sol-gel process is a type of chemical synthesis that is typically used to manufacture ceramics, glasses, and composites at temperatures far lower than those required by more conventional techniques like hot pressing and sintering.[29] The sol-gel technology has been used since the middle of the 1800s, and over the past three decades, a rapid expansion of sol-gel applications has been achieved.[76,77]

Sol-gel HA coating is produced by combining precursors for calcium and phosphorus and preparing the sol with the addition of solvents, often ethanol and water.[78] Ca and P precursors' chemical nature is crucial for the production of HA.[79,80] Due to its predominant role in osseointegration, the surface chemistry of sol-gel-produced HA coatings has drawn a lot of attention as one of the most significant factors for the coating in a clinical application.[29,81] A deposition method like dip coating, spin coating, or spray coating can be conveniently coupled to the sol-gel process. The sol-gel dip-coating technique comes out as the most effective of these. The benefits of adopting a sol-gel dip-coating method are its independence from substrate shape and its ability to establish good control over surface attributes like composition, thickness, and topography.[48] Sol-gel processing is a versatile and appealing method since it may be used to produce ceramic coatings from chemically based solutions. Table 16.3 shows the variation in the properties of several microcoatings and nanocoatings with sol-gel process. Figure 16.6 shows the feature of sol-gel.

Table 16.3 Micro- and nano-hydroxyapatite coatings by using sol-gel spraying technique

S. no.	Author	Coating material	Observation
1.	Chai et al.[82]	Nano-HA	It was found that to induce the formation of a coating that is monophasic hydroxyapatite (HA); solutions should be aged for 1±7 days prior to application. Solutions were observed for wet substrates well and produced crystalline calcium phosphate coatings with some CaO also present when used immediately after solution preparation.
2.	Jaafar et al.[83]	HA	The main drawback of HA to Ti surface is poor adhesion and low corrosion resistance. However, thick interlayers act as a barrier against corrosion while decreasing the bonding strength of the multilayer due to the mismatched coefficients of thermal expansion (CTE) in intermediate layers and Ti substrate's.
3.	Kim et al.[84]	HA and FHA	When compared to pure HA, the fluor-hydroxyapatite (FHA) layer demonstrated a substantially lower rate of degradation. When compared to pure Ti, HA- and FHA-coated Ti showed higher levels of alkaline phosphate expression, demonstrating that the coatings improved the activity and functionality of the cells on the substrate.
4.	Kim et al.[85]	HA/TiO_2 double layer coating	The TiO_2 layer greatly enhanced the HA layer's ability to adhere to the Ti substrate. When compared to the HA single coating, the strength of the double layer coating increased by around 60%, reaching a maximum strength of 55 MPa. However, TiO_2 layer at 200 nm considerably increased the bonding strength and corrosion resistance.
5.	Hadad et al.[86]	HA	The size of the crystallites and the level of crystallinity of the coatings made from HA sol-gel are significantly influenced by the heat treatment that was employed. The results showed excellent bioactivity after 15 days of immersion of coating in simulated body fluid (SBF).

(Continued)

Table 16.3 (Continued) Micro- and nano-hydroxyapatite coatings by using sol-gel spraying technique

S. no.	Author	Coating material	Observation
6.	Ramires et al.[87]	HA+TiO$_2$	HA-TiO$_2$ coatings made by sol-gel manufacturing were reported to be non-cytotoxic and also promoted the growth of cells like human osteoblasts.
7.	Kim et al.[88]	HA+TiO$_2$	From a mechanical and biological viewpoint, the HA/TiO$_2$ composite sol-gel coatings produced have good characteristics for hard tissue applications.
8.	Vijayalakshmi et al.[89]	Nano-HA	At a temperature of 600°C, it was shown that HA nanoparticles partially crystallized and remained thermally stable up to 1,200°C without experiencing secondary phase formation. The implant's surface had a nanostructured sol-gel HA coating that improved the implant's corrosion resistance and adhesion strength, showing its suitability for long-term orthopaedic usage.
9.	Guo et al.[90]	Nano-HA	With the rise in firing temperature, the mean crystallite size increased and microstrain significantly decreased. The firing temperature was between 400°C and 600°C.
10.	Jafari et al.[91]	Nano-HA	The corrosion performance of the HA-coated specimens significantly improved after sintering. In addition, sintering at 700°C maintained the coated alloy's increased bioactivity compared to the uncoated alloy.
11.	Rojaee et al.[92]	Nano-HA	The researchers came to the conclusion that n-HA-coated AZ91 alloy would be a promising biodegradable implant material for biomedical purposes.
12.	Guo et al.[93]	HA+TiO$_2$	It is observed that Mg alloy's in vitro cytocompatibility and antibacterial characteristics are improved by HA/TiO$_2$ composite coatings.
13.	Kaur et al.[94]	HA	It was discovered that samples coated with HA exhibited superior corrosion resistance and improved implant characteristics than untreated 316L stainless steel.
14.	Mohammed et al.[95]	Nano-HA+TiO$_2$	The researchers discovered that the HA/TiO$_2$ film has excellent wear and corrosion capabilities and can be regarded as a possible thin film that is suitable for use on hard tissues because of its advanced crystallization and homogenous nanoscale structure.

Figure 16.6 Features of the sol-gel coatings.

It has been shown that the nanocrystalline grain structure of the coatings created by the sol-gel technique leads to superior biological and mechanical capabilities. The sol-gel process is also very simple to carry out and may coat complicated shapes.

16.6.1 Advantage of using nano-hydroxyapatite sol-gel coating

- Multifunctional HA nanocoating is expected to speed up the bone healing and increase the useful life of orthopaedic implants.
- Nano-HA sol-gel coatings exhibit superior bioactivity and osseointegration, which is due to their significant surface area and resemblance in size to HA crystal found in bone.

- The corrosion resistance and adhesion strength are improved by sol-gel technique with nano-HA.
- Excellent bioactivity and an antibacterial effect are produced when metal-doped and nanocomposite HA coatings are made using the sol-gel method.
- Using the sol-gel coating process, nano-HA+TiO$_2$ is quite effective in improving the bond strength and wear resistance.

Producing thin, uniform HA coatings on metallic implants is an approach that can be achieved via the sol-gel method. In reality, nanoscale coatings lower the temperature at which ceramics sinter, which reduces the possibility that the coating may degrade due to a mismatch in the thermal conductivities of the metal and ceramic. Sol-gel HA coating with the correct amount of Ag and Zn can be used to build implants with excellent biocompatibility and effective microbial resistance.[96] Recently, interest has increased in sol-gel coatings as corrosion protection layers for various metallic subtracts, such as galvanized steel, carbon steel, stainless steel, and aluminium and its alloys.[97] Utilizing the dip-coating process, sol-gel-generated nanostructured hydroxyapatite (n-HA) was applied to AZ91 alloy to stabilize alkalization behaviour and increase corrosion resistance.[92] According to the results, sol-gel coatings enhance the substrate's biocompatibility, wear resistance, and corrosion resistance.

Strong adherence was observed at the coating/alloy contact by HA sol-gel-derived coatings. The Ti6Al4V alloy's ion release to the simulated body fluid (SBF) was prevented by the HA coating/Ti6Al4V systems; effective corrosion protection was performed after thermal treatment in the 600°C–800°C range.[86]

According to the discussion above, various qualities in biomedical implants can be improved with post heat treatment, i.e., by annealing. As a consequence, it drove our desire to learn more about how post heat treatment in various coating techniques affect implant surface characteristics.

16.7 EFFECT OF POST HEAT TREATMENT ON HYDROXYAPATITE BY USING VARIOUS COATING TECHNIQUES

Many researchers have examined the effects of heat treatment, particularly annealing treatment, on the performance of HA coating and have suggested that it may be an effective method for improving the coating's crystallinity, microstructure, and corrosion resistance. Table 16.4 shows the Post heat treatment of hydroxyapatite by various coating techniques.

Fernández et al.[98] measured the adhesion strength of the coatings by analysing the dissolution/precipitation behaviour and degradation of HA coatings brought on by the immersion test. Bond strength reduced for the

Table 16.4 Post heat treatment of hydroxyapatite by various coating techniques

S. no.	Author	Coatings material	Coating techniques	Optimized annealing temperature and time
1	Fernandez et al.[98]	HA	HVOF	700°C for 1 hour
2	Liu et al.[99]	HA	Micro-plasma-spray coating	650°C for 3 hours
3	Zhao et al.[100]	HA	Micro-plasma-spray coating	760°C for 2 hours
4	Singh et al.[101]	HA	Vacuum plasma-spray coating	700°C for 1 hour
5	Tiwari et al.[102]	HA	LVOF coating	600°C for 2 hours
6	Li et al.[103]	HA	HVOF coating	750°C for 1 hour
7	Liu et al.[104]	HA	Sol-gel spray coating	400°C for 20 minutes
8	Xu et al.[105]	HA+TiO$_2$	Sol-gel coating	780°C for 20 minutes
9	Rocha et al.[106]	HA+TiO$_2$	Plasma spraying	750°C for 1 hour

as-sprayed coatings after immersion in simulated bodily fluid (SBF), while it remained unchanged for the heat-treated coatings. Tensile adhesion was reported to be 44.2 8.3 MPa for heat-treated coating, whereas it was 37.5 4.8 MPa for coating without heat treatment. Li et al.[103] studied that the heat treatment at 750°C for 1 hour after HVOF HA coatings increases the bond strength.

Liu et al.[99] observed that TTCP, TCP, and ACP were transformed into HA after being heated at 650°C for 3 hours. Tiwari et al.[102] and Singh et al.[101] studied that post heat treatment significantly increased the corrosion resistance of coatings because it caused the microstructure of the coatings to become denser, further reducing the active sites for dissolution. The post-coating heat treatment at 700°C for 1 hour fully eliminated the amorphous phase that was present in the coatings when they were being sprayed. Liu et al.[104] observed that apatite coatings can be made dense and adhesive through water-based sol-gel technology and short-term annealing at about 400°C in air. The coatings' bonding strength, 44 MPa, is greater than that of those created by plasma-spraying deposition.

According to the findings of Xu et al.,[105] the HA phase began to crystallize at a heat treatment temperature of 580°C, while the crystallinity significantly increased at a temperature of 780°C. The HA film displayed a porous structure and a thickness of 5–7 mm after being heated to 780°C. Rocha et al.[106] investigated the result of plasma spraying on Ti-alloy and they demonstrate that the specimen's crystallinity was observed to be 55.6% without heat treatment and 75.6% after heat treatment. As a result, the composite's crystallinity increased by around 36% because of heat treatment. Roughness and porosity showed no noticeable fluctuation, proving that the heat treatment had no impact on these properties. The composite had a HV_{100} hardness of 874±26 and adhesion strength of 30.0±2.0 MPa after the heat treatment.

Therefore, optimize parameter (i.e., annealing temperature and time) is one of the most important factors in order to get the desired optimal properties in the coating.

16.7.1 Advantage of post heat treatment on HA coating

- Improvement of the crystallinity of the HA coating.
- Nanograins developed on the coated surface as a result of heat treatment.
- Tensile adhesion strength of the coating is increased.
- Microcracks and pores have been significantly decreased in quantity by post heat treatment.
- Heat treatment successfully converted the foreign phases into apatite.

16.8 MARKET EVALUATION AND FUTURE PERSPECTIVES OF BIOMEDICAL INDUSTRY

Approximately 126.6 million Americans, according to the "American Academy of Orthopaedic Surgeons," had a musculoskeletal disorder in 2012. To be more precise, arthritis affects 50% of adults in the United States aged 65 and above (approx. 51.8 million). The number of Americans with arthritis is expected to rise to 67 million among the country's older populations. In America, 10 million people have osteoporosis, and another 19 million people (mainly women) are at risk of developing it. One in two women and one in four men over the age of 50 may experience an osteoporosis-related fracture, and 20% of hip fracture patients over the age of 50 will pass away within a year of their injury. Treatment for a musculoskeletal disorder cost $7,800 on average per person per year. The cost of treating and replacing lost wages due to musculoskeletal disorders in the United States in 2011 was $213 billion, or 1.4% of the nation's gross domestic product (GDP).[107]

Around 16% of the country's population in India suffers from musculoskeleton disorder (MSD), while orthopaedic items imported at high cost satisfy 70% of the country's entire local demand. Indian orthopaedic devices now have a massive market of USD 375 million, and it is estimated that this market would grow by 20% annually until it reaches USD 2.5 billion by 2030.[108]

Based on the techniques currently employed to produce biomedical implants, additive manufacturing has gained a lot of interest in the sector, which is primarily divided into four categories: laser or electron beam melting, paste extrusion technology, friction stir processing, and photo-polymerization.

Due to its simplicity in creating customized implants, additive manufacturing is one of the most growing processes used in the developing health care sector. Because of its advanced features, this approach is vital in the bioprinting of complicated organs because the implants made using it are more biocompatible than those made using traditional methods. In the future, it may be possible to collect stem cells from a baby's teeth and use them as a toolkit throughout his or her life to repair missing organs and tissues using additive manufacturing.

As the advancements increase in the field of orthopaedic surgery, the number of people who require implants and who are dependent on several different medical devices is increasing quickly. These implants are being used more often now, and this trend is anticipated to grow significantly over the next few decades as the population ages and medical care becomes more prudent.[109,110]

16.9 CONCLUSION

The present work elaborated the significance of biotribology. The work also stressed upon the type of infections occurring in different implants even after successful surgeries. This chapter also brought forth the influence of postoperative, delayed, and late infections on the replacement or sometimes removal of implants, which invariably makes a patient to bear additional costs besides being traumatized in the process. The formation of bacteria in the implants has been discussed and further the possibilities to reduce or eliminate such bacteria have also been mentioned.

The study also brought forth the significance of coatings with nanoparticles of HA as compared to its microparticles. The nanoparticles led to the reduction of cytotoxicity, enhancement of osteoblast adhesion and tensile strength, improved fixing and stability, and promoted bone development and osteointegration. Nano-TiO_2 reinforced with HA are considered to exhibit the antimicrobial property and also addition of TiO_2 in HA increases fracture toughness, tensile strength, and osteoblast adhesion.

The authors have presented the usefulness and limitations of different coating processes mainly sol-gel and HVOF process. Though sol-gel is a time-consuming process, it has a number of benefits such as low processing temperature, simplicity of manufacturing homogenous stoichiometric coating, control of coating chemical and microstructure, and strong adhesion strength. Sol-gel processing is a low-cost and simple method, which allows to achieve molecular-level mixing of the HA. The alteration of the surface of the implants may be carried out at reasonable cost with sol-gel coating. HVOF is less time consuming, but the high-temperature process results in the formation of oxides, thus reducing the bioactivity in the implants.

However, thickness and porosity are high in the HVOF coating process, but they are low to medium in the sol-gel. Based on the previous studies, the authors suggested the use of HVOF as a best alternative to sol-gel and plasma coating techniques. Despite high temperature, it is possible to obtain a homogenous, crystalline, and oxide-free coating by optimizing the HVOF coating's process parameters. The authors also made a mention that optimized post-annealing time and temperature for the sol-gel coating are less as compared to the plasma and HVOF coatings. However, among all the three-coating technique, it is observed that crystallinity increases after the post heat treatment.

REFERENCES

1. Cacopardo. Biomaterials and biocompatibility. In *Human Orthopaedic Biomechanics: Fundamentals, Devices and Applications.* 2022:341–359. Editor: Bernardo Innocenti, Fabio Galbusera. Academic Press. Science direct doi:10.1016/B978-0-12-824481-4.00038-X
2. Sharma M, Khurana SMP. Biomedical engineering: The recent trends. In Barh and Azevedo (eds.), *Omics Technologies and Bio-Engineering: Volume 2: Towards Improving Quality of Life.* 2018:323–336. Elsevier. doi:10.1016/B978-0-12-815870-8.00016-6
3. Zhou Z, Jin ZM. Biotribology: Recent progresses and future perspectives. *Biosurface and Biotribology.* 2015;1(1):3–24. Accessed August 3, 2022, https://www.sciencedirect.com/science/article/pii/S2405451815000082
4. Affatato S, Traina F. Bio and medical tribology. *Tribology for Engineers.* 2011:243–286. Accessed August 4, 2022, https://www.sciencedirect.com/science/article/pii/B9780857091147500062
5. Chen J, Shi W, Norman AJ, Ilavarasan P. Electrical impact of high-speed bus crossing plane split. *IEEE International Symposium on Electromagnetic Compatibility.* 2002;2:861–865. doi:10.1109/ISEMC.2002.1032709
6. The Biomedical Engineering Handbook—Google Books. Accessed June 12, 2022, https://www.google.co.in/books/edition/_/T2UIoAxcFdIC?hl=en&gbpv=1&pg=PA3&dq=7.%09Bronzino,+J.+D.+2000.+The+Biomedical+Engineering+Handbook,+2nd+edn.+Boca+Raton,+FL:+CRC+Press
7. Love BJ. *Biomaterials: A systems approach to engineering concepts.* Elsevier/Academic Press. 2016:1–387. ISBN: ISBN-13: 978-0128094785 https://www.researchgate.net/publication/319547299_Biomaterials_A_Systems_Approach_to_Engineering_Concepts.
8. Biomaterials Science—Google Books. Accessed June 12, 2022, https://www.google.co.in/books/edition/Biomaterials_Science/Uzmrq7LO7loC?hl=en&gbpv=1&dq=An+Introduction+to+Materials+in+Medicine.+San+Diego,+CA:+Academic+Press.+&pg=PR5&printsec=frontcover
9. Special Feature: Consistency in Postoperative Education Prog... : Topics in Geriatric Rehabilitation. Accessed June 12, 2022, https://journals.lww.com/topicsingeriatricrehabilitation/Abstract/2000/06000/Special_Feature__Consistency_in_Postoperative.8.aspx

10. Christenson EM, Anseth KS, van den Beucken JJJP, et al. Nanobiomaterial applications in orthopedics. *Journal of Orthopaedic Research*. 2007;25(1):11–22. doi:10.1002/JOR.20305
11. Arciola CR, Campoccia D, Montanaro L. Implant... - Google Scholar. Accessed June 12, 2022, https://scholar.google.com/scholar?hl=en&as_sdt =0%2C5&q=Arciola+CR%2C+Campoccia+D%2C+Montanaro+L+%2820 18%29+Implant+infections%3A+adhesion%2C+biofilm+formation+and+im mune+evasion.+Nat+Rev+Microbiol+16%287%29%3A397%E2%80%9340 9&btnG=
12. Panteli M, Giannoudis PV. Chronic osteomyelitis: What the surgeon needs to know. *EFORT Open Reviews*. 2016;1(5):128–135. doi:10.1302/2058-5241. 1.000016
13. Vester H, Wildemann B, Schmidmaier G, Stöckle U, Lucke M. Gentamycin delivered from a PDLLA coating of metallic implants: In vivo and in vitro characterisation for local prophylaxis of implant-related osteomyelitis. *Injury*. 2010;41(10):1053–1059. doi:10.1016/J.INJURY.2010.05.010
14. Edmiston CE, Spencer M, Lewis BD, et al. Reducing the risk of surgical site infections: Did we really think SCIP was going to lead us to the promised land? Surgical Infections (Larchmt). 2011;12(3):169–167. doi:10.1089/ SUR.2011.036
15. Widmer AF. New developments in diagnosis and treatment of infection in orthopedic implants. *Clinical Infectious Diseases*. 2001;33(2):S94–S106. https://academic.oup.com/cid/article-abstract/33/Supplement_2/S94/528013
16. Khatoon Z, McTiernan CD, Suuronen EJ, Mah TF, Alarcon EI. Bacterial biofilm formation on implantable devices and approaches to its treatment and prevention. *Heliyon*. 2018;4(12). doi:10.1016/J.HELIYON.2018.E01067
17. Prasad R, Siddhardha B, Dyavaiah M, eds. *Nanostructures for antimicrobial and antibiofilm applications*. Springer. 2020. https://link.springer.com/ book/10.1007/978-3-030-40337-9. doi:10.1007/978-3-030-40337-9
18. Blecher K, Nasir A, Friedman A. The growing role of nanotechnology in combating infectious disease. *Virulence*. 2(5):395–401. doi:10.4161/ viru.2.5.16035
19. Palanikumar L, Ramasamy SN, Balachandran C. Size-dependent antimicrobial response of zinc oxide nanoparticles. *IET Nanobiotechnology*. 2014;8(2):111–116. doi:10.1049/IET-NBT.2012.0008
20. Hench LL. Biomaterials: A forecast for the future. *Biomaterials*. 1998;19(16):1419–1423. doi:10.1016/S0142-9612(98)00133-1
21. Poitout DG, eds. *Biomechanics and biomaterials in orthopedics*. Springer. 2004. https://link.springer.com/book/10.1007/978-1-4471-3774-0. doi:10. 1007/978-1-4471-3774-0
22. Cvijović-Alagić I, Cvijović Z, Rakin M. Damage behavior of orthopedic titanium alloys with martensitic microstructure during sliding wear in physiological solution. *International Journal of Damage Mechanics*. 2019;28(8):1228–1247. doi:10.1167/1056789518823049
23. ISO-ISO/TS 80004-1:2010–Nanotechnologies—Vocabulary—Part 1: Core terms. Accessed June 12, 2022, https://www.iso.org/standard/51240.html
24. ISO-ISO/TS 27687:2008–Nanotechnologies—Terminology and definitions for nano-objects—Nanoparticle, nanofibre and nanoplate. Accessed June 12, 2022, https://www.iso.org/standard/44278.html

25. Mansoori GA, Soelaiman TAF. Nanotechnology—An introduction for the standards community. *Journal of ASTM International*. 2005;2(6). Accessed June 12, 2022, www.astm.org
26. Prasad R, Jha AK, Prasad K, eds. *Exploring the realms of nature for nanosynthesis*. https://link.springer.com/book/10.1007/978-3-319-99570-0 Springer. 2018. doi:10.1007/978-3-319-99570-0
27. Goudarzi MR, Bagherzadeh M, Fazilati M, Riahi F, Salavati H, Esfahani S. Evaluation of antibacterial property of hydroxyapatite and zirconium oxide-modificated magnetic nanoparticles against Staphylococcus aureus and Escherichia coli. *IET Nanobiotechnology Research Article*. 2019;17(4):86–95. doi:10.1049/iet-nbt.2018.5029
28. Kon M, Hirakata LM, Miyamoto Y, Kawano F, Asaoka K. Surface-layer modification of hydroxyapatite ceramic with acid and heat treatments. *Dental Materials Journal*. 2002;21(2):160–180. doi:10.4012/DMJ.21.160
29. Hench LL. *An introduction to bioceramics*. World Scientific. 1993:386. https://doi.org/10.1142/p884 | June 2013. https://www.worldscientific.com/worldscibooks/10.1142/p884#t=aboutBook
30. Kay M, Young R, Nature AP. Crystal structure of hydroxyapatite. *Nature*. 1964;204:1050–1052. Accessed June 12, 2022, https://www.nature.com/articles/2041050a0
31. Rogowska-Tylman J, Locs J, Salma I, et al. In vivo and in vitro study of a novel nanohydroxyapatite sonocoated scaffolds for enhanced bone regeneration. *Materials Science and Engineering: C*. 2019;99:669–684. doi:10.1016/J.MSEC.2019.01.084
32. Johansson P, Jimbo R, Kozai Y, et al. Nanosized hydroxyapatite coating on PEEK implants enhances early bone formation: A histological and three-dimensional investigation in rabbit bone. *Materials*. 2015;8(7):3815–3830. doi:10.3390/MA8073815
33. Geuli O, Metoki N, Zada T, Reches M, Eliaz N, Mandler D. Synthesis, coating, and drug-release of hydroxyapatite nanoparticles loaded with antibiotics. *Journal of Materials Chemistry B*. 2017;5:7819–7830. Accessed June 12, 2022, https://pubs.rsc.org/en/content/articlehtml/2016/tb/c7tb02105d
34. Huang D, Zuo Y, Zou Q, et al. Antibacterial chitosan coating on nano-hydroxyapatite/polyamide66 porous bone scaffold for drug delivery. *Journal of Biomaterials Science, Polymer Edition*. 2011;22(7):931–944. doi:10.1163/092050610X496576
35. Campbell AA, Song L, Li XS, Nelson BJ, Bottoni C, Brooks DE, DeJong ES. Development, characterization, and anti-microbial efficacy of hydroxyapatite-chlorhexidine coatings produced by surface-induced mineralization. *Journal of Biomedical Materials Research*. 2000;53(4):400–407. Accessed June 12, 2022, https://onlinelibrary.wiley.com/doi/epdf/10.1002/1097-4636%282000%2953%3A4%3C400%3A%3AAID-JBM14%3E3.0.CO%3B2-Z
36. Wang X, Li Y, Wei J, de Groot K. Development of biomimetic nano-hydroxyapatite/poly (hexamethylene adipamide) composites. *Biomaterials*. 2002;23(24):4787–4791. Accessed June 18, 2022, https://www.sciencedirect.com/science/article/pii/S0142961202002296?casa_token=Qj7P3d3mvMMAAAAA:H2ULVFl_vUS5WemuF3sOtQ4lycdI_2krVzAdg9JsM5w3DEDDxo4TIXSwtF3Gh3XObN3K8djv9VY

37. Aksakal B, Kom M, Tosun HB, Demirel M. Influence of micro-and nanohydroxyapatite coatings on the osteointegration of metallic (Ti6Al4V) and bioabsorbable interference screws: An in vivo study. *European Journal of Orthopaedic Surgery & Traumatology.* 2014;24(5):813–819. doi:10.1007/s00590-013-1236-8
38. Webster TJ, Ergun C, Doremus RH, Siegel RW, Bizios R. Enhanced functions of osteoblasts on nanophase ceramics. *Biomaterials.* 2000;21(16):1803–1810. doi:10.1016/S0142-9612(00)00075-2
39. Ergun C, Liu H, Webster TJ, Olcay E, Yilmaz Ş, Sahin FC. Increased osteoblast adhesion on nanoparticulate calcium phosphates with higher Ca/P ratios. *Journal of Biomedical Materials Research—Part A.* 2008;85(1):236–241. doi:10.1002/JBM.A.31555
40. Nelson M, Balasundaram G, Webster TJ. Increased osteoblast adhesion on nanoparticulate crystalline hydroxyapatite functionalized with KRSR. *International Journal of Nanomedicine.* 2006;1(3):339–349. Accessed June 12, 2022, https://www.ncbi.nlm.nih.gov/pmc/articles/pmc2426799/
41. Cai Y, Liu Y, Yan. Role of hydroxyapatite nanoparticle size in bone cell proliferation. *Journal of Materials Chemistry.* 2007;16(36):3780–3787. doi:10.1039/B705129H
42. Zhou H, Lee J. Nanoscale hydroxyapatite particles for bone tissue engineering. *Acta Biomaterialia.* 2011;7(7):2769–2781. doi:10.1016/J.ACTBIO.2011.03.019
43. Oka S, Fung YC. Biorheology. *Journal of Biomechanical Engineering.* 1985;107(1):87. doi:10.1115/1.3138527
44. Vedantam R, Ruddlesdin C. The fully hydroxyapatite-coated total hip implant: Clinical and roentgenographic results. *Journal of Arthroplasty.* 1996;11(5):534–542. doi:10.1016/S0883-5403(96)80106-9
45. Li H, Khor KA, Cheang P. Titanium dioxide reinforced hydroxyapatite coatings deposited by high velocity oxy-fuel (HVOF) spray. *Biomaterials.* 2002;23(1):85–91. doi:10.1016/S0142-9612(01)00082-5
46. Li H, Khor KA, Cheang P. Impact formation and microstructure characterization of thermal sprayed hydroxyapatite/titania composite coatings. *Biomaterials.* 2003;24(6):949–957. doi:10.1016/S0142-9612(02)00431-3
47. Murphy W, Black J, Hastings G. *Handbook of biomaterial properties.* Springer. 1998. https://link.springer.com/book/10.1007/978-1-4939-3305-1. doi:10.1007/978-1-4615-5801-9
48. Liu JX, Yang DZ, Shi F, Cai YJ. Sol-gel deposited TiO2 film on NiTi surgical alloy for biocompatibility improvement. *Thin Solid Films.* 2003;429(1–2):225–230. doi:10.1016/S0040-6090(03)00146-9
49. Wu S, Weng Z, Liu X, Yeung KWK, Chu PK. Functionalized TiO2 based nanomaterials for biomedical applications. *Advanced Functional Materials.* 2014;24(35):5464–5481. doi:10.1002/ADFM.201400706
50. Koch CC, Morris DG, Lu K, Inoue A. Ductility of nanostructured materials. *MRS Bulletin.* 1999;24(2):54–58. doi:10.1557/S0883769400051551
51. Vaßen R, Stöver D. Processing and properties of nanophase ceramics. *Journal of Materials Processing Technology.* 1999;92–93:77–84. Accessed June 12, 2022, https://www.sciencedirect.com/science/article/pii/S0924013699002186

52. Gell M, Jordan EH, Sohn YH, Goberman D, Shaw L, Xiao TD. Development and implementation of plasma sprayed nanostructured ceramic coatings. *Surface and Coatings Technology*. 2001;146–147:48–54. doi:10.1016/S0257-8972(01)01470-0
53. US6835449B2—Nanostructured titania coated titanium—Google Patents. Accessed June 12, 2022, https://patents.google.com/patent/US6835449B2/en
54. Abrasion behavior of nanostructured and conventional...—Google Scholar. Accessed June 12, 2022, https://scholar.google.com/scholar?hl=en&as_sdt=0%2C5&q=+Abrasion+behavior+of+nanostructured+and+conventional+titania+coatings+thermally+sprayed+via+APS%2C+VPS+and+HVOF&btnG=
55. Zhang S. *Biological and biomedical coatings handbook: Applications*. Routledge. 2016:1–489. https://doi.org/10.1201/b10871. https://www.taylorfrancis.com/books/edit/10.1201/b10871/biological-biomedical-coatings-handbook-sam-zhang
56. Pawłowski L. *The science and engineering of thermal spray coatings*. Wiley. 2008:626. DOI:10.1002/9780470754085. https://onlinelibrary.wiley.com/doi/book/10.1002/9780470754085
57. Kweh SWK, Khor KA, Cheang P. An in vitro investigation of plasma sprayed hydroxyapatite (HA) coatings produced with flame-spheroidized feedstock. *Biomaterials*. 2002;23(3):775–785. doi:10.1016/S0142-9612(01)00183-1
58. Sun L, Berndt CC, Grey CP. Phase, structural and microstructural investigations of plasma sprayed hydroxyapatite coatings. *Materials Science and Engineering A*. 2003;360(1–2):70–84. doi:10.1016/S0921-5093(03)00439-8
59. Ferrer A, García I, Fernández J, Guilemany JM. Study of adhesion relationship of hydroxyapatite-titania coating obtained by HVOF. *Materials Science Forum*. 2010;636–637:82–88. doi:10.4028/WWW.SCIENTIFIC.NET/MSF.636-637.82
60. Gadow R, Killinger A, Stiegler N. Hydroxyapatite coatings for biomedical applications deposited by different thermal spray techniques. *Surface and Coatings Technology*. 2010;205(4):1157–1164. Accessed June 13, 2022, https://www.sciencedirect.com/science/article/pii/S0257897210002628
61. Khor KA, Li H, Cheang P. Significance of melt-fraction in HVOF sprayed hydroxyapatite particles, splats and coatings. *Biomaterials*. 2004;25(7–8):1167–1186. doi:10.1016/J.BIOMATERIALS.2003.08.008
62. Fukumoto M, Yamaguchi T, Yamada M, Yasui T. Splash splat to disk splat transition behavior in plasma-sprayed metallic materials. *Journal of Thermal Spray Technology*. 2007;16(5–6):905–912. doi:10.1007/S11666-007-9083-Y
63. Li H, Khor KA, Cheang P. Effect of steam treatment during plasma spraying on the microstructure of hydroxyapatite splats and coatings. *Journal of Thermal Spray Technology*. 2006;15:610–616. doi:10.1361/105996306X146938
64. Fauchais P, Fukumoto M, Vardelle A, Vardelle M. Knowledge concerning splat formation: An invited review. *Journal of Thermal Spray Technology*. 2004;13:337–360. doi:10.1361/10599630419670
65. Fukumoto M, Huang Y. Flattening mechanism in thermal sprayed nickel particle impinging on flat substrate surface. *Journal of Thermal Spray Technology*. 1999;8:427–432.

66. Gross K, Berndt CC. Thermal processing of hydroxyapatite for coating production. *Journal of Biomedical Materials Research*. 1998;39(4):580–587. Accessed August 4, 2022, https://onlinelibrary.wiley.com/doi/abs/10.1002/(SICI)1097-4636(19980315)39:4%3C580::AID-JBM12%3E3.0.CO;2-B
67. Gu YW, Khor KA, Cheang P. In vitro studies of plasma-sprayed hydroxyapatite/Ti-6Al-4V composite coatings in simulated body fluid (SBF). *Biomaterials*. 2003;24(9):1603–1611. doi:10.1016/S0142-9612(02)00573-2
68. Hermann-Muñoz J, Rincón-López JA, Clavijo-Mejíaa GA, et al. Influence of HVOF parameters on HAp coating generation: An integrated approach using process maps. *Surface and Coatings Technology*. 2019;358:299–307. Accessed June 13, 2022, https://www.sciencedirect.com/science/article/pii/S0257897218312398
69. Gaona M, Lima R, Marple BR. Nanostructured titania/hydroxyapatite composite coatings deposited by high velocity oxy-fuel (HVOF) spraying. *Materials Science and Engineering: A*. 2007;458:141–149. doi:10.1016/j.msea.2006.12.090
70. Henao J, Cruz-Bautista M, Hincapie-Bedoya J, et al. HVOF hydroxyapatite/titania-graded coatings: Microstructural, mechanical, and in vitro characterization. *Journal of Thermal Spray Technology*. 2018;27:1302–1321. doi:10.1007/s11666-018-0811-2
71. Melero HC, Sakai RT, Vignatti CA, et al. Corrosion resistance evaluation of HVOF produced hydroxyapatite and TiO2-hydroxyapatite coatings in Hanks' solution. *Materials Research*. 2018;21(2):20160210. doi:10.1590/1980-5373-MR-2016-0210
72. Lima RS, Khor KA, Li H, Cheang P, Marple BR, Lima S. HVOF spraying of nanostructured hydroxyapatite for biomedical applications. *Materials Science and Engineering: A*. 2005;396:181–187. doi:10.1016/j.msea.2005.01.037
73. Kulpetchdara K, Limpichaipanit A, Rujijanagul G, Randorn C, Chokethawai K. Influence of the nano hydroxyapatite powder on thermally sprayed HA coatings onto stainless steel. *Surface and Coatings Technology*. 2016;306:181–186. doi:10.1016/J.SURFCOAT.2016.05.069
74. Karlsson M, Pålsgård E, Wilshaw PR, di Silvio L. Initial in vitro interaction of osteoblasts with nano-porous alumina. *Biomaterials*. 2003;24(18):3039–3046. doi:10.1016/S0142-9612(03)00146-7
75. Jagadeeshanayaka N, Awasthi S, Jambagi SC, Srivastava C. Bioactive surface modifications through thermally sprayed hydroxyapatite composite coatings: A review over selective reinforcements. *Biomaterials Science*. 2022;10:2484–2523. Accessed August 4, 2022, https://pubs.rsc.org/en/content/articlehtml/2022/bm/d2bm00039c
76. Gupta R, Kumar A. Bioactive materials for biomedical applications using sol-gel technology. *Biomedical Materials*. 2008;3(3). doi:10.1088/1648-6041/3/3/034005
77. Wang D, Bierwagen GP. Review. *Progress in Organic Coatings*. 2009;4(64):327–338. doi:10.1016/J.PORGCOAT.2008.08.010
78. Costa DO, Dixon SJ, Rizkalla AS. One- and three-dimensional growth of hydroxyapatite nanowires during sol-gel-hydrothermal synthesis. *ACS Applied Materials and Interfaces*. 2012;4(3):1490–1499. doi:10.1021/AM201635K/SUPPL_FILE/AM201635K_SI_001.PDF

79. Alves CD, Jansen JA, Leeuwenburgh SCG. Synthesis and application of nanostructured calcium phosphate ceramics for bone regeneration. *Journal of Biomedical Materials Research Part B-Applied Biomaterials*. 2012;100:2316–2326. Accessed June 13, 2022, https://repository.ubn.ru.nl/handle/2066/109661
80. Shadanbaz S, Dias GJ. Calcium phosphate coatings on magnesium alloys for biomedical applications: A review. *Acta Biomaterialia*. 2012;8(1):20–30. doi:10.1016/J.ACTBIO.2011.10.016
81. Kačiulis S, Mattogno G, Napoli A, Bemporad E, Ferraric F, Montenero A, Gnappi G. Surface analysis of biocompatible coatings on titanium. *Journal of Electron Spectroscopy and Related Phenomena*. 1998;95(1):61–69. Accessed June 13, 2022, https://www.sciencedirect.com/science/article/pii/S0368204898002023
82. Chai CS, Ben-Nissan B. Bioactive nanocrystalline sol-gel hydroxyapatite coatings. *Journal of Materials Science: Materials in Medicine*. 1999;10(8):465–469. doi:10.1023/A:1008992807888
83. Jaafar A, Hecker C, Árki P, Joseph Y. Sol-gel derived hydroxyapatite coatings for titanium implants: A review. *Bioengineering*. 2020;7(4):127. doi:10.3390/bioengineering7040127
84. Kim HW, Kim HE, Knowles JC. Fluor-hydroxyapatite sol-gel coating on titanium substrate for hard tissue implants. *Biomaterials*. 2004;25(16):3351–3358. doi:10.1016/J.BIOMATERIALS.2003.09.104
85. Kim HW, Koh YH, Li LH, Lee S, Kim HE. Hydroxyapatite coating on titanium substrate with titania buffer layer processed by sol-gel method. *Biomaterials*. 2004;25(13):2533–2538. doi:10.1016/J.BIOMATERIALS.2003.09.041
86. El Hadad AA, Peón E, García-Galván FR, et al. Biocompatibility and corrosion protection behaviour of hydroxyapatite sol-gel-derived coatings on Ti6Al4V alloy. *Materials*. 2017;10(2):94. doi:10.3390/ma10020094
87. Ramires PA, Romito A, Cosentino F, Milella E. The influence of titania/hydroxyapatite composite coatings on in vitro osteoblasts behaviour. *Biomaterials*. 2001;22(12):1467–1474. doi:10.1016/S0142-9612(00)00269-6
88. Kim HW, Kim HE, Salih V, Knowles JC. Hydroxyapatite and titania sol-gel composite coatings on titanium for hard tissue implants; mechanical and in vitro biological performance. *Journal of Biomedical Materials Research*. 2004;72:1–8. doi:10.1002/jbm.b.30073
89. Vijayalakshmi U, Rajeswari S. Influence of process parameters on the sol-gel synthesis of nano hydroxyapatite using various phosphorus precursors. *Journal of Sol-Gel Science and Technology*. 2012;63:45–55. doi:10.1007/s10971-012-2762-2
90. Guo L, Li H. Fabrication and characterization of thin nano-hydroxyapatite coatings on titanium. *Surface and Coatings Technology*. 2004;185(2–3):268–274. doi:10.1016/J.SURFCOAT.2004.01.013
91. Jafari H, Hessam H, Morteza S, et al. Characterizing sintered nano-hydroxyapatite sol-gel coating deposited on a biomedical Ti-Zr-Nb alloy. *Journal of Materials Engineering and Performance*. 1944;25:901–909. doi:10.1007/s11665-016-1944-4
92. Rojaee R, Fathi M, Raeissi K. Controlling the degradation rate of AZ91 magnesium alloy via sol-gel derived nanostructured hydroxyapatite coating. *Materials Science and Engineering C*. 2013;33(7):3816–3825. doi:10.1016/J.MSEC.2013.05.014

93. Guo Y, Su Y, Jia S, et al. Hydroxyapatite/titania composite coatings on biodegradable magnesium alloy for enhanced corrosion resistance, cytocompatibility and antibacterial properties. *Journal of the Electrochemical Society.* 2018;165(14):C962–C972. doi:10.1149/2.1161814JES
94. Kaur S, Bala N, Khosla C. Characterization of hydroxyapatite coating on 316L stainless steel by sol-gel technique. *Surface Engineering and Applied Electrochemistry.* 2019;55(3):357–366. doi:10.3103/S1068375519030104
95. Mohammed MT, Hussein SM. Synthesis and characterization of nanocoatings derived by sol-gel onto a new surgical titanium surface. *Materials Research Express.* 2019;6(7). doi:10.1088/2053-1591/AB1622
96. Chung RJ, Hsieh MF, Huang CW, Perng LH, Wen HW, Chin TS. Antimicrobial effects and human gingival biocompatibility of hydroxyapatite sol-gel coatings. *Journal of Biomedical Materials Research—Part B Applied Biomaterials.* 2006;76(1):169–168. doi:10.1002/JBM.B.30365
97. Figueira RB. Hybrid sol-gel coatings for corrosion mitigation: A critical review. *Polymers.* 2020;12(3):689. doi:10.3390/polym12030689
98. Fernández J, Gaona M, Guilemany JM. Effect of heat treatments on hvof hydroxyapatite coatings. *Journal of Thermal Spray Technology.* 2007;16:220–228. doi:10.1007/s11666-007-9034-7
99. Liu X, He D, Zhou Z, et al. Effect of post-heat treatment on the microstructure of micro-plasma sprayed hydroxyapatite coatings. *Surface and Coatings Technology.* 2019;367:225–230. Accessed June 13, 2022, https://www.sciencedirect.com/science/article/pii/S0257897219303329
100. Zhao G, Wen G, Wu K. Influence of processing parameters and heat treatment on phase composition and microstructure of plasma sprayed hydroxyapatite coatings. *Transactions of Nonferrous Metals Society of China.* 2009;19(2):S463–S469. Accessed June 13, 2022, https://www.sciencedirect.com/science/article/pii/S1003632610600908
101. Singh A, Singh G, Chawla V. Influence of post coating heat treatment on microstructural, mechanical and electrochemical corrosion behaviour of vacuum plasma sprayed reinforced hydroxyapatite. *Journal of the Mechanical Behavior of Biomedical Materials.* 2018;85:20–36. Accessed June 13, 2022, https://pubmed.ncbi.nlm.nih.gov/29843093/
102. Tiwari S, Mishra SB. Post annealing effect on corrosion behavior, bacterial adhesion, and bioactivity of LVOF sprayed hydroxyapatite coating. *Surface and Coatings Technology.* 2021;405:126500. Accessed June 13, 2022, https://www.sciencedirect.com/science/article/pii/S0257897220311695
103. Li H, Khor K, Cheang P. Effect of the powders' melting state on the properties of HVOF sprayed hydroxyapatite coatings. *Materials Science and Engineering: A.* 2000;293:71–80. Accessed June 13, 2022, https://www.sciencedirect.com/science/article/pii/S0921509300012454
104. Liu DM, Yang Q, Troczynski T. Sol-gel hydroxyapatite coatings on stainless steel substrates. *Biomaterials.* 2002;23(3):691–698. doi:10.1016/S0142-9612(01)00157-0
105. Xu W, Hu W, Li M, Wen C. Sol–gel derived hydroxyapatite/titania biocoatings on titanium substrate. *Materials Letters.* 2006;60(13–14):1575–1578. Accessed June 14, 2022, https://www.sciencedirect.com/science/article/pii/S0167577X05012048

106. Rocha RC, de Sousa Galdino AG, da Silva SN, Machado MLP. Surface, microstructural, and adhesion strength investigations of a bioactive hydroxyapatite-titanium oxide ceramic coating applied to Ti-6Al-4V alloys by plasma thermal spraying. *Materials Research*. 2018;21(4):20161144. doi:10.1590/1980-5373-MR-2016-1144
107. One in two Americans have a musculoskeletal condition: New report outlines the prevalence, scope, cost and projected growth of musculoskeletal disorders in the U.S. Accessed June 13, 2022, https://www.sciencedaily.com/releases/2016/03/160301114116.htm
108. Haralson RH, Zuckerman JD. Prevalence, health care expenditures, and orthopedic surgery workforce for musculoskeletal conditions. *JAMA*. 2009;302(14):1586–1587. doi:10.1001/JAMA.2009.1489
109. Holzwarth U, Cotogno G. *Total hip arthroplasty : State of the art, prospects and challenges*. Luxembourg: Publications Office of the European Union. 2012. doi:10.2788/31286
110. Iorio R, Robb WJ, Healy WL, et al. Orthopaedic surgeon workforce and volume assessment for total hip and knee replacement in the United States: Preparing for an epidemic. *Journal of Bone and Joint Surgery*. 2008;90(7):1598–1605. doi:10.2106/JBJS.H.00067

Index

accumulative roll bonding 31
aluminium alloys 35

biomedical implants 261
biotribology 258

ceramic nanocoatings 210
chemical vapour deposition 174, 206
cold spray 171
copper composites 35
critical mass 249

damping coefficients 247
disintegrated melt deposition 29

electrodeposition 208
electroplating 173

friction stir processing 31

in situ processing 55

laser cladding 208
liquid processing 54
lubricants 11

magnesium alloys 35
metallic nanocoatings 209
multifunctional anti-corrosive nanocoatings 211

nickel-graphite composites 36

one-dimensional nanomaterials 3

physical vapour deposition (PVD) 207
plasma spray 170
polymeric nanocoatings 210

sedimentation technique 107
semi-solid casting 29
semisolid materials 13, 15
semi-solid processing 54
sol-gel 207
solid processing 55
spin coatings 175
spray coating 207
sputtering 172
stability of nanofluids 102
stiffness coefficients 247
stir-casting 28

three-dimensional nanomaterials 4
threshold speed 249
turbidimetry 106
two-dimensional nanomaterials 3

ultraviolet–visible spectroscopy 107

whirl frequency ratio 249

zero-dimensional nanomaterials 3
zeta-potential 104

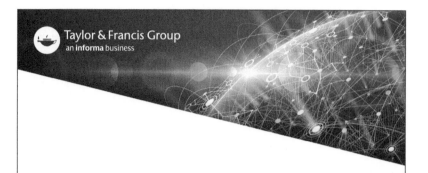